Math & YOU™

The Power & Use of Mathematics

Ron Larson

andYOU.com
1762 Norcross Road
Erie, PA 16510-3838
USA

For product information and customer support, contact andYOU.com
at 1-866-690-ANDU or visit us at andYOU.com.

Connect with us on Facebook and Twitter:

 http://www.facebook.com/andyoucom

http://www.twitter.com/andyoucom

Distributed by W.H. Freeman.

Printed in the U.S.A.

ISBN 13: 978-1-60840-602-9
ISBN 10: 1-60840-602-4

2 3 4 5 6 7 8 9 10-WEB-15 14 13 12 11

A Word from the Author

Dr. Ron Larson has written over 200 mathematics books and considers this book to be his most important. He says, "Math & YOU will liberate people from the feeling of being inept at mathematics and help them feel at ease in using math to make important decisions."

I started writing Math & YOU in the Spring of 2010. It didn't take long before the spirit and potential of the book became evident. The more I worked on the book, the more enthused I became.

All of a sudden I had a career altering realization: There is a terrible imbalance between the mathematics that is used to run our country's households, businesses, and governments, and the mathematics that we teach.

After 6th or 7th grade, the mathematics that we *teach* is almost all centered in algebra. But, whether at home, in business, or in government, the mathematics we *use* consists of little, if any, algebra!

Math & YOU changes that. In this book you will add, subtract, multiply, divide, take percents, use significant digits, and represent and interpret data. As such, you should find that much of the mathematics in this book is easy.

What is not so easy is the interpretation of the results you obtain with the mathematics.

As you read this book, remember that additional supporting material is online at *math.andyou.com*. I hope you love this book as much as I do.

Ron Larson

Ron Larson, Ph.D
Professor of Mathematics
Penn State University at Erie
www.RonLarson.com
odx@psu.edu

Contents

The Mathematics of Calculation

The Mathematics of Consumption

The Mathematics of Logic & the Media

The Mathematics of Inflation & Depreciation

The Mathematics of Taxation

The Mathematics of Borrowing & Saving

The Mathematics of Patterns & Nature

The Mathematics of Likelihood

The Mathematics of Description

The Mathematics of Fitness & Sports

Easy Access Social Textbook Online

View full-size printable PDFs of every page.

Download spreadsheets to make real-life decisions.

Navigate to any page.

Complete Student Textbook

- All examples and exercises are available.
- Browse by topic, contents, and index.

Additional Online Resources

- Math Help, Consumer Suggestions, and Checkpoint Solutions
- Worked-out solutions for odd-numbered exercises
- Discussions for every lesson and exercise
- Shareable via Facebook and Twitter

Access a glossary of math terms, online calculators, and other useful tools.

Search [   ] GO   TOC & Index   ABC   Follow us on:

Math Help

As is true of the terms "markup" and "markup percent," the terms "discount" and "discount percent" are technical terms that are used in business.

Discount: Technically this term only applies to retail (or reselling) businesses. An item is discounted when its normal retail price is lowered. The discount is the difference between the "Regular Price" and the "Discounted Price." You can see the potential for abuse with this definition. It is common that a retailer artificially inflates the "Regular Price" for the sake of being able to claim a larger discount.

Discount Percent: When using the term discount percent, be sure that the numerator is the discount and the denominator is the regular retail price.

Additional instruction for every lesson.

Read tips on how to become a better consumer.

+ **Consumer Suggestion**
+ **Checkpoint Solution**
+ **Comments (0)**

View worked-out solutions for every Checkpoint.

Read and share comments with other students.

Catch up with us
on our blog:
andYOU.com/blog

Program Overview

Student Resources

AVAILABLE IN SIX FORMATS

- Online access to full textbook
- Chapter-by-chapter PDF download
- eBook download
- Loose-leaf, 3-hole punch
- Paperback
- Hardcover

Check out the options at *math.andyou.com.*

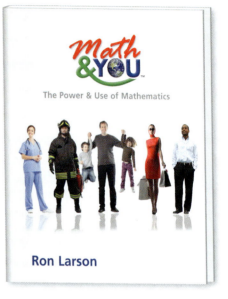

STUDENT SOLUTIONS GUIDE

Worked-out solutions to all odd-numbered exercises are available online.

MATH HELP FOR EVERY EXAMPLE

Additional instruction, written by Ron Larson, for every example in the book is available at "Math Help" online.

CHECKPOINT SOLUTIONS

All examples are followed by a Checkpoint to assess your understanding. The solutions to these Checkpoints are available, step-by-step, online.

AUTOGRADED SELF ASSESSMENTS

Each chapter in the book has 2 quizzes and 1 chapter review. Autograded versions are available online.

SOCIAL MEDIA INTERACTIONS

All examples and exercises have live interactive comments online, which are monitored by full-time social media learning experts.

INTERACTIVE GLOSSARY & SPREADSHEETS

A comprehensive math glossary is available online. All data sets are available online in spreadsheet form.

Find these free resources online at *math.andyou.com.*

Instructor Resources

AVAILABLE IN SIX FORMATS
• Online access to full textbook
• Chapter-by-chapter PDF download
• eBook download
• Loose-leaf, 3-hole punch
• Paperback
• Hardcover

Contact us for a complimentary instructor copy at *math.andyou.com.*

INSTRUCTOR GUIDE
• Pacing guides
• Summary of each chapter
• Teaching suggestions for all examples
• Solutions for all exercises, quiz questions, chapter review exercises, and Checkpoint questions

The Instructor Guide is available in print and online.

For a complimentary print copy or an instructor password, contact us at *math.andyou.com.*

ASSESSMENT BY ExamView
Assessment Suite

WHITEBOARD COMPATIBLE

1 The Mathematics of Calculation

1.1 Order of Operations & Formulas

▶ Use the order of operations to evaluate a numerical expression.

▶ Use a calculator to evaluate a numerical expression.

▶ Use the order of operations to evaluate a formula.

1.2 Rounding & Calculators

▶ Round numbers in a real-life context.

▶ Read large and small numbers.

▶ Understand the concept of "garbage in, garbage out."

1.3 Using Percent

▶ Understand and find a percent of a number.

▶ Determine what percent one number is of another number.

▶ Use percent to represent change.

1.4 Units & Conversions

▶ Use unit analysis to "balance" both sides of a formula.

▶ Convert within a given system of measure.

▶ Convert between different systems of measure.

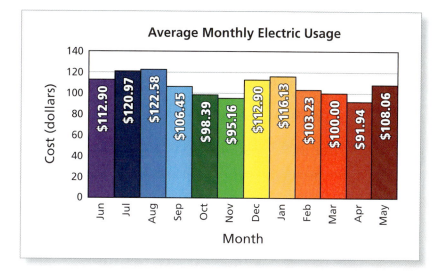

Average Monthly Electric Usage

Cost (dollars) vs. Month

- Jun: $112.90
- Jul: $120.97
- Aug: $122.58
- Sep: $106.45
- Oct: $98.39
- Nov: $95.16
- Dec: $112.90
- Jan: $116.13
- Feb: $103.23
- Mar: $100.00
- Apr: $91.94
- May: $108.06

Example 1 on page 2 shows how you can calculate your reconciliation payment when you pay your electric bill on an equal-payment plan. In May, do you owe more or will you get a refund?

1.1 Order of Operations & Formulas

▶ Use the order of operations to evaluate a numerical expression.
▶ Use a calculator to evaluate a numerical expression.
▶ Use the order of operations to evaluate a formula.

Order of Operations

Mathematics is a language. In this language, numbers are the nouns. The operations $+$, $-$, $\times$, $\div$, and exponentiation are the verbs. The **order of operations** is a set of rules that tells you which operations have priority.

Study Tip

A mnemonic for remembering the order of operations is "**P**lease **E**xcuse **M**y **D**ear **A**unt **S**ally."

Order of Operations

1. Perform operations in **P**arentheses.
2. Evaluate numbers with **E**xponents.
3. **M**ultiply or **D**ivide from left to right.
4. **A**dd or **S**ubtract from left to right.

EXAMPLE 1   Using Order of Operations

From June through April, on an equal-payment plan, you pay \$110 each month on your electric bill. How much do you owe in May?

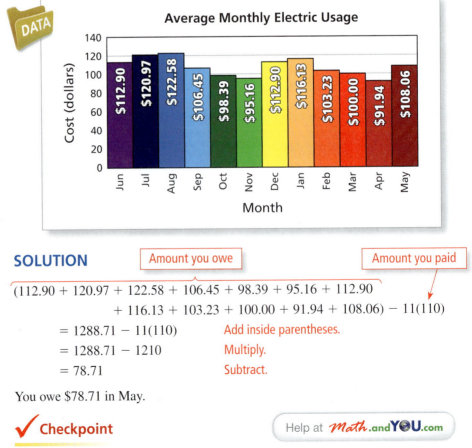

Average Monthly Electric Usage

In 2009, the greatest consumption of electricity in the United States occurred in August. The least occurred in April.

SOLUTION

Amount you owe   Amount you paid

$(112.90 + 120.97 + 122.58 + 106.45 + 98.39 + 95.16 + 112.90$
$\qquad + 116.13 + 103.23 + 100.00 + 91.94 + 108.06) - 11(110)$

$= 1288.71 - 11(110)$ Add inside parentheses.

$= 1288.71 - 1210$ Multiply.

$= 78.71$ Subtract.

You owe \$78.71 in May.

✓ **Checkpoint**   Help at Math.andYOU.com

Suppose you paid \$95 per month. How much would you owe in May?

EXAMPLE 2   **Using Order of Operations**

There are 1950 calories in a cup of salad oil, 55 calories in an egg yolk, and 16 calories in a teaspoon of sugar. The other ingredients in the mayonnaise recipe are essentially calorie free. How many calories are in the mayonnaise recipe?

Homemade Mayonnaise

* 2 egg yolks
* ½ tsp powdered mustard
* Pinch cayenne pepper
* 1½ cups salad oil
* ¾ tsp salt
* ¼ tsp sugar
* 4½ tsp white vinegar
* 4 tsp hot water

Preparation:

Beat yolks, salt, mustard, sugar, pepper, and 1 teaspoon vinegar until thick and pale yellow. Add about ¼ cup oil, drop by drop, beating vigorously. Beat in 1 teaspoon each vinegar and hot water. Add another ¼ cup oil, a few drops at a time, beating vigorously. Beat in another teaspoon each vinegar and water. Add ½ cup oil in a fine steady stream, beating constantly. Mix in remaining vinegar and water. Slowly beat in remaining oil. Cover and refrigerate until needed. Do not keep longer than 1 week.

Some historians say that mayonnaise originated in Bayonne, France, and was originally called bayonnaise.

SOLUTION

| 1.5 cups of salad oil | 2 egg yolks | Quarter teaspoon of sugar |

$$\text{Total calories} = 1.5(1950) + 2(55) + 0.25(16)$$
$$= 2925 + 110 + 4 \qquad \text{Multiply.}$$
$$= 3039 \qquad \text{Add.}$$

There are 3039 calories in the recipe.

✓ **Checkpoint**   Help at *Math*.and**Y⊙U**.com

The total number of teaspoons in the mayonnaise recipe is given by the expression below.

3 tsp in an egg yolk | Mustard | 48 tsp in a cup | Salt | Sugar | Vinegar | Water

$$2(3) + 0.5 + 1.5(48) + 0.75 + 0.25 + 4.5 + 4$$

a. How many teaspoons are in the recipe?

b. How many calories are in a teaspoon of mayonnaise?

Calculators and Order of Operations

Some calculators use the standard order of operations and some do not. Rather than relying on your calculator to perform the operations in the correct order, you should use a calculator that has parentheses. Also, for this book, you will need a calculator with an exponent key.

Recommended Calculator Keys

Study Tip

Here is an example of the exponent key. There are 12^3 cubic inches in 1 cubic foot.

EXAMPLE 3   Using a Calculator

You are taking a 3-credit evening course. The cost is $150 for registration and $219 for each credit. Which of the following keystroke sequences is better for finding the total cost? Explain your reasoning.

a.   1 5 0 + (3 × 2 1 9) =

b.   1 5 0 + 3 × 2 1 9 =

SOLUTION

a. This sequence is better. You are forcing the calculator to multiply 3 by 219 before adding 150.

$$\text{Total cost} = 150 + 3(219)$$
$$= 150 + 657 \qquad \text{Multiply.}$$
$$= 807 \qquad \text{Add.}$$

The correct total cost is $807.

b. On some calculators, this keystroke sequence gives an incorrect total because it adds 150 and 3 to get 153 and then multiplies by 219 to get a total of $33,507, which is an unreasonable answer.

✓ **Checkpoint**

Help at *Math*.andY☺U.com

Your cell phone bills for 3 months are $50, $62, and $73. Which of the following keystroke sequences is better for finding your average monthly bill? Explain your reasoning.

c.   5 0 + 6 2 + 7 3 ÷ 3 =

d.   (5 0 + 6 2 + 7 3) ÷ 3 =

Using and Evaluating Formulas

A **formula** is an equation that relates one quantity to one or more other quantities. For instance, the area of a rectangle is given by the formula

$$\text{Area} = (\text{base})(\text{height}) \quad \text{or} \quad A = bh.$$

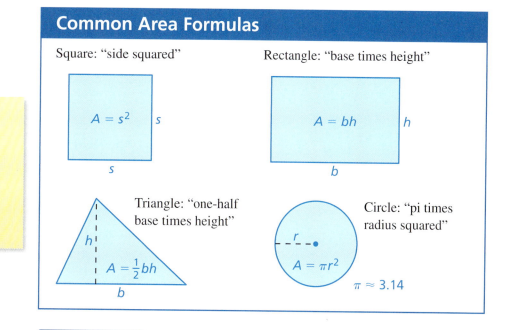

Common Area Formulas

Square: "side squared"

$A = s^2$ s
s

Rectangle: "base times height"

$A = bh$ h
b

Triangle: "one-half base times height"

h
$A = \frac{1}{2}bh$
b

Circle: "pi times radius squared"

r
$A = \pi r^2$
$\pi \approx 3.14$

> **Study Tip**
>
> Another formula for the area of a rectangle is "length times width."
>
> $A = \ell w$

EXAMPLE 4   Buying Floor Tiles

12 in.

12 in.

Ceramic tile has been around for at least 4000 years. Tiled surfaces have been found in the ruins of Egypt, Babylon, and Greece.

You are buying tiles for your kitchen floor. The room is 8 feet by 12 feet. The tiles are 12-inch squares that cost $8.97 each. The tiles come in boxes of 15. You should order an extra 20 tiles to allow for waste.

a. How many boxes should you order?

b. What is the total cost of your order?

SOLUTION

a. The area of the room is 8×12, or 96 square feet. So, you need 96 tiles plus an extra 20 tiles for waste. This is a total of 116 tiles. Because there are 15 tiles in a box, you should order 8 boxes, giving you 120 tiles.

b. The total cost of your order is

Number of tiles   Cost per tile

$120 \times 8.97 = \$1076.40.$

✔ **Checkpoint**

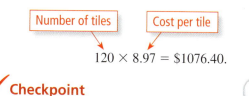

You are buying tiles for 2 bathroom floors. One room is 8.5 square feet. The other room is 9 feet by 10 feet. The tiles are 12-inch squares that cost $7.65 each. The tiles come in boxes of 12. You should order an extra 20 tiles to allow for waste.

c. How many boxes should you order?

d. What is the total cost of your order?

Formulas for Distance, Rate, and Time

Distance = d	Rate = r	Time = t
$d = rt$	$r = \dfrac{d}{t}$	$t = \dfrac{d}{r}$
Distance equals rate times time.	Rate equals distance divided by time.	Time equals distance divided by rate.

You can access a distance, rate, and time calculator at *Math.andYou.com*.

EXAMPLE 5   **Using the Distance Formula**

Each morning, you run 7.5 miles in 45 minutes. You weigh 155 pounds. How many calories do you burn running each morning?

Calories Burned During 1 Hour of Activity

Activity	Weight 130 lb	155 lb	190 lb
Bicycling, 12 mph	472	563	690
Billiards	148	176	216
Fishing from boat	148	176	216
Golfing	236	281	345
Running, 10 mph	944	1126	1380
Running, 8 mph	797	950	1165
Swimming laps	590	704	863
Volleyball at beach	472	563	690

SOLUTION

You are given the time and the distance. Your rate is given by

$$r = \frac{d}{t}$$

$$= \frac{7.5 \text{ miles}}{0.75 \text{ hour}} \qquad \text{Write 45 minutes as 0.75 hour.}$$

$$= 10 \frac{\text{miles}}{\text{hour}}.$$

Because 45 minutes is $\frac{3}{4}$ of an hour, you burn $\frac{3}{4}$ of 1126 calories.

Calories burned = $0.75(1126) = 844.5$

So, you burn about 850 calories each morning.

According to the Centers for Disease Control and Prevention, people who exercise regularly lower their risk of heart disease, stroke, and colon cancer.

✓ **Checkpoint**   Help at *Math.andYou.com*

For lunch, you eat a third-pound hamburger with 590 calories, a medium cola with 210 calories, and a large order of French fries with 510 calories. How long do you have to run at 10 miles per hour to burn the calories you eat at lunch?

Formulas for Earnings, Rate, and Time

Earnings = E	Rate = r	Time = t
$E = rt$	$r = \dfrac{E}{t}$	$t = \dfrac{E}{r}$
Earnings equal rate times time.	Rate equals earnings divided by time.	Time equals earnings divided by rate.

You can access an earnings, rate, and time calculator at *Math.andYou.com*.

Study Tip

Notice the similarity between these formulas and the ones for distance, rate, and time.

EXAMPLE 6   Comparing Job Offers

Which job offer has the better total compensation? Explain your reasoning.

a.

Salary Rate:	$30 per hour
401(k):	5% matching
Health Insurance:	$600 per month

b.

Salary Rate:	$59,000 per year
401(k):	6% matching
Health Insurance:	$900 per month
Profit Sharing:	$0–$20,000 per year

SOLUTION

a. There are 52 weeks in a year. At 40 hours a week, you work 2080 hours in a year. Your yearly earnings are

$$E = \left(30\,\frac{\$}{\text{hr}}\right)(2080\,\text{hr}) = \$62{,}400.$$

5% of this is $3120 (see Section 1.3). So, your total compensation is

$$62{,}400 + 3120 + 12(600) = \$72{,}720.$$

b. 6% of $59,000 is $3540 (see Section 1.3). Your total compensation is

$$59{,}000 + 3540 + 12(900) = \$73{,}340.$$

So, even without profit sharing, this total compensation is better. If the company has a profitable year, the total compensation could be *much* better.

✓ **Checkpoint**   Help at *Math.andYOU.com*

c. You get a third job offer. How does it compare with the other two offers?

Salary Rate:	$4800 per month
401(k):	4% matching
Health Insurance:	$1200 per month
Sales Commission:	$0–$3000 per month

1.1 Exercises

 Equal-Payment Plan In Exercises 1 and 2, use the graph. *(See Example 1.)*

1. From October through August, on an equal-payment plan, you pay $50 each month on your gas bill. How much do you owe in September?

2. You are on an equal-payment plan. What is the least amount that you can pay each month from October through August to owe nothing in September?

Actual Monthly Gas Usage

Cost (dollars) vs. Month

Month	Cost
Oct	$42.58
Nov	$52.16
Dec	$69.47
Jan	$84.61
Feb	$77.06
Mar	$62.83
Apr	$49.16
May	$40.45
Jun	$28.17
Jul	$29.54
Aug	$26.87
Sep	$34.18

 Equal-Payment Plan In Exercises 3 and 4, use the table. *(See Example 1.)*

Actual Monthly Electric Usage											
Jan	**Feb**	**Mar**	**Apr**	**May**	**Jun**	**Jul**	**Aug**	**Sep**	**Oct**	**Nov**	**Dec**
$84.02	$86.59	$77.65	$72.10	$68.26	$62.06	$63.47	$65.19	$60.12	$67.61	$75.36	$81.98

3. From January through November, on an equal-payment plan, you pay $75 each month on your electric bill. How much do you owe in December?

4. You are on an equal-payment plan. What is the least amount that you can pay each month from January through November to owe nothing in December?

Old-Fashioned Pink Lemonade

* 1 ½ cups white sugar
* 6 ¾ cups water
* 1 ½ cups fresh lemon juice
* ¾ cup cranberry juice

5. Pink Lemonade There are 774 calories in a cup of sugar, 65 calories in a cup of fresh lemon juice, and 137 calories in a cup of cranberry juice. *(See Example 2.)*

a. How many calories are in the old-fashioned pink lemonade recipe?

b. This recipe makes nine servings. How many calories are in one serving?

---

6. Fudge There are 99 calories in an ounce of cream cheese, 12 calories in a teaspoon of vanilla extract, 389 calories in a cup of powdered sugar, and 145 calories in a square of chocolate. Salt has 0 calories. *(See Example 2.)*

a. How many calories are in the fudge recipe?

b. This recipe uses an 8-inch by 8-inch dish. How many calories are in a 1-inch by 1-inch piece of fudge?

Homemade Fudge

* 6 oz cream cheese, softened
* ⅛ tsp salt
* ½ tsp vanilla extract
* 4 cups powdered sugar, sifted
* 4 squares unsweetened chocolate

7. **Parking** You buy a campus parking permit for $250. During the year, you get four $15 parking tickets. Which of the following keystroke sequences is better for finding your total parking cost? Explain your reasoning. *(See Example 3.)*

   a. (2 5 0 + 4) × 1 5 =

   b. 2 5 0 + (4 × 1 5) =

8. **Books** Your books for this semester cost $95, $120, and $115. Which of the following keystroke sequences is better for finding your average book cost? Explain your reasoning. *(See Example 3.)*

   a. (9 5 + 1 2 0 + 1 1 5) ÷ 3 =

   b. (9 5 + 1 2 0) + 1 1 5 ÷ 3 =

9. **Pizza Party** You buy 3 pizzas for $10 each and 5 bottles of cola for $2 each. Which of the following keystroke sequences is better for finding the total cost? Explain your reasoning. *(See Example 3.)*

   a. (3 × 1 0) + (5 × 2) =

   b. (3 × 1 0) + 5 × 2 =

10. **Depreciation** A new car worth $25,000 depreciates at a rate of 50% every 3 years. The value of the car after 6 years is $25{,}000(0.5^2)$. Which of the following keystroke sequences is better for finding the value of the car? Explain your reasoning. *(See Example 3.)*

    a. (2 5 0 0 0 × . 5) ^ 2 =

    b. 2 5 0 0 0 × (. 5 ^ 2) =

Football Field In Exercises 11 and 12, use the diagram of the football field. *(See Example 4.)*

11. You are buying lawn seed for the football field. A 10-pound bag of lawn seed covers 2000 square feet.

    a. How many 10-pound bags of lawn seed should you buy?

    b. A 10-pound bag of lawn seed costs $27.79. What is the total cost of your purchase?

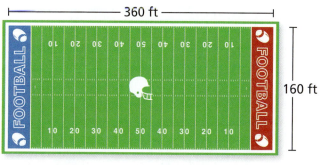

12. Latex paint is used to paint the end zones of the football field. Each gallon of paint must be diluted with 1 gallon of water. One gallon of diluted latex paint covers 100 square feet.

    a. Each end zone is 30 feet long. How many gallons of undiluted latex paint are needed to paint both end zones?

    b. Why do you think the paint is diluted?

Solar System In Exercises 13–16, use the table. The speed of light is about 299,792 kilometers per second. *(See Examples 5 and 6.)*

Planet	Mean Distance from the Sun (km)
Mercury	57,909,175
Venus	108,208,930
Earth	149,597,890
Mars	227,936,640
Jupiter	778,412,020
Saturn	1,426,725,400
Uranus	2,870,972,200
Neptune	4,498,252,900

Mercury Earth
Venus Mars
Jupiter Saturn
Uranus Neptune
Sun

Saturn is the only planet in the Solar System that is less dense than water. It would float in a bathtub if you could build a bathtub big enough.

13. How long does it take sunlight to reach Earth?

14. How long does it take sunlight to reach Neptune?

15. In the book *Men Are from Mars, Women Are from Venus*, the author John Gray suggests that communication is a key component to successful relationships among couples. Data transmission through space occurs at the speed of light. How long would it take for a text message sent from Venus to reach Mars?

16. A meteor streaks through Jupiter's atmosphere, causing a bright flash. How long before the flash is seen from Earth?

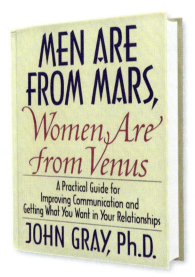

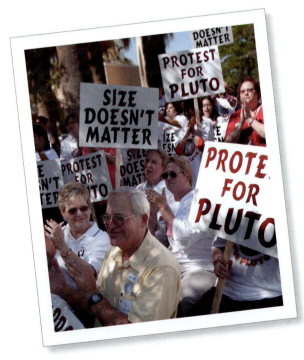

17. **Pluto** From its discovery in 1930 until 2006, Pluto was considered the Solar System's ninth planet. In 2006, Pluto was reclassified as a dwarf planet because of its low mass and its similarity to other dwarf planets beyond Neptune in the Kuiper Belt. To better understand Pluto, NASA planned a mission to Pluto called New Horizons. The spacecraft launched on January 19, 2006, and is supposed to reach Pluto 9.5 years (83,220 hours) later after traveling about 3 billion miles. What is the average speed of the spacecraft in miles per hour? *(See Example 5.)*

18. **Light-Year** What is a light-year? How is the term *light-year* used?

▶ Extending Concepts

DATA **19. Equal-Payment Plan** From May through March, on an equal-payment plan, you pay $90 each month on your electric bill.

a. How much do you owe in April?

b. Interpret your answer to part (a).

c. What should you pay each month to have 12 equal monthly payments?

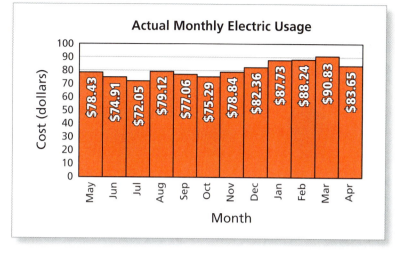

Actual Monthly Electric Usage

Month	Cost
May	$78.43
Jun	$74.91
Jul	$72.05
Aug	$79.12
Sep	$77.06
Oct	$75.29
Nov	$78.84
Dec	$82.36
Jan	$87.73
Feb	$88.24
Mar	$90.83
Apr	$83.65

20. Ink Cartridge A combo pack has 2^3 ink cartridges. A shipment has 6^2 combo packs. Which of the following keystroke sequences is better for finding the total number of ink cartridges in a shipment? Explain your reasoning.

a. [2] [^] [3] [( ] [6] [^] [2] [ )] [=]

b. [( ] [2] [^] [3] [ )] [( ] [6] [^] [2] [ )] [=]

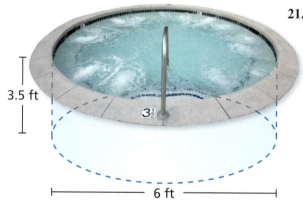

3.5 ft

$3\frac{1}{2}$

6 ft

21. Hot Tub The inside of a circular hot tub is lined with 1-inch by 1-inch tiles.

a. Explain how to estimate the number of tiles needed for the hot tub.

b. The hot tub is 3.5 feet deep and has a diameter of 6 feet. What is the minimum number of tiles needed? (The formula for the circumference of a circle is $C = \pi d$.)

Job Offers In Exercises 22–24, use the job offers shown.

Offer A

Salary Rate:	$3200 per month
401(k):	3% matching
Health Insurance:	$850 per month
Sales Commission:	$0–2500 per month

Offer B

Salary Rate:	$25 per hour
401(k):	5% matching
Health Insurance:	$650 per month
Profit Sharing:	$0–15,000 per year

22. What are the minimum and maximum yearly total compensations for each job offer?

23. Which job offer do you think is better? Explain your reasoning.

24. You decide to make a counteroffer to the company you did not choose in Exercise 23. How does your counteroffer differ from the original? Explain your reasoning.

1.2 Rounding & Calculators

▶ Round numbers in a real-life context.
▶ Read large and small numbers.
▶ Understand the concept of "garbage in, garbage out."

Rounding Numbers in a Real-Life Context

People tend to think that mathematics gives exact answers. In a real-life context, however, listing "exact" calculator displays is often unreasonable. It is usually reasonable to **round** the calculator display according to the context of the real-life application.

Rounding Numbers

Determine your rounding digit and look at the digit to the right of it.
1. **Rounding Down:** If the digit is 0, 1, 2, 3, or 4, do not change the rounding digit. All digits to the right of the rounding digit become 0.
2. **Rounding Up:** If the digit is 5, 6, 7, 8, or 9, add 1 to the rounding digit. All digits to the right of the rounding digit become 0.

EXAMPLE 1   Rounding Numbers in Context

Round the defense spending per person for each country so it is reasonable for the context.

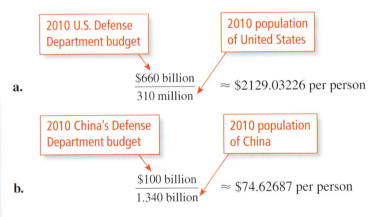

a. $\dfrac{\$660 \text{ billion}}{310 \text{ million}} \approx \$2129.03226$ per person

> 2010 U.S. Defense Department budget
> 2010 population of United States

b. $\dfrac{\$100 \text{ billion}}{1.340 \text{ billion}} \approx \$74.62687$ per person

> 2010 China's Defense Department budget
> 2010 population of China

SOLUTION

None of the numbers given in the problem are exact. So, rounding is reasonable.

a. A reasonable answer is that defense spending in the United States is about $2130 per person.

b. A reasonable answer is that defense spending in China is about $75 per person.

✓ **Checkpoint**   Help at *Math*.and**YOU**.com

In 2010, the population of the United Kingdom was about 62 million. Estimate its defense spending per person.

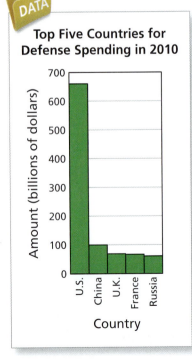

DATA

Top Five Countries for Defense Spending in 2010

Country

EXAMPLE 2   **Comparing Annual Fuel Cost**

Assume that you drive 15,000 miles in a year. Use the information given to estimate the annual fuel cost based on a fuel price of $2.68 per gallon for 87 octane and $2.92 per gallon for 93 octane.

	a. 2011 Toyota Prius	b. 2011 Infiniti EX35
Miles on a Tank*	536 miles	360 miles
Tank Size	11.9 gal	20.0 gal
Recommended Octane	87	93

*Based on 45% highway and 55% city driving.

SOLUTION

a. This car can travel 536 miles on a tank of gas. To drive 15,000 miles, you need the following number of tanks.

$$\frac{15{,}000 \text{ miles}}{536 \text{ miles per tank}} \approx 27.985 \text{ tanks}$$ See page 34.

Each tank contains 11.9 gallons, so the number of gallons you use is

$$(27.985 \text{ tanks})\left(11.9 \frac{\text{gallons}}{\text{tank}}\right) = 333.0215 \text{ gallons.}$$

At $2.68 per gallon, your annual fuel cost is

$$(333.0215 \text{ gallons})\left(2.68 \frac{\text{dollars}}{\text{gallon}}\right) = \$892.49762.$$

Your annual fuel cost is about $900. In the context of the question, it is misleading to specify an annual cost of $892.50. This would lead your reader to believe that you know more about the context than you do.

b. This car can travel 360 miles on a tank of gas. To drive 15,000 miles, you need the following number of tanks.

$$\frac{15{,}000 \text{ miles}}{360 \text{ miles per tank}} \approx 41.667 \text{ tanks}$$ See page 34.

Each tank contains 20 gallons, so the number of gallons you use is

$$(41.667 \text{ tanks})\left(20 \frac{\text{gallons}}{\text{tank}}\right) = 833.34 \text{ gallons.}$$

At $2.92 per gallon, your annual fuel cost is

$$(833.34 \text{ gallons})\left(2.92 \frac{\text{dollars}}{\text{gallon}}\right) = \$2433.3528.$$

Your annual fuel cost is about $2400 or $2500.

✓ **Checkpoint**   Help at *Math*.and**Y⊕U**.com

How many miles per gallon does each car get?

c. A 2011 Toyota Prius   **d.** A 2011 Infiniti EX35

Study Tip

Notice that listing the units, such as gallons per tank, helps determine the units of the answer.

The United States uses about 140 billion gallons of gasoline in a year. Many countries place high taxes on gasoline. Some people have suggested a tax of at least $1 per gallon in the United States.

Reading Large and Small Numbers

When numbers are too large or too small to be conveniently written in standard decimal notation, most calculators switch to *scientific* or *exponential notation.*

Exponential Notation

In **exponential notation,** numbers are written as *a* times a power of 10,

$$a \times 10^b$$

where *a* is at least 1 and less than 10, and *b* is an integer. Here are two examples.

Standard Decimal Notation	Exponential Notation
6,830,000,000	6.83×10^9
0.0000000000683	6.83×10^{-11}

EXAMPLE 3   Describing Large and Small Numbers

Describe the numbers in the article about bacteria.

It is estimated that 500 to 1000 species of bacteria live in the human digestive system, and a roughly similar number live on the skin. Bacteria cells are much smaller than human cells (typically 3×10^{-6} meter in length), and there are at least 10 times as many bacteria as human cells in the body $\left(\text{approximately } 10^{14} \text{ versus } 10^{13}\right)$. There are approximately 5×10^{30} bacteria on Earth.

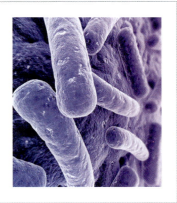

SOLUTION

Length of a bacteria cell:

3×10^{-6} meter $= 0.000003$ meter   3 millionths

Number of bacteria in a human:

10^{14} bacteria $= 100,000,000,000,000$ bacteria   100 trillion

Number of human cells in a human:

10^{13} cells $= 10,000,000,000,000$ cells   10 trillion

Number of bacteria on Earth:

5×10^{30} bacteria

$= 5,000,000,000,000,000,000,000,000,000,000$ bacteria

✓ Checkpoint   Help at *Math*.andY☺U.com

The diameter of a virus is less than 3×10^{-8} meter. Write this number in standard decimal notation and describe it in words. Which is larger, a bacteria or a virus?

EXAMPLE 4   Comparing Hard Drive Storage

You are buying a computer. For an additional $80, you can get 1 terabyte of storage instead of 100 gigabytes of storage.

a. How much more storage is that?

b. A typical movie uses 1 gigabyte of storage. How many movies could you store with 1 terabyte of storage?

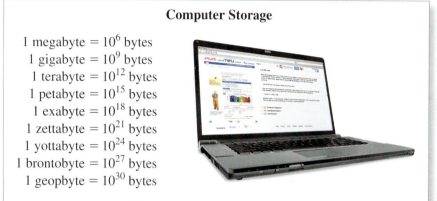

Computer Storage

$$1 \text{ megabyte} = 10^6 \text{ bytes}$$
$$1 \text{ gigabyte} = 10^9 \text{ bytes}$$
$$1 \text{ terabyte} = 10^{12} \text{ bytes}$$
$$1 \text{ petabyte} = 10^{15} \text{ bytes}$$
$$1 \text{ exabyte} = 10^{18} \text{ bytes}$$
$$1 \text{ zettabyte} = 10^{21} \text{ bytes}$$
$$1 \text{ yottabyte} = 10^{24} \text{ bytes}$$
$$1 \text{ brontobyte} = 10^{27} \text{ bytes}$$
$$1 \text{ geopbyte} = 10^{30} \text{ bytes}$$

SOLUTION

a. 100 gigabytes of storage is

$$100 \times 10^9 = 100{,}000{,}000{,}000 \text{ bytes.} \qquad \text{100 billion bytes}$$

1 terabyte of storage is

$$10^{12} = 1{,}000{,}000{,}000{,}000 \text{ bytes.} \qquad \text{1 trillion bytes}$$

1 trillion is 10 times more than 100 billion. So, the additional $80 will give you 10 times the amount of storage.

b. A terabyte is 1000 gigabytes. So, you could store about 1000 movies with 1 terabyte of storage.

✓ Checkpoint

Help at Math.andYOU.com

Use the information in the paragraph below to find the storage of an Apple iPad.

Hello Zettabytes

The so-called digital universe has grown to 800,000 petabytes, or 0.8 zettabyte. A petabyte is a million gigabytes. That is equivalent to all the information that can be stored on 50 billion Apple iPads.

Understanding Garbage In, Garbage Out

Humans tend to have unwarranted faith in numbers that are generated by calculators or by computers. When stating a number as an answer to a question, you should remember that the accuracy of the output is only as good as the accuracy of the input.

Study Tip

The best way to communicate measurements is to tell the reader how you obtained your numbers and to give some indication of the accuracy of the measurements.

Significant Digit Rule

A general rule for writing numbers is to not list more **significant digits** in the output than you can guarantee in the input.

Examples: The number 1.26 has 3 significant digits. The number 0.034 has 2 significant digits. The number 25,400 has 3 significant digits.

EXAMPLE 5 **Estimating the Accuracy of Measurements**

Rewrite each statement using the number of significant digits that seems appropriate in the context of the statement. Explain your reasoning.

a. In 2009, the population of Florida was 18,537,969 people.

b. The total amount of a loan is $3,546.28.

c. The area of an apartment is 1891 square feet.

SOLUTION

a. Taking a population census of a large region has many difficulties. Moreover, because 2009 is 9 years after an official census, this statement would be better phrased as, "In 2009, the population of Florida was about 18.5 million people." This number has 3 significant digits.

b. It is fair to assume that numbers dealing with banking or payroll are exact. So, this statement can remain as it is. All of the digits are significant.

c. Without knowing who is making this statement, it is difficult to determine the accuracy. Even so, specifying the area of an apartment exactly within 1 square foot seems unrealistic. Suppose the apartment was measured to be a rectangle that is 31 feet by 61 feet. This would yield an area of $31 \times 61 = 1891$ square feet. If the measurements were off by only 0.1 foot, the actual area could range between

$$30.9 \times 60.9 = 1881.81 \quad \text{and} \quad 31.1 \times 61.1 = 1900.21.$$

It would be better to say that the area of the apartment is about 1900 square feet. This number has 2 significant digits.

✓ **Checkpoint**

Help at *Math*.and**Y♥U**.com

Rewrite each statement using the number of significant digits that seems appropriate in the context of the statement. Explain your reasoning.

d. The weight of an athlete is 213.6 pounds.

e. The record time for a 100-meter dash is 9.58 seconds.

f. The distance between Earth and the Sun is 92,955,819 miles.

EXAMPLE 6   **Writing Significant Digits**

Here is a description of the Vietnam Veterans Memorial in Washington, D.C.

> The Wall consists of the East Wall and the West Wall. The triangular walls are each 246.75 feet long and 10.1 feet tall where they meet at a 125-degree angle. The West Wall points to the Lincoln Memorial, and the East Wall points to the Washington Monument. Each Wall consists of 72 panels: 70 with names and 2 very small, blank panels at each end. There are 58,267 names listed on the Memorial. Approximately 1200 of these names are listed as missing.

Use the information in this description to estimate (a) the area of the Wall and (b) the size of the lettering used on the Wall. In each case, write your answer with only as many significant digits as you think are reasonable.

SOLUTION

a. Find the area of each triangle.

$$\text{Area of each triangle} = \frac{1}{2}bh$$

$$= 0.5(246.75)(10.1)$$

Base = 246.75 ft
Height = 10.1 ft

$$= 1246.0875 \text{ ft}^2$$

Doubling this amount gives a total of 2492.175 square feet. Because the height is given with only 3 significant digits, it is best to round the answer to the same. So, the most accurate you can be with the given description is to say that the Wall has an area of about 2490 square feet.

b. Using an area of 2490 square feet, the average area per name is

$$\frac{58,267 \text{ names}}{2490 \text{ ft}^2} \approx 23.4 \text{ names per square foot.}$$

A square foot contains 144 square inches. So, you would have at most 6 square inches of space to carve each name. One reasonable solution is that the letters might be 1/2 inch high. This would leave 12 inches for the length of each name.

✓ **Checkpoint**

Help at ***Math*.and*YOU*.com**

Suppose you were planning the Wall. You want to list each name as

First Name   Middle Initial   Last Name.

How could you estimate the average number of characters used in the names? What is your estimate?

1.2 Exercises

DATA **Electricity Consumption** In Exercises 1–4, use the graph. Round your answer so it is reasonable for the context. *(See Example 1.)*

1. Estimate the electricity consumption of the United States.

2. Estimate the total electricity consumption of Japan, Russia, India, and Canada. Is it greater than or less than the electricity consumption of the United States?

3. In 2008, the population of the United States was about 304 million. Estimate the amount of electricity consumed per person.

Top Six Countries for Electricity Consumption in 2008

Electricity consumption (billions of kilowatt-hours) vs. Country (United States, China, Japan, Russia, India, Canada), y-axis from 0 to 4500 in increments of 500.

4. In 2008, the population of Canada was about 33 million. Estimate the amount of electricity consumed per person. Which country consumed more electricity per person, Canada or the United States?

Refrigerator Costs Two refrigerator models and their annual electricity consumptions are shown. In Exercises 5–8, assume the price of electricity is $0.1202 per kilowatt-hour (kWh). *(See Example 2.)*

5. You are buying a refrigerator.

   a. Estimate the annual electricity cost of each model.

   b. How much will you save in electric bills each year by buying the top-freezer model instead of the side-by-side model?

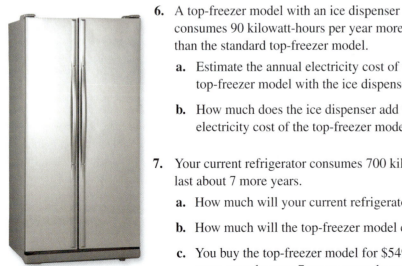

6. A top-freezer model with an ice dispenser consumes 90 kilowatt-hours per year more than the standard top-freezer model.

   a. Estimate the annual electricity cost of the top-freezer model with the ice dispenser.

   b. How much does the ice dispenser add to the annual electricity cost of the top-freezer model?

Top-freezer refrigerator
529 kWh/yr

7. Your current refrigerator consumes 700 kilowatt-hours per year and should last about 7 more years.

   a. How much will your current refrigerator cost you over the next 7 years?

   b. How much will the top-freezer model cost you over the same time period?

   c. You buy the top-freezer model for $549.99. Will the difference in electricity costs over the next 7 years cover the price of the new refrigerator? Explain.

Side-by-side refrigerator
634 kWh/yr

8. You own the top-freezer model. Suppose the price of electricity decreases to $0.12 per kilowatt-hour. How much will you save in electric bills each year?

Plankton In Exercises 9 and 10, write the length of the plankton in exponential notation. Then use the chart to classify the plankton. *(See Examples 3 and 4.)*

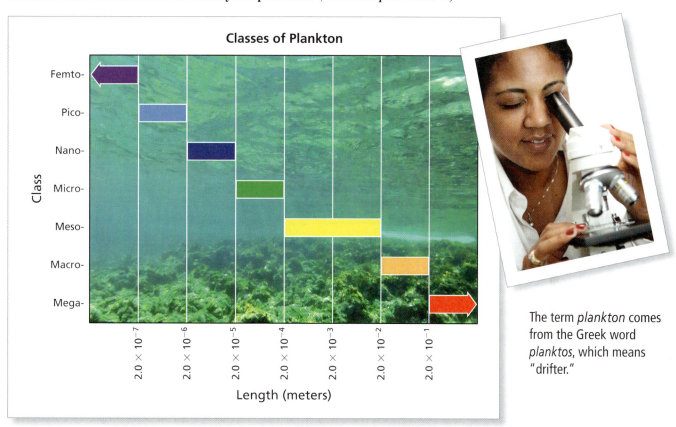

Classes of Plankton

The term *plankton* comes from the Greek word *planktos*, which means "drifter."

9. Copepod: 0.0012 meter

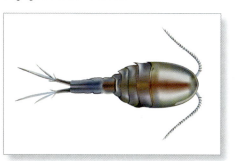

10. Golden algae: 0.0000091 meter

11. Marine Virus Marine viruses are a type of plankton. The length of a marine virus is less than 2.0×10^{-7} meter. Write this number in standard decimal notation and describe it in words. Use the chart above to classify the virus. *(See Examples 3 and 4.)*

12. Blue Whale A blue whale eats about 5000 pounds of plankton per day during a 120-day feeding period. How many pounds of plankton does the whale eat in one feeding period? Write your answer in standard decimal notation and in exponential notation. *(See Examples 3 and 4.)*

Significant Digits In Exercises 13–16, rewrite each statement using the number of significant digits that seems appropriate in the context of the statement. Explain your reasoning. *(See Example 5.)*

13. In 2009, the population of Texas was 24,782,302 people.

14. The price of a house is $239,900.

15. The weight of a bag of sugar is 5.01 pounds.

16. The number of calories in a sandwich is 297.

Home Improvement In Exercises 17 and 18, use the floor plan. Write your answer with only as many significant digits as you think are reasonable. *(See Example 6.)*

17. You are carpeting the living room.

   a. How many square feet of carpet should you buy?

   b. The cost of the carpet including installation is $3.49 per square foot. How much does it cost to carpet the living room?

18. You are replacing the linoleum in the kitchen.

   a. How many square feet of linoleum should you buy? Should you include the area of the two counters in your estimate? Explain.

   b. The cost of the linoleum including installation is $1.25 per square foot. How much does it cost to replace the linoleum in the kitchen?

▶ Extending Concepts

Grocery Bill In Exercises 19 and 20, use rounding to decide whether the total is reasonable. Explain your reasoning.

19.

Bananas	$2.79
Water	$9.89
Pizza	$5.99
Cereal	$3.08
Lettuce	$3.39
Tofu	$6.48
	$31.62

20.

Oranges	$1.89
Pie	$3.49
Ice cream	$3.22
Carrots	$2.79
Dressing	$2.96
Ketchup	$2.46
	$23.11

Maps In Exercises 21–25, use a ruler and the scale on the map of Wyoming.

21. Estimate the distance from Cheyenne to Gillette.

22. You are planning a trip from Riverton to Gillette. You want to stop in Douglas along the way. Estimate the total distance of the trip.

23. Your car can travel 274 miles on a tank of gas. Each tank contains 12.7 gallons. Estimate the fuel cost for a trip from Laramie to Evanston. Use a gas price of $2.92 per gallon.

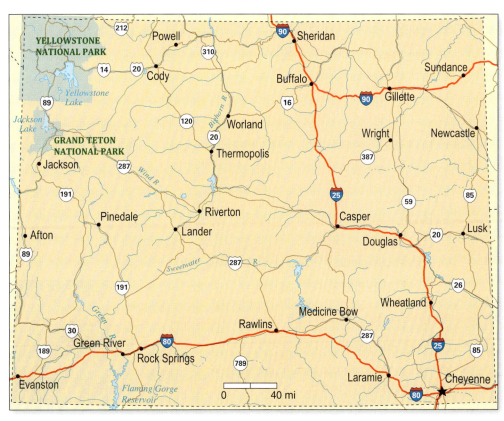

24. Estimate the area of the portion of Yellowstone National Park that is in Wyoming.

25. Describe different ways to find the distance between two cities.

1.1–1.2 Quiz

Patio **You are expanding a patio as shown. In Exercises 1 and 2, use the diagram.**

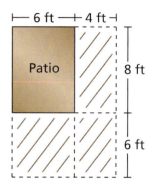

1. Which of the following keystroke sequences is better for finding the area that you are adding to the patio? Explain your reasoning.

a.

b.

2. You are laying 1-foot by 1-foot tiles. You estimate that it will take you 1 hour to lay 45 tiles. Estimate how long it will take you to tile the area that you are adding to the patio.

Population Density **In Exercises 3–6, use the information given in the table.**

3. Population density is the number of people per unit of area.

   a. Use the total area of Canada to estimate the country's population density.

   b. Land area is equal to total area minus water area. Use the land area of Canada to estimate the country's population density.

   c. Do you think population density should be defined in terms of the total area of a country or in terms of the land area? Explain your reasoning.

4. Excluding Canada's three northern territories, the population of Canada is about 34.0 million people, and the land area of Canada is about 5.5 million square kilometers.

   a. Estimate the population density of Canada, excluding the three northern territories.

   b. How does excluding the territories change your estimate of the population density? Explain your reasoning.

5. The population of Toronto is about 2.5 million people, and the land area of Toronto is about 630 square kilometers.

   a. Estimate the population density of Toronto.

   b. How does your estimate for Toronto compare to your estimates for Canada?

Canada	
Population (2010)	34,108,800
Total area	9,984,670 km²
Water area	891,163 km²

6. Do you think that population density is an accurate measure of a country's population distribution? Explain your reasoning.

Math & Lunar Eclipses

PROJECT: Finding Patterns in Nature

Total Eclipse
Partial Eclipse

Cross section
of penumbra

Cross section
of umbra

Sun

1. Use the *Lunar Eclipse Computer** at *Math.andYou.com* to find information about the next total eclipse of the moon in your area.

Eclipse:

2010 December 21 (Total) ▾

City or Town Name:

Chicago

State or Territory:

Illinois ▾

⁞ Get Data

Begin by entering your location information in the calculator. Then, press "Get Data" to obtain the information, as shown below.

			h m	Moon's Azimuth °	Moon's Altitude °
Moonrise	2010 Dec 20		15:51	56.9	----
Moon enters penumbra	2010 Dec 20		23:27.7	169.3	71.5
Moon enters umbra	2010 Dec 21		00:32.3	212.5	69.1
Moon enters totality	2010 Dec 21		01:40.4	241.3	59.9
Middle of eclipse	2010 Dec 21		02:17.0	251.2	53.8
Moon leaves totality	2010 Dec 21		02:53.6	259.1	47.4
Moon leaves umbra	2010 Dec 21		04:01.7	271.0	35.1
Moon leaves penumbra	2010 Dec 21		05:06.1	280.6	23.5
Moonset	2010 Dec 21		07:32	302.1	----

2. How long does a total lunar eclipse last?

3. How often does a total lunar eclipse occur? Explain your reasoning.

4. How has the time between lunar eclipses or between full moons influenced life on Earth? Include a discussion of tides, folklore, calendars, literature, and language.

**Provided by U.S. Naval Observatory

1.3 Using Percent

▶ Understand and find a percent of a number.

▶ Determine what percent one number is of another number.

▶ Use percent to represent change.

Finding a Percent of a Number

"Cent" implies 100, as in the word *century*. A **percent** is the number of parts per one hundred.

$$25\% = 0.25 = \frac{25}{100}$$

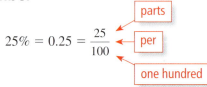

parts

per

one hundred

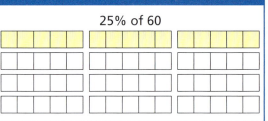

Study Tip

To rewrite a percent as a decimal, move the decimal point 2 places to the left.

$$25\% = 0.25$$

Finding a Percent of a Number

To find a percent of a number, write the percent as a decimal and multiply by the number.

Example: 25% of 60

$$0.25 \times 60 = 15$$

25% of 60

EXAMPLE 1　**Finding a Percent of a Number**

The map shows the percent of land in each state that is owned by the federal government. The land area of Nevada is 109,826 square miles. How many square miles of Nevada are owned by the federal government?

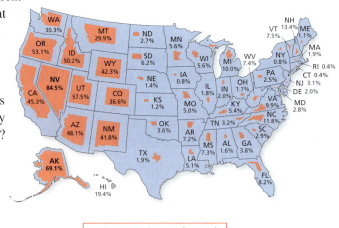

Percent of Land Owned by the Federal Government

1. Nevada　　84.5%
2. Alaska　　69.1%
3. Utah　　57.5%
4. Oregon　　53.1%
5. Idaho　　50.2%
6. Arizona　　48.1%
7. California　　45.3%
8. Wyoming　　42.3%
9. New Mexico　　41.8%
10. Colorado　　36.6%

SOLUTION

Write 84.5% as a decimal.

$$0.845 \times 109,826 = 92,802.97$$

The federal government owns about 92,800 square miles of Nevada.

✔ **Checkpoint**　　　　　　Help at *Math*.and**YOU**.com

The land area of Alaska is 571,951 square miles. How many square miles of Alaska are owned by the federal government? Does the federal government own more land in Alaska or in Nevada?

EXAMPLE 2   **Finding Percents**

There are about 117 million households in the United States. Use the circle graph (also called a pie chart or pie graph) to estimate the number of pet dogs in the United States.

Dog Ownership in U.S. Households

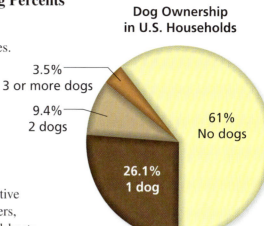

3.5%
3 or more dogs

9.4%
2 dogs

61%
No dogs

26.1%
1 dog

SOLUTION

Problems like this, with repetitive steps and sums of large numbers, lend themselves well to spreadsheets.

$117{,}000{,}000 \times 0.094 \times 2 = 21{,}996{,}000$

	A Households	B Number of Dogs	C Percent	D Dogs
1	Households	Number of Dogs	Percent	Dogs
2	117,000,000	0	61.0%	0
3	117,000,000	1	26.1%	30,537,000
4	117,000,000	2	9.4%	21,996,000
5	117,000,000	5	3.5%	20,475,000
6			100.0%	73,008,000
7				
8				

Total percent

Total number of dogs

Notice that it was assumed that households with 3 or more dogs have an average of 5 dogs. This assumption affects the final answer. With this assumption and with the context of the problem, you cannot assume much accuracy in the answer. Perhaps an answer of "about 70 million dogs" is reasonable.

✔ **Checkpoint**

Help at *Math*.and**Y♥U**.com

There are about 117 million households in the United States. Thirty-three percent of the households own at least one cat. On average, these households have 2.45 cats. Estimate the number of pet cats in the United States.

Cat Ownership in U.S. Households

67%
No cats

33%
At least one cat

Using Percent to Compare Two Numbers

Comparing Two Numbers Using Percent

To find what percent a part is of a base, divide the part by the base.

16 is 20% or $\frac{1}{5}$ of 80

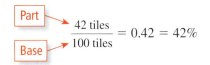

Example:

Part $\rightarrow$
$\frac{16}{80} = 0.2 = 20\%$
Base $\rightarrow$

EXAMPLE 3   Comparing Two Numbers Using Percent

The number of tiles of each letter in SCRABBLE depends on the language. There are 100 letters in the traditional version of SCRABBLE. The distribution of the letters in the English version is shown at the right. In this version, what percent of the tiles are vowels?

SOLUTION

There are 9 A's, 12 E's, 9 I's, 8 O's, and 4 U's. That is a total of 42 vowels.

Part $\rightarrow$
$\frac{42 \text{ tiles}}{100 \text{ tiles}} = 0.42 = 42\%$
Base $\rightarrow$

42% of the tiles in the English version of SCRABBLE are vowels.

✓ **Checkpoint**   Help at

The Spanish version of SCRABBLE has the following tiles.

2 blank tiles, 11 A's, 3 B's, 4 C's, 4 D's, 11 E's, 2 F's, 2 G's, 2 H's,

6 I's, 2 J's, 1 K, 4 L's, 1 LL, 3 M's, 5 N's, 1 Ñ, 8 O's, 2 P's, 1 Q,

4 R's, 1 RR, 7 S's, 4 T's, 6 U's, 2 V's, 1 W, 1 X, 1 Y, and 1 Z

What percent of these tiles are vowels?

| EXAMPLE 4 | **Comparing Two Numbers Using Percent** |

A person's body fat percentage is the weight of the person's fat divided by the person's weight. The following ideal body fat percentages are from the American Council on Exercise.

Description	Women	Men
Athletes	14–20%	6–13%
Fitness	21–24%	14–17%
Average	25–31%	18–24%
Obese	32%+	25%+

Find the body fat percentage for the following people.

a. A man who weighs 210 pounds with 44 pounds of fat

b. A woman who weighs 145 pounds with 38 pounds of fat

SOLUTION

a. A man who weighs 210 pounds with 44 pounds of fat:

Fat weight → Total weight
$$\frac{44 \text{ lb}}{210 \text{ lb}} \approx 0.2095 \approx 20.1\%$$

The man's body fat percentage is about 20%.

b. A woman who weighs 145 pounds with 38 pounds of fat:

Fat weight → Total weight
$$\frac{38 \text{ lb}}{145 \text{ lb}} \approx 0.2621 \approx 26.2\%$$

The woman's body fat percentage is about 26%.

> **Study Tip**
>
> When you find a percent by dividing, the numerator and the denominator should have the same units.

✓ Checkpoint

Help at *Math*.and*YOU*.com

The circle graph represents a typical 180-pound man. Find the percent of each type of material.

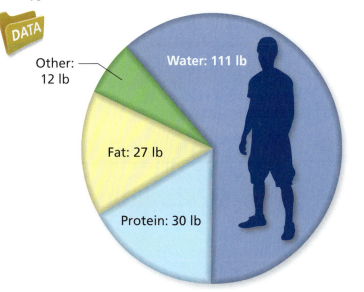

DATA

Other: 12 lb
Water: 111 lb
Fat: 27 lb
Protein: 30 lb

Study Tip

In this section, we will only look at quantities that increase over time. In Chapter 4, we will look at quantities that increase or decrease over time.

Using Percent to Represent Change

In the first four examples of this section, percent is used to represent a part of a whole.

Example 1: Federally owned land is part of the *total* state land.

Example 2: Households with dogs are part of *all* households.

Example 3: Tiles with vowels are part of *all* tiles in SCRABBLE.

Example 4: Fat weight is part of the *total* weight of a person.

The last two examples in this section look at another common use of percent—that is, to describe the amount that a quantity changes over time.

EXAMPLE 5   Using Percent to Describe an Increase

Reword the statement so it uses a percent that is greater than 100%.

> *Life expectancy at birth in the United States in 1901 was 49 years. At the end of the century, it was 77 years, an increase of more than 50%.*

SOLUTION

Compare 77 years to 49 years using division.

Increased amount
Original amount

$$\frac{77 \text{ years}}{49 \text{ years}} \approx 1.57 = 157\%$$

So, one way to rephrase the statement is as follows.

> *Life expectancy at birth in the United States in 1901 was 49 years. The life expectancy of 77 years in 2000 was 157% of the 1901 life expectancy.*

✔ **Checkpoint**   Help at *Math*.and**Y�**U.com

Use the line graph to write two paragraphs describing the change in women's life expectancy at birth from 1900 to 2000.

a. In the first paragraph, describe the change in life expectancy.

b. In the second paragraph, compare the newer life expectancy to the older one.

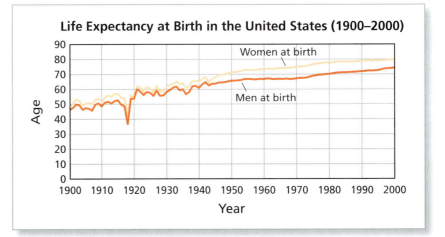

EXAMPLE 6   **Using Percent to Describe an Increase**

The line graph shows the daily high price of gold per ounce. Use the graph to describe the change in the high price of gold (a) from 1980 to 2010 and (b) from 2000 to 2010.

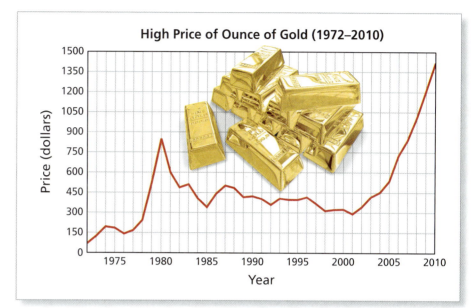

High Price of Ounce of Gold (1972–2010)

SOLUTION

a. Suppose you bought gold in 1980 for $800 per ounce and sold it in 2010 for $1200 per ounce.

$$\frac{\$1200 \quad \text{2010 price}}{\$800 \quad \text{1980 price}} = 1.5 = 150\%$$

The 2010 price was 150% of the 1980 price. Another way of saying this is that your investment increased by 50%.

b. Suppose you bought gold in 2000 for $250 per ounce and sold it in 2010 for $1250 per ounce.

$$\frac{\$1250 \quad \text{2010 price}}{\$250 \quad \text{2000 price}} = 5.0 = 500\%$$

The 2010 price was 500% of the 2000 price. Another way of saying this is that your investment increased by 400%. Be sure you see that the 500% is a result of comparing the two prices. If you compare only the *increase* in price to the original price, you obtain 400%.

$$\frac{\$1000 \quad \text{Price increase}}{\$250 \quad \text{2000 price}} = 4.0 = 400\%$$

All the gold that has been mined in human history would fit into a cube that is 65.5 feet on each side.

65.5 ft
65.5 ft
65.5 ft

✓ **Checkpoint**

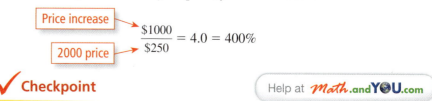
Help at *Math*.and**Y©U**.com

Use the graph to describe the change in the high price of gold from 1975 to 2010.

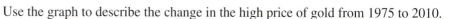

1.3 Exercises

Stem Cell Research **The bar graph shows the results of a poll of adult Americans about federal government funding of stem cell research. In Exercises 1–4, use the bar graph.** *(See Example 1.)*

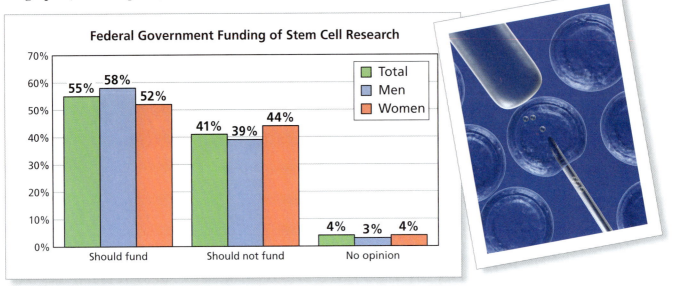

Federal Government Funding of Stem Cell Research

Legend: Total (green), Men (blue), Women (orange)

Should fund: 55%, 58%, 52%
Should not fund: 41%, 39%, 44%
No opinion: 4%, 3%, 4%

1. There are about 117,000,000 adult American men. About how many of these men think the federal government should fund stem cell research?

2. There are about 123,000,000 adult American women. About how many of these women think the federal government should fund stem cell research?

3. Describe two ways to find how many American adults have no opinion about federal government funding of stem cell research. Do both ways result in the same answer? Explain.

4. Are the results for men significantly different than the results for women? Explain.

Stem Cell Research **The circle graph shows the results of a poll of adult Americans about the moral acceptability of stem cell research. In Exercises 5–8, use the circle graph.** *(See Example 2.)*

5. In a city, 412,230 adults think that stem cell research is morally wrong. Estimate the population of the city.

6. How many adults from the city in Exercise 5 think stem cell research is morally acceptable?

7. Conduct a poll on stem cell research by asking at least 30 people for their opinion. Compare your results to the bar graph above and the circle graph.

8. Summarize the results of your poll in Exercise 7 graphically. Explain your choice of graph.

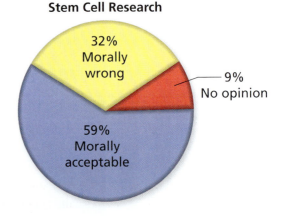

Stem Cell Research

32% Morally wrong
9% No opinion
59% Morally acceptable

DATA **Skeletal System** In Exercises 9–14, use the information below. *(See Examples 3 and 4.)*

The stapes (pronounced stay-peas) bone is the smallest bone in the body. It is sometimes called the stirrup bone because of its shape. It is located in the middle ear and is about the size of a grain of rice.

Bones in the Human Body

- Head – 29
- Hands – 38
- Wrists – 16
- Trunk – 51
- Feet – 38
- Ankles – 14
- Shoulders and arms – 10
- Hips and legs – 10

9. Of the hand bones, 28 are finger bones (14 phalanges in each hand). What percent of the hand bones are finger bones?

Distal phalanges

Intermediate phalanges

Proximal phalanges

Metacarpals

Carpals

10. Of the head bones, 14 are facial bones. What percent of the head bones are facial bones?

11. What percent of the bones in the human body are in the hips and legs?

12. What percent of the bones in the human body are in the trunk?

13. The skeletal system is a complex work of nature, with the hands and feet being the most intricate. What percent of the bones are in the hands, wrists, feet, and ankles?

14. The longest bone in the body is the thigh bone (femur). The average adult male has a femur that is 48 centimeters long. Use estimation and percent to compare the length of the stapes to the length of the femur in an average adult male.

15. **Skeleton** The skeleton of a 160-pound man weighs about 30 pounds. What percent of the man's weight is his skeleton? *(See Examples 3 and 4.)*

16. **Weight Gain** The man in Exercise 15 gains 30 pounds. What percent of the man's new weight is his skeleton? *(See Examples 3 and 4.)*

Diamond Rings The graph shows the price ranges for five weights of diamond rings at a jewelry store. In Exercises 17–24, use the graph. *(See Examples 5 and 6.)*

Diamond Rings

Price (dollars) vs. Weight (carats)

- 0.30: $971
- 0.50: $1055
- 1.00: $4866
- 2.00: $9200
- 3.00: $13,304

17. The jewelry store has a diamond ring listed at $11,499. What is the carat weight of the diamond ring? Explain your reasoning.

18. What is the price range of a 1.00-carat diamond ring?

19. How much more does a 2.00-carat diamond ring cost than a 1.00-carat diamond ring?

20. Reword the statement so that it uses a percent that is greater than 200%.

A 0.50-carat diamond ring costs $2026, which is about 108.7% more than the cost of a 0.30-carat diamond ring.

21. Is it true that a 0.50-carat diamond is more than 60% larger than a 0.30-carat diamond? Explain your reasoning.

22. Use percent to describe the change in maximum price.
    a. from a 0.50-carat diamond ring to a 1.00-carat diamond ring
    b. from a 1.00-carat diamond ring to a 2.00-carat diamond ring
    c. from a 2.00-carat diamond ring to a 3.00-carat diamond ring

23. Use percent to describe the change in carat weight.
    a. from a 0.50-carat diamond ring to a 1.00-carat diamond ring
    b. from a 1.00-carat diamond ring to a 2.00-carat diamond ring
    c. from a 2.00-carat diamond ring to a 3.00-carat diamond ring

24. Describe the relationship between the carat weight and the price of a diamond ring.

▶ Extending Concepts

True or False In Exercises 25–28, decide whether the statement is true or false. Explain your reasoning.

25. You invest $1250 in the stock market and lose 20% during the first year. To get back to your original investment, your portfolio needs to gain 20% during the second year.

26. At work, you receive a cost-of-living adjustment (COLA) of 3% each year. Over a 5-year period, your pay will increase by 15%.

27. Your property taxes increase by 5% one year and decrease by 1% the next year. Overall, your property taxes are up by 4% over 2 years.

28. Everything in a store is on sale for 40% off. You have a coupon for 10% off the purchase price. When you go to the counter to pay, you will receive 50% off the original price.

The cost of living is the cost of maintaining a certain standard of living. It is usually tied to a cost-of-living index such as the consumer price index (CPI).

29. Savings Account You open a savings account that has a 2.0% simple annual interest rate. At the end of the first year, you have earned $30 in interest. No other transactions were posted to the account. What is the balance of the account at the end of the first year?

30. Political Platform As part of a political platform, a politician promises to cut spending by 110%. Is this possible? Explain your reasoning.

DATA **31. Test Scores** Four exams are used to determine your final grade in one of your classes. The bar graph shows your scores on the first three exams. What percent do you need on the last exam to receive an 85% in the class?

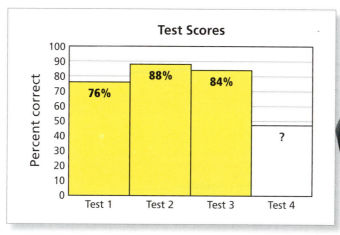

32. Profit Sharing Your company currently allocates 5% of the company profits each year to a profit sharing plan for the employees. The board of directors decides to increase the amount allocated by 10%. Does this mean the new policy is to allocate 5.5% or 15% to the profit sharing plan each year? Explain your reasoning.

1.4 Units & Conversions

▶ Use unit analysis to "balance" both sides of a formula.
▶ Convert within a given system of measure.
▶ Convert between different systems of measure.

Using Unit Analysis when Evaluating Formulas

Everything you measure has a number with some sort of **unit of measure** attached. The unit can be feet, pounds, dollars per hour, people, degrees, or any of countless other ways in which real-life objects are measured.

Study Tip

Many people have found that unit analysis is the most useful "trick" they were ever taught in math. Using it and understanding it can open your eyes to all sorts of math that appears mysterious without it.

Unit Analysis

Unit analysis is the process of changing from one unit of measure to another. Change occurs when units are multiplied or divided. Addition and subtraction always preserve units.

Examples:

$2 \text{ ft} + 3 \text{ ft} = 5 \text{ ft}$ Addition preserves units.

$8 \text{ oz} - 2 \text{ oz} = 6 \text{ oz}$ Subtraction preserves units.

$\left(18 \frac{\$}{\cancel{hr}}\right)(10 \cancel{hr}) = \$180$ Multiplication changes units.

$\dfrac{320 \text{ mi}}{8 \text{ gal}} = 40 \dfrac{\text{mi}}{\text{gal}}$ Division changes units.

EXAMPLE 1   Using Unit Analysis

In 2010, the average hourly billing rate for attorneys in the United States was $413 per hour for litigation, $302 per hour for labor or employee cases, and $294 per hour for real estate cases.

a. What would you expect to pay for 1200 hours of litigation?

b. You paid $2850 for 9.5 hours of legal work on a mortgage. What hourly rate did your attorney charge?

c. Your attorney represented you for 25% contingency on a personal injury case in which you were awarded $600,000. Your attorney spent 1000 hours on the case. What was the hourly billing rate?

SOLUTION

a. $\left(413 \frac{\$}{\cancel{hr}}\right)(1200 \cancel{hr}) = \$495{,}600$ **b.** $\dfrac{\$2850}{9.5 \text{ hr}} = 300 \dfrac{\$}{\text{hr}}$

c. $\dfrac{25\% \text{ of } \$600{,}000}{1000 \text{ hr}} = \dfrac{\$150{,}000}{1000 \text{ hr}} = 150 \dfrac{\$}{\text{hr}}$

✓ **Checkpoint**   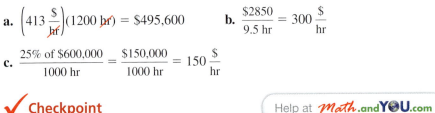   Help at *Math*.and**Y©U**.com

An attorney at a law firm billed an average of 55 hours in litigation fees per week for a year. Estimate the amount of revenue the attorney brought into the law firm during the year.

EXAMPLE 2　**Using Unit Analysis**

It is estimated that about 125 species of birds and 60 species of mammals have become extinct since 1600. There are approximately 1000 species of birds and mammals that are currently facing extinction. Including all types of plant and animal species, the number of endangered species is around 20,000.

a. Estimate the number of species of birds that will become extinct in the next 100 years.

b. Estimate the number of years in which 40 more species of mammals will have become extinct.

The Endangered Species Coalition works to protect endangered species and their habitats.

The giant panda was listed as endangered in 1990.

SOLUTION

This estimate is, of course, quite primitive because it is based only on the rate of extinction during the 410 years from 1600 to 2010.

a. $\left(\dfrac{125\ \text{species}}{410\ \text{years}}\right)(100\ \text{years}) \approx 30.49\ \text{species}$

So, you could estimate that 30 species of birds will become extinct during the next 100 years.

b. This is the most difficult type of question in unit analysis. You need to perform an operation so that the resulting units are years. To do this, you can use the *Invert and Multiply Rule* for dividing by a fraction.

$$\dfrac{40\ \text{species}}{\dfrac{60\ \text{species}}{410\ \text{years}}} = (40\ \text{species})\left(\dfrac{410\ \text{years}}{60\ \text{species}}\right) \approx 273.33\ \text{years}$$

Invert the denominator and multiply.

So, you could estimate that it might take about 270 years for an additional 40 species of mammals to become extinct.

Here is a simple example of dividing by a fraction. How many one-eighth inches are in 2 inches?

$$\dfrac{2\ \text{in.}}{\dfrac{1}{8}\ \text{in.}} = 2\left(\dfrac{8}{1}\right) = 16$$

Invert the denominator and multiply.

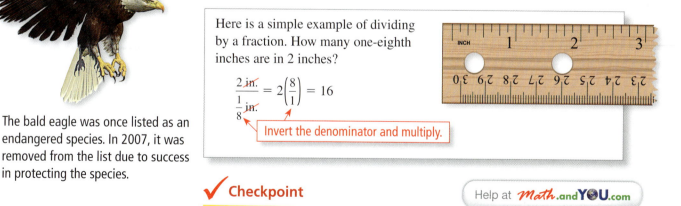

The bald eagle was once listed as an endangered species. In 2007, it was removed from the list due to success in protecting the species.

✓ Checkpoint

Help at *Math*.and**YOU**.com

c. Estimate the number of species of mammals that will become extinct in the next 100 years.

d. Estimate the number of years in which 40 more species of birds will have become extinct.

Converting Units within a Given System of Measure

Converting Units

To **convert** from one type of unit to another, multiply by a convenient form of the number one.

One

Example: 2 ft = 2 ft $\left(\dfrac{12\ \text{in.}}{1\ \text{ft}}\right)$ = 24 in.

Common Unit Conversions

	U.S. Customary	Metric
Distance:	1 mi = 5280 ft	1 km = 1000 m
	1 yd = 3 ft	1 m = 100 cm
	1 ft = 12 in.	1 cm = 10 mm
Volume/Capacity:	1 gal = 4 qt	1 L = 1000 mL
	1 qt = 4 cups	
	1 cup = 8 fl oz	
Weight/Mass:	1 lb = 16 oz	1 kg = 1000 g

EXAMPLE 3   Converting Units

How many seconds are in a calendar year of 365 days?

SOLUTION

At first, this question might seem overwhelming. But you can find the answer by simply multiplying by three carefully chosen forms of the number one.

One

$$365 \text{ days} = (365 \text{ days})\left(\frac{24 \text{ hr}}{1 \text{ day}}\right)\left(\frac{60 \text{ min}}{1 \text{ hr}}\right)\left(\frac{60 \text{ sec}}{1 \text{ min}}\right)$$

$$= 31{,}536{,}000 \text{ sec}$$

There are 31,536,000 seconds in a calendar year of 365 days.

Study Tip

A day is the length of time it takes Earth to complete one revolution about its axis. The fact that a solar year is not a whole number of days has perplexed calendar makers for many centuries.

✓ **Checkpoint**   Help at *Math*.and**YOU**.com

A *solar year* is the time it takes Earth to complete one revolution of its orbit about the Sun. A solar year is 365.24218967 days. How many more seconds are in a solar year than in a calendar year?

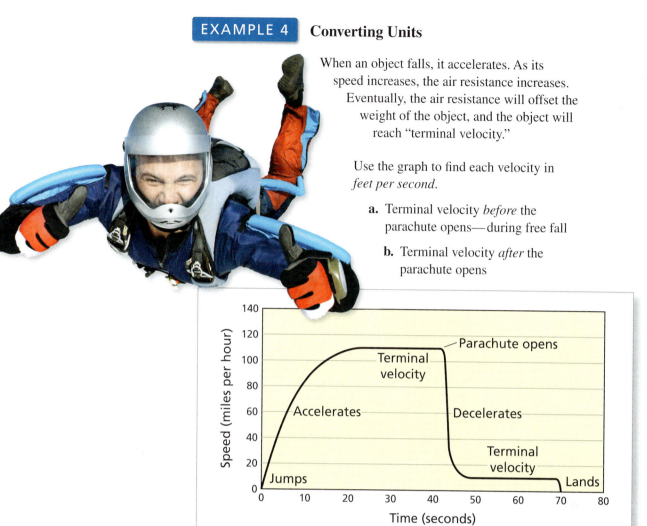

EXAMPLE 4   **Converting Units**

When an object falls, it accelerates. As its speed increases, the air resistance increases. Eventually, the air resistance will offset the weight of the object, and the object will reach "terminal velocity."

Use the graph to find each velocity in *feet per second*.

a. Terminal velocity *before* the parachute opens—during free fall

b. Terminal velocity *after* the parachute opens

SOLUTION

a. From the graph, the free-fall terminal velocity is about 110 miles per hour.

$$110\,\frac{\text{mi}}{\text{hr}} = \left(110\,\frac{\text{mi}}{\text{hr}}\right)\left(\frac{5280\text{ ft}}{1\text{ mi}}\right)\left(\frac{1\text{ hr}}{60\text{ min}}\right)\left(\frac{1\text{ min}}{60\text{ sec}}\right) \approx 161.33\,\frac{\text{ft}}{\text{sec}}$$

The free-fall terminal velocity is about 161 feet per second.

b. The parachute terminal velocity is about 10 miles per hour.

$$10\,\frac{\text{mi}}{\text{hr}} = \left(10\,\frac{\text{mi}}{\text{hr}}\right)\left(\frac{5280\text{ ft}}{1\text{ mi}}\right)\left(\frac{1\text{ hr}}{60\text{ min}}\right)\left(\frac{1\text{ min}}{60\text{ sec}}\right) \approx 14.66\,\frac{\text{ft}}{\text{sec}}$$

The parachute terminal velocity is about 15 feet per second.

Study Tip

Notice the difference in the "form of one" used in the solutions.

Example 3: $\dfrac{60\text{ sec}}{1\text{ min}}$

Example 4: $\dfrac{1\text{ min}}{60\text{ sec}}$

The form used depends on the units you are trying to obtain in your answer.

✓ **Checkpoint**   Help at

c. Estimate the time the parachutist spends in free fall. Explain your reasoning.

d. Estimate the time the parachutist spends with the parachute open. Explain your reasoning.

Converting Units Between Different Systems

Study Tip
Quick approximations:

$1 \text{ mi} \approx \frac{8}{5} \text{ km}$

$F \approx 2C + 30$

$1 \text{ gal} \approx 4 \text{ L}$

$1 \text{ lb} \approx \frac{2}{5} \text{ kg}$

Common Unit Conversions

Distance:

$1 \text{ mi} \approx 1.61 \text{ km}$

$1 \text{ in.} = 2.54 \text{ cm}$

Volume/Capacity:

$1 \text{ gal} \approx 3.79 \text{ L}$

Temperature:

$F = \frac{9}{5}C + 32$

Weight/Mass:

$1 \text{ lb} \approx 0.45 \text{ kg}$

EXAMPLE 5   Converting Units

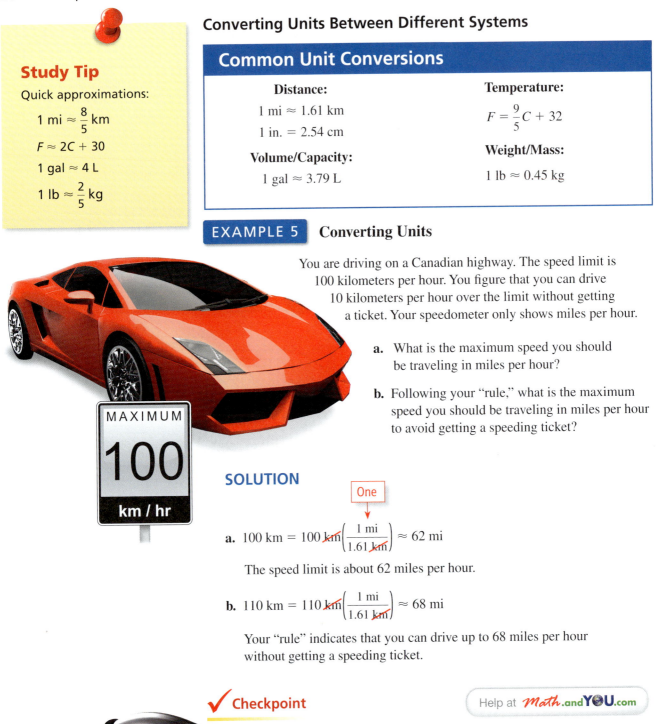

You are driving on a Canadian highway. The speed limit is 100 kilometers per hour. You figure that you can drive 10 kilometers per hour over the limit without getting a ticket. Your speedometer only shows miles per hour.

a. What is the maximum speed you should be traveling in miles per hour?

b. Following your "rule," what is the maximum speed you should be traveling in miles per hour to avoid getting a speeding ticket?

SOLUTION

One

a. $100 \text{ km} = 100 \text{ km} \left(\dfrac{1 \text{ mi}}{1.61 \text{ km}} \right) \approx 62 \text{ mi}$

The speed limit is about 62 miles per hour.

b. $110 \text{ km} = 110 \text{ km} \left(\dfrac{1 \text{ mi}}{1.61 \text{ km}} \right) \approx 68 \text{ mi}$

Your "rule" indicates that you can drive up to 68 miles per hour without getting a speeding ticket.

✓ Checkpoint

Help at *Math*.and**Y☺U**.com

While in Canada, you stop at a gas station. You have heard that gas prices in Canada are considerably more expensive than in the United States.

c. From the sign at the left, does this seem to be true?

d. To answer part (c), what conversion should you consider in addition to the conversion between gallons and liters?

EXAMPLE 6   **Converting Temperature Units**

You remember hearing that there is one temperature that has the same degree measure in the Fahrenheit and Celsius scales. You cannot remember what the temperature is. How can you find it?

SOLUTION

One way is to use algebra. You could let $C = F$ and solve for F in the equation

$$F = \frac{9}{5}F + 32.$$

$$0 = \frac{9}{5}F - F + 32 \qquad \text{Subtract } F \text{ from each side.}$$

$$0 = \frac{4}{5}F + 32 \qquad \text{Combine like terms.}$$

$$-32 = \frac{4}{5}F \qquad \text{Subtract 32 from each side.}$$

$$-160 = 4F \qquad \text{Multiply each side by 5.}$$

$$-40 = F \qquad \text{Divide each side by 4.}$$

But you may have forgotten how to solve this equation. Another instructive way to find the temperature is to use a spreadsheet.

	A	B	C
1	**Celsius**	**Fahrenheit**	
2	100	212	
3	90	194	(9/5)*A2+32
4	80	176	
5	70	158	
6	60	140	
7	50	122	
8	40	104	
9	30	86	
10	20	68	
11	10	50	
12	0	32	
13	-10	14	
14	-20	-4	
15	-30	-22	
16	-40	-40	
17	-50	-58	

The temperature -40 degrees is the same on both scales.

✓ **Checkpoint**

Help at ***Math*.andYOU.com**

You are staying at a hotel in Canada. Your room feels cold, and you notice that the temperature is set at 20°C.

a. What is the room temperature in degrees Fahrenheit?

b. What should you set the temperature at to obtain a temperature of 77°F?

1.4 Exercises

Online Shopping The table shows an online store's shipping rates. In Exercises 1–6, use the table. *(See Examples 1 and 2.)*

Delivery Time	Shipping Rate
3–5 days	$0.49/lb
2 days	$0.89/lb
1 day	$1.89/lb

1. You order a 5-pound textbook with 3–5 day shipping. What is the shipping fee?

2. You order a 330-pound weight set with 2-day shipping. What is the shipping fee?

3. You order a 26-pound television. Suppose you use 3–5 day shipping instead of 2-day shipping. How much will you save?

4. You order a 119-pound couch. Suppose you use 1-day shipping instead of 3–5 day shipping. How much more will you pay?

5. You wait until the last minute to buy a 159-pound treadmill for a birthday present. The online price of the treadmill is $110 less than the price of the same treadmill at a local fitness store. You need the treadmill by tomorrow, so you will have to pay for 1-day shipping.

  a. Estimate the shipping fee for the treadmill.

  b. Will it cost more to order the treadmill online with 1-day shipping or to buy the treadmill at the local fitness store?

  c. Does your answer to part (b) change if you had an extra day to place your order? What if you had an extra week?

6. You recently purchased a pair of headphones online. The shipping fee was $12.24. You believe that you were overcharged for 3–5 day shipping, but the weight of the headphones is not listed on the receipt.

  a. How much do the headphones have to weigh for the fee to be correct? Is the weight reasonable?

  b. If the weight is not reasonable, write an email to the store explaining the error and asking for a refund.

Travel Time The road sign shows your distance to several nearby cities in miles.
In Exercises 7 and 8, use the road sign. *(See Example 3.)*

7. You are driving at 45 miles per hour. Which of the following expressions should you use to determine how long it will take to get to Ventura? How long will it take?

a. $54 \text{ mi} \times \dfrac{45 \text{ mi}}{1 \text{ hr}} \times \dfrac{60 \text{ min}}{1 \text{ hr}} = $

b. $54 \text{ mi} \times \dfrac{1 \text{ hr}}{45 \text{ mi}} \times \dfrac{60 \text{ min}}{1 \text{ hr}} = $

c. $54 \text{ mi} \times \dfrac{45 \text{ mi}}{1 \text{ hr}} \times \dfrac{1 \text{ hr}}{60 \text{ min}} = $

d. $54 \text{ mi} \times \dfrac{1 \text{ hr}}{45 \text{ mi}} \times \dfrac{1 \text{ hr}}{60 \text{ min}} = $

Santa Barbara   26
Ventura   54
Los Angeles   122

8. You are driving at 65 miles per hour. How long will it take to get to Los Angeles?

A human heart weighs about 10 ounces. On average, it beats about 70 times per minute and pumps about 2.5 fluid ounces of blood with each beat.

Human Heart In Exercises 9–12, use the information. *(See Examples 3 and 4.)*

9. How much does a heart weigh in pounds?

10. A human brain weighs about 3 pounds. How many ounces heavier is a brain than a heart?

11. How many quarts of blood does a heart pump in 1 minute?

12. How many gallons of blood does a heart pump in 1 day?

Heart Rate A patient's heart rate can be determined from an electrocardiogram (EKG). The time between two peaks on the EKG represents one heartbeat. In Exercises 13 and 14, use the EKG. *(See Example 4.)*

13. Ringo Starr sang a song that had the words "In a heartbeat, I'll be by your side." What is the length of one heartbeat?

14. What is the heart rate of the person whose EKG is shown? Is this heart rate normal?

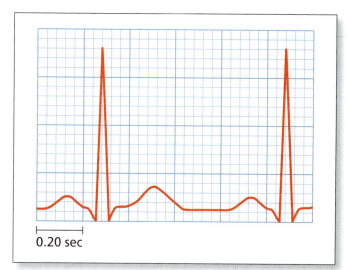

0.20 sec

Seven Summits The map shows the height of the highest mountain on each of Earth's seven continents. Climbing all seven mountains, also known as the Seven Summits, is considered a challenge in mountaineering. In Exercises 15–20, use the information on the map. *(See Examples 5 and 6.)*

Mt. McKinley
6194 m

Mt. Elbrus
5.642 km

Mt. Everest
9676 yd

Mt. Carstensz
16,024 ft

Mt. Kilimanjaro
5.895 km

Mt. Aconcagua
7614 yd

Mt. Vinson
16,067 ft

15. Determine the height of Mount McKinley in feet.

16. Order the heights of the Seven Summits from highest to lowest.

17. You are climbing Mount Everest. After 1 hour, you have climbed 183 meters. Find your rate of ascent in feet per minute.

18. The original Seven Summits included Mount Kosciuszko in mainland Australia instead of Mount Carstensz. Mount Kosciuszko is about 2228 meters high. Is it higher or lower than Mount Carstensz? Explain.

19. At the summit of Mount Everest, water boils at 69°C.

   a. Convert this temperature to degrees Fahrenheit.

   b. Could you boil a potato at the summit until it is cooked? Explain.

20. For every 1000 feet of elevation, the air temperature drops about 6.5°C. Assume the temperature at the foot of Mount Kilimanjaro is 23°C. Estimate the temperature at the summit in degrees Fahrenheit.

The Seven Summits on the map are also known as the Messner List, named after mountaineer Reinhold Messner.

▶ Extending Concepts

Exchange Rates The table shows a sample of currency exchange rates for November 2010. In Exercises 21–24, use the table.

Currency	From U.S. Dollars to Foreign Currency	From Foreign Currency to U.S. Dollars
British pound	$1 = £0.6325	£1 = $1.5813
European euro	$1 = €0.7465	€1 = $1.3396
Canadian dollar	$1 = C$1.0239	C$1 = $0.9767
Mexican peso	$1 = Mex$12.4823	Mex$1 = $0.0801
Japanese yen	$1 = ¥83.065	¥1 = $0.012
Indian rupee	$1 = ₹45.81	₹1 = $0.0218

21. You order a product from a British website. The list price is £223. How much does the product cost in U.S. dollars?

22. In Germany, an airline charges an extra €10 for each additional kilogram of carry-on luggage over 10 kilograms. Your carry-on luggage weighs 26.5 pounds. How much extra can you expect to pay in U.S. dollars?

23. A gas station in Mexico City charges Mex$7.73 for a liter of gasoline. Your car's 23-gallon tank is almost empty. How much will it cost in U.S. dollars to fill the tank?

24. You are in the United States and plan to travel to several other countries. You have $3000 in cash.

    a. Your first destination is Canada. Convert your cash to Canadian dollars.

    b. You leave Canada for Great Britain with C$2800. Convert this to British pounds.

    c. You depart from Great Britain for Spain with 1400 British pounds. Convert this to European euros.

    d. Your next stop is India. After your time in Spain, you have 1000 euros. Convert this to Indian rupees.

    e. From India, you leave for Japan. You have 50,000 rupees after your time in India. Convert this to Japanese yen.

    f. You return to the United States with 80,000 Japanese yen. Convert this to U.S. dollars. How much money did you spend total?

1.3–1.4   Quiz

Disabilities The circle graph shows the age categories of people with disabilities in the United States. In Exercises 1–4, use the circle graph.

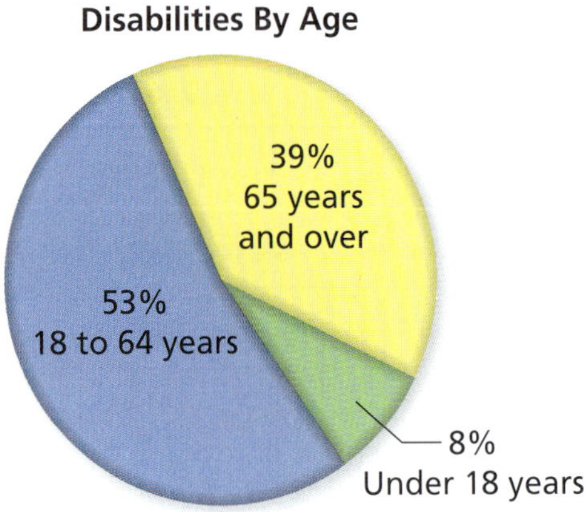

1. There are about 36 million people with disabilities in the United States. How many of the people with disabilities are 65 years old or older?

2. The population of the United States is about 310 million. Estimate the percent of the general population that is disabled.

3. People who are 65 years old and over only make up about 13% of the general population. Compare this to the percent of the disabled population for this group. Why do you think there is a discrepancy?

4. Compare the percent for people under 18 years old to the percents for the other categories. Why do you think the percent for people under 18 years old is so small?

Disabilities By Age

39%
65 years
and over

53%
18 to 64 years

8%
Under 18 years

5. **Horse Height** The height of a horse is measured from the ground to its withers (shoulders). The traditional unit of measure is "hands," where 1 hand is equal to 4 inches. A horse is 15 hands high. What is its height in inches?

6. **Top Speed** American Quarter Horses are the fastest horses in the world. They can run 55 miles per hour. How fast is this in feet per second?

7. **Belmont Stakes** The length of a horse race is often given in furlongs, where 1 furlong is equal to 0.125 mile. The Belmont Stakes is a 1.5-mile horse race that takes place every June at Belmont Park in Elmont, New York. How long is the Belmont Stakes in furlongs?

8. **Record High** The record high temperature in June in Elmont, New York, is 101°F. What is the record high temperature in degrees Celsius?

Chapter 1   Summary

Section Objectives	How does it apply to you?

Section 1

Use the order of operations to evaluate a numerical expression.	→	You need to know the order of operations to do calculations properly in your daily life. *(See Examples 1 and 2.)*
Use a calculator to evaluate a numerical expression.	→	To get an accurate answer, you must use the correct keystroke sequence when using a calculator. *(See Example 3.)*
Use the order of operations to evaluate a formula.	→	You will use common formulas many times throughout your life. *(See Examples 4, 5, and 6.)*

Section 2

Round numbers in a real-life context.	→	Using "exact" numbers is often unreasonable. It is usually reasonable to round numbers according to the context. *(See Examples 1 and 2.)*
Read large and small numbers.	→	Large and small numbers are often written in *scientific* or *exponential notation*. *(See Examples 3 and 4.)*
Understand the concept of "garbage in, garbage out."	→	When answering a question, remember that the accuracy of the output is only as good as the accuracy of the input. *(See Examples 5 and 6.)*

Section 3

Understand and find a percent of a number.	→	Percents are everywhere. You need to have a firm grasp of percents to make sense of today's world. *(See Examples 1 and 2.)*
Determine what percent one number is of another number.	→	Percent can be used to compare two numbers in the same context. *(See Examples 3 and 4.)*
Use percent to represent change.	→	Percent can be used to describe the amount that a quantity changes over time. *(See Examples 5 and 6.)*

Section 4

Use unit analysis to "balance" both sides of a formula.	→	Everything you measure has a number with some sort of unit of measure attached. *(See Examples 1 and 2.)*
Convert within a given system of measure.	→	You need to know how to compare units. *(See Examples 3 and 4.)*
Convert between different systems of measure.	→	Two systems of measure are used in the United States. You need to know how to compare units. *(See Examples 5 and 6.)*

Chapter 1   Review Exercises

Section 1.1

1. **Train Fare** A family of six takes a one-way trip on a train. The cost is $60 per adult and $45 per child. Which of the following keystroke sequences is better for finding the total cost? Explain your reasoning.

    a.

    b. [2] [×] [(] [6] [0] [+] [4] [×] [4] [5] [)] [=]

2. **Seating** The seating area of a passenger car on a high-speed train is in the shape of a rectangle. It has a length of 70 feet and a width of 9 feet. Estimate how many seats can fit in the passenger car. Explain your reasoning.

High-speed Rail In Exercises 3–6, use the facts about high-speed rail in China.

- In 2010, a Chinese passenger train hit a record speed of 302 miles per hour during a test run.

- It reached the top speed on a segment of the 824-mile-long line.

- The line is expected to open in 2012 and will halve the current travel time between Beijing and Shanghai to five hours.

- China already has the world's longest high-speed rail network, and it plans to cover 8125 miles by 2012 and 10,000 miles by 2020.

3. At its top speed, about how long does it take the passenger train to travel 1050 miles?

4. How long did it take to travel between Beijing and Shanghai before 2012?

5. From 2012 to 2020, about how many miles per year does China plan to expand its high-speed rail network?

6. The track distance between Beijing and Shanghai is 819 miles. Why do you think it takes five hours to make this trip?

Section 1.2

Oil In Exercises 7–10, use the pictograph.

Top Five Oil-consuming Countries in 2008

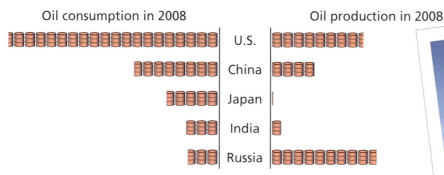

Oil consumption in 2008   Oil production in 2008

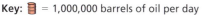

U.S.
China
Japan
India
Russia

Key: = 1,000,000 barrels of oil per day

7. Consider the consumption side of the pictograph.

   a. Estimate the number of barrels of oil consumed by the United States per day. Write your answer in standard decimal notation.

   b. For how many years do you think the United States and the rest of the world can continue consuming oil at this rate? Explain your reasoning.

8. Estimate the number of barrels of oil produced by Russia per day. Write your answer in exponential notation.

9. A barrel of oil contains 42 gallons of oil.

   a. In 2008, the population of China was about 1.3 billion. Estimate the number of gallons of oil consumed by China per person per day.

   b. In 2008, the population of the United States was about 304 million. Compare your estimate in part (a) with the number of gallons of oil consumed by the United States per person per day.

10. Use the formula below to estimate and compare the net exports for Russia and Japan. What does a negative value indicate?

    Net exports = oil produced − oil consumed

11. **Iran** Use the information in the doughnut graph to estimate the total number of barrels of oil Japan imported from Iran in 2008. Looking at the pictograph above, why do you think that Japan is one of Iran's top importers?

12. **U.S. Sanctions** The United States does not import any oil from Iran because of economic sanctions against the Middle Eastern country. It is estimated that lifting the sanctions could save the United States $76 billion per year. Suppose sanctions were lifted. Estimate how much the United States would save per day.

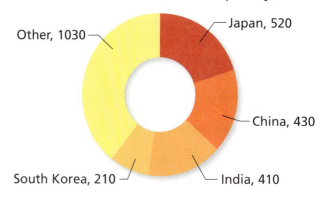

**Destination of Oil Exports from Iran in 2008
(in thousands of barrels of oil per day)**

Other, 1030
Japan, 520
China, 430
India, 410
South Korea, 210

Section 1.3

Earth's Water Distribution In Exercises 13–20, use the graph.

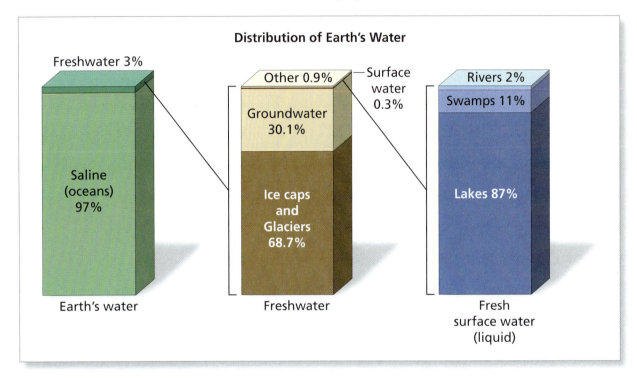

Distribution of Earth's Water

Freshwater 3%

Saline (oceans) 97%

Earth's water

Other 0.9%

Groundwater 30.1%

Surface water 0.3%

Ice caps and Glaciers 68.7%

Freshwater

Rivers 2%

Swamps 11%

Lakes 87%

Fresh surface water (liquid)

13. There are about 332.5 million cubic miles of water on Earth. How many cubic miles of Earth's water are freshwater?

14. How much freshwater is frozen in ice caps and glaciers?

15. The lakes on Earth hold 42,320 cubic miles of water. Of this, 20,490 cubic miles are saline. The rest are freshwater. What percent of the water in lakes is freshwater?

16. Groundwater accounts for about 5.6 million cubic miles of water. What percent of Earth's water is underground?

17. Suppose climate change causes some of the ice caps and glaciers to melt, adding 6000 cubic miles of freshwater to the surface water. Use percent to describe the increase in surface water.

18. What effect would a significant increase in surface water have on Earth and its inhabitants?

19. One of the categories in the graph is "Other." Where else do you think you can find freshwater besides the categories shown?

20. How do you think scientists estimated that there are about 332.5 million cubic miles of water on Earth?

Section 1.4

Nautical Units   In Exercises 21–26, use the table.

1 fathom = 2 yards
1 nautical mile ≈ 1013 fathoms
1 nautical mile ≈ 1.15 miles
1 league = 3 nautical miles
1 league ≈ 3.45 miles
1 knot = 1 nautical mile/hr

21. During the U.S. Civil War, the Confederate ironclad CSS *Virginia* clashed with the Union ironclad USS *Monitor* in the Battle of Hampton Roads. The speed of the *Monitor* was about 9 miles per hour, and the speed of the *Virginia* was about 6 knots. Which ship was faster?

22. The phrase "deep six" means to throw something away. It was originally used by boaters to mean 6 fathoms. Convert this depth to feet.

23. Territorial waters around the United States used to be defined by the reach of a cannonball fired from shore, or a "cannon shot." One cannon shot was defined as 3 nautical miles.

   a. Convert 1 cannon shot to miles.

   b. Compare this definition of territorial waters to the modern definition.

24. Jules Verne wrote a famous novel called *20,000 Leagues Under the Sea*.

   a. Convert 20,000 leagues to miles.

   b. At its deepest point, the ocean is about 7 miles deep. What do you think the title of Verne's novel means? Explain your reasoning.

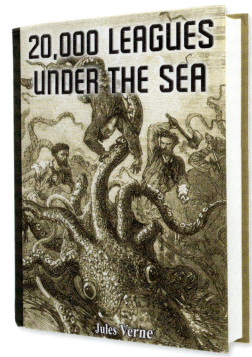

25. Below is an excerpt from William Shakespeare's *The Tempest*.

> *Full fathom five thy father lies;*
> *Of his bones are coral made;*
> *Those are pearls that were his eyes;*
> *Nothing of him that doth fade*
> *But doth suffer a sea-change*
> *Into something rich and strange.*

Shakespeare writes that the man lies 5 fathoms deep. Convert this depth to feet.

26. Mark Twain was the pen name of Samuel Clemens, who spent several years as a steamboat pilot. He chose the name because the leadsman on a steamboat would call out "mark twain" to indicate that the water depth was 12 feet and safe for the boat to pass. Convert this depth to fathoms.

2 The Mathematics of Consumption

2.1 Unit Prices

▶ Find the unit price of an item.

▶ Compare the unit prices of two or more items.

▶ Find the annual cost of an item.

2.2 Markup & Discount

▶ Find the markup on an item.

▶ Find the discount on an item.

▶ Find the final price after multiple discounts.

2.3 Consumption Taxes

▶ Find the sales tax on an item.

▶ Find the excise tax on an item.

▶ Find the value-added tax on an item.

2.4 Budgeting

▶ Create and balance a monthly budget.

▶ Write checks and balance a checkbook.

▶ Analyze a budget.

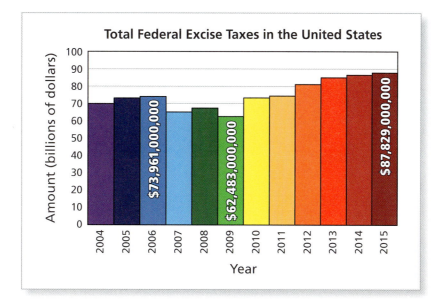

Total Federal Excise Taxes in the United States

$73,961,000,000

$62,483,000,000

$87,829,000,000

Amount (billions of dollars)

Year

Example 4 on page 77 describes the changes in federal excise tax revenues. Why did the excise tax revenue decline from 2006 to 2009?

2.1 Unit Prices

▶ Find the unit price of an item.
▶ Compare the unit prices of two or more items.
▶ Find the annual cost of an item.

Unit Prices

The **unit price** of an item is the cost per pound, quart, or other unit of weight or volume. It is often posted on the shelf tag below the item, along with the item price.

North Coast Co-op — Item price — **$2.75**
ARROWHEAD MILLS — Total units — 24 OZ
CORN GRITS - Organic Yellow
07433338525 **00179** 12 — Unit price — 0.11/OZ

Finding a Unit Price

To find a unit price, divide the item price by the total number of units.

$$\text{Unit price} = \frac{\text{item price}}{\text{total units}}$$

EXAMPLE 1 **Finding a Unit Price**

Check that the unit price shown above is correct.

SOLUTION

$$\text{Unit price} = \frac{\text{item price}}{\text{total units}}$$

$$= \frac{\$2.75}{24 \text{ oz}} \approx \$0.1146$$

Rounded to the nearest cent, the unit price is $0.11 per ounce.

✓ **Checkpoint**   Help at *Math*.and**Y⊙U**.com

Find the unit price of each item.

a.

b.

c.

$853.65 for 8 oz $119 for 1.7 fl oz $10.50 for 8 oz

EXAMPLE 2   **Analyzing a Unit Price**

Dairy farmers are paid for every 100 pounds of milk produced.

a. In 2010, how much did a dairy farmer earn per gallon of milk produced? (A gallon of whole milk weighs about 8.6 pounds.)

b. In 2010, the price of a gallon of milk was about $3.25. What percent of the price did the dairy farmer earn?

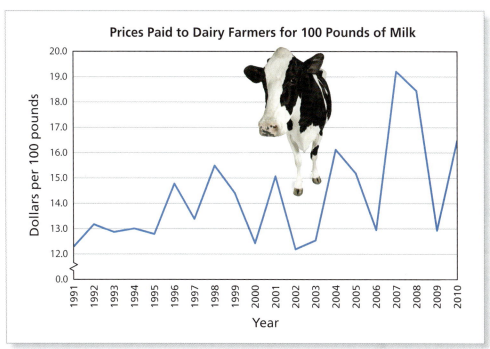

Prices Paid to Dairy Farmers for 100 Pounds of Milk

Study Tip

An old saying is "A pint is a pound the world around." The actual weight of a pint of water is about 1.04 pounds. There are 8 pints in a gallon, so a gallon of water weighs about 8.32 pounds.

SOLUTION

a. In 2010, a dairy farmer earned about $16.50 for every 100 pounds of milk produced.

$$\text{Unit price} = \frac{\$16.50}{100 \text{ lb}} = \frac{\$16.50}{100 \text{ lb}}\left(\frac{8.6 \text{ lb}}{1 \text{ gal}}\right) = \frac{\$1.419}{\text{gal}}$$

A dairy farmer earned about $1.42 per gallon of milk produced.

b.

Amount earned by farmer

Price per gallon

$$\frac{\$1.42}{\$3.25} \approx 0.44$$

The dairy farmer earned about 44% of the price.

For milk to be labeled "organic," the cows must meet four criteria.

● Not treated with growth hormone
● Not given antibiotics while in a herd
● Not fed feed treated with pesticides
● Must have access to pasture

✓ **Checkpoint**   Help at

In 2010, a dairy farmer earned about $25 for every 100 pounds of milk produced organically. The price of a gallon of organic milk was about $5.25.

c. How much did a dairy farmer earn for a gallon of organic milk in 2010?

d. What percent of the price did the dairy farmer earn?

Comparing Unit Prices

EXAMPLE 3   **Comparing Unit Prices**

Compare the unit prices of the three laundry detergents.

a. Brand A — 100 fl oz $12.99

b. Brand B — 2 gal $17.99

c. Brand C — 50 fl oz $7.99

Study Tip

To compare unit prices, you need the same units. That is why a conversion factor is used to calculate the unit price of brand B.

SOLUTION

a. Unit price $= \dfrac{\text{item price}}{\text{total units}} = \dfrac{\$12.99}{100 \text{ fl oz}} \approx \$0.13$ per fl oz

b. Unit price $= \dfrac{\text{item price}}{\text{total units}} = \dfrac{\$17.99}{2 \text{ gal}}\left(\dfrac{1 \text{ gal}}{128 \text{ fl oz}}\right) \approx \$0.07$ per fl oz

c. Unit price $= \dfrac{\text{item price}}{\text{total units}} = \dfrac{\$7.99}{50 \text{ fl oz}} \approx \$0.16$ per fl oz

Brands A and C are comparable, with brand A being a little less per fluid ounce. Brand B has a considerably lower unit price.

✓ **Checkpoint**   Help at *Math*.and**Y☉U**.com

Each of the above detergents recommends using 2 fluid ounces per load. Compare the cost per load of brand B with the cost per load of homemade laundry soap.

Homemade Laundry Soap (Use 1/2 cup per load)
∗ ⅓ bar Fels-Naptha Soap ($0.40)   ∗ ½ cup washing soda ($0.17)   ∗ ½ cup borax powder ($0.14)
Preparation: Grate the soap and put it in a cooking pan. Add 6 cups water and heat until the soap melts. Add the washing soda and borax and stir until dissolved. Pour 4 cups hot water into a bucket. Add the soap mixture and stir. Then add 1 gallon plus 6 cups water and stir. Let the soap mixture stand for about 24 hours and it will gel.

EXAMPLE 4   **Comparing Unit Prices**

Compare the unit prices of the different sizes of pizza.

a. Personal (10-inch diameter):
$5.99

b. Small (12-inch diameter):
$9.99

c. Medium (16-inch diameter):
$13.99

d. Large (20-inch diameter):
$18.99

$r = 5$ in.
$A \approx 78.5$ in.^{2}

$r = 6$ in.
$A \approx 113$ in.^{2}

$r = 8$ in.
$A \approx 201$ in.^{2}

$r = 10$ in.
$A \approx 314$ in.^{2}

SOLUTION

Use the formula for the area of a circle to find the area of each pizza. Remember that the formula for the area of a circle is $A = \pi r^2$, where π is approximately equal to 3.14 and the radius r is half the diameter.

a. Area $\approx 3.14(5^2) = 78.5$ in.^{2}

$$\text{Unit price} \approx \frac{\$5.99}{78.5 \text{ in.}^2} \approx \$0.076 \text{ per sq in.} \qquad \text{Personal}$$

b. Area $\approx 3.14(6^2) \approx 113$ in.^{2}

$$\text{Unit price} \approx \frac{\$9.99}{113 \text{ in.}^2} \approx \$0.088 \text{ per sq in.} \qquad \text{Small}$$

c. Area $\approx 3.14(8^2) \approx 201$ in.^{2}

$$\text{Unit price} \approx \frac{\$13.99}{201 \text{ in.}^2} \approx \$0.070 \text{ per sq in.} \qquad \text{Medium}$$

d. Area $\approx 3.14(10^2) = 314$ in.^{2}

$$\text{Unit price} \approx \frac{\$18.99}{314 \text{ in.}^2} \approx \$0.060 \text{ per sq in.} \qquad \text{Large}$$

The small pizza has the greatest unit price. The large pizza has the least unit price.

Notice that when the radius of the pizza doubles, the area is four times greater.

✓ **Checkpoint**

Find the unit price of a jumbo pizza.

e. Jumbo (24-inch diameter): $24.99

Using Unit Prices to Find Annual Savings

Because unit prices often differ by small amounts, people often ignore the differences and think, "It's just a few cents different." However, you can get a better idea of the differences when you compare them on the basis of annual consumption.

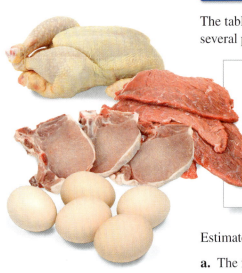

EXAMPLE 5 **Finding the Annual Savings**

The table shows the average annual consumption by Americans for several products.

Average Annual Consumption by Americans					
Beef	61.2 lb	Chicken	58.8 lb	Pork	46.0 lb
Turkey	13.9 lb	Margarine	4.2 lb	Eggs	247
Whole milk	6.1 gal	Tea	8.0 gal	Butter	5.0 lb
Cheese	32.4 lb	Wheat flour	136.6 lb	Sugar	65.7 lb

Estimate the annual savings for a family of four people for the following.

a. The family purchases all of its whole milk at a discount store in which the whole milk averages $1.20 less per gallon than at a supermarket.

b. The family purchases all of its meat at a discount store in which the meat averages $0.85 less per pound than at a supermarket.

SOLUTION

a. The total amount of whole milk consumed (per person) is 6.1 gallons.

For a family of 4, this amounts to

$$4(6.1) = 24.4 \text{ gallons.}$$

If the family was able to save $1.20 per gallon, the annual savings would be $1.20(24.4) = \$29.28$.

b. The total amount of meat consumed (per person) is

$$61.2 + 58.8 + 46.0 + 13.9 = 179.9 \text{ pounds.}$$

For a family of 4, this amounts to

$$4(179.9) = 719.6 \text{ pounds.}$$

If the family was able to save $0.85 per pound, the annual savings would be $0.85(719.6) = \$611.66$.

✓ **Checkpoint**   Help at

Estimate the annual savings for a family of 4 people when the family purchases all of its wheat flour and sugar at a discount store at an average savings of $0.67 per pound.

EXAMPLE 6   Eating In or Eating Out?

The list shows the ingredients for six cheeseburgers.

$2.84	1 1/2 lb hamburger
$2.99	6 hamburger buns
$0.90	2 tomatoes
$0.35	1 onion
$2.00	6 cheese slices
$0.59	1 head lettuce

a. Estimate the cost per cheeseburger.

b. You can buy a quarter-pound cheeseburger for $3.49. How much will you save by making six cheeseburgers instead of buying six?

c. Suppose that once a week you make six cheeseburgers instead of buying six. How much will you save in a year?

SOLUTION

a. The cost of the ingredients is

$$2.84 + 2.99 + 0.90 + 0.35 + 2.00 + 0.59 = \$9.67.$$

The unit price per cheeseburger is

$$\text{Unit price} = \frac{\$9.67}{6 \text{ cheeseburgers}} \approx \$1.61.$$

b. The cost of buying 6 cheeseburgers is

$$6(3.49) = \$20.94.$$

You will save $20.94 - 9.67 = \$11.27$.

c. By saving the amount in part (b) once a week for a year, you will save

$$(11.27)(52) = \$586.04.$$

✓ Checkpoint

Help at *Math*.and**YOU**.com

You are preparing spaghetti for four people.

$1.59	1 box spaghetti
$1.89	1 lb hamburger
$3.49	1 jar spaghetti sauce
$2.89	1 package parmesan cheese
$0.35	1 onion
$0.79	1 can sliced mushrooms

d. What is the unit price per person?

e. How much more would you pay to buy 4 spaghetti meals at a restaurant for $7.95 each?

2.1 Exercises

Groceries In Exercises 1–4, find the unit price of the item. *(See Example 1.)*

1. Peanuts

$2.79 for 16 oz

2. Ramen

$4.29 for 72 oz

3. Grape juice

$3.39 for 64 fl oz

4. Ice cream

$4.49 for 1.5 qt

Strawberries In Exercises 5 and 6, use the graph. *(See Example 2.)*

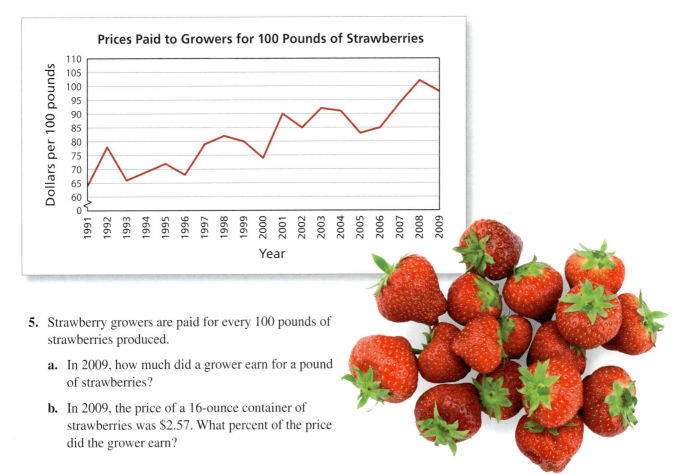

5. Strawberry growers are paid for every 100 pounds of strawberries produced.

  a. In 2009, how much did a grower earn for a pound of strawberries?

  b. In 2009, the price of a 16-ounce container of strawberries was $2.57. What percent of the price did the grower earn?

6. In 1991, the average retail price of a pound of strawberries was $1.32.

  a. In 1991, how much did a grower earn for a pound of strawberries?

  b. In 1991, what percent of the average retail price did the grower earn?

  c. Compare the percent in part (b) with the percent from Exercise 5. Did growers earn a larger portion of the average retail price in 1991 or in 2009? Explain your reasoning.

Ketchup **In Exercises 7–10, use the information below.** *(See Examples 3 and 4.)*

Brand A   **Brand B**   **Brand C**

20 oz
$1.89

4 lb
$3.46

40 oz
$2.40

7. Compare the unit prices of the three brands of ketchup. Which is the best buy? Explain your reasoning.

8. Brand D contains 114 ounces and costs $8.38.

   a. Compare the unit price of brand D with the unit prices of the three brands above. Is the largest product necessarily the best buy? Explain your reasoning.

   b. You use less than 20 ounces of ketchup per year. Does it make sense for you to buy the brand with the lowest unit price? Explain your reasoning.

9. A 40-ounce bottle of ketchup contains about 4.2 cups of ketchup. (One cup equals 8 fluid ounces.)

   a. What is the unit price of brand C in dollars per fluid ounce?

   b. Why do you think a manufacturer would prefer to use ounces in the unit rate?

10. Below is a recipe for homemade ketchup.

Homemade Ketchup (Makes 48 ounces)

* 4 lb tomatoes	($7.99)	* 1½ onions	($0.37)
* 1 tsp ground cloves	($0.65)	* 1 tsp salt	($0.01)
* 1 tsp allspice	($0.28)	* 1 cup vinegar	($0.24)

Preparation: Place the tomatoes in a boiling pot of water and leave them in until the skins split. Strain the tomatoes and allow them to cool. Peel the skins, slice the tomatoes open, and remove the seeds. Dice the tomatoes and onions and put them in a saucepan. Simmer for 10 minutes. Pour into a blender. After blending, pour back into saucepan and add the other ingredients. Stir the mixture. Simmer for 1½ hours. Stir often. After one-third has burned off, turn heat off. Allow time to cool.

   a. Compare the unit price of homemade ketchup with the unit price of brand A above.

   b. Your neighbor gives you 4 pounds of tomatoes from his garden. What is the unit price of homemade ketchup when you do not have to pay for the tomatoes?

Beverage Consumption In Exercises 11–14, use the graph. *(See Examples 5 and 6.)*

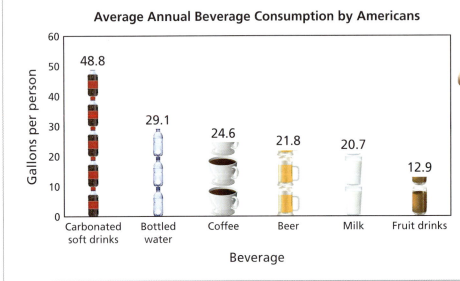

Average Annual Beverage Consumption by Americans

Gallons per person (y-axis, 0 to 60)

- Carbonated soft drinks: 48.8
- Bottled water: 29.1
- Coffee: 24.6
- Beer: 21.8
- Milk: 20.7
- Fruit drinks: 12.9

Beverage (x-axis)

Some brands of bottled water indicate on the label that the water is from a public water source. This means that the bottle contains purified tap water.

11. At a discount store, fruit drinks average $1.08 less per gallon than at a supermarket. A family of four people purchases all of its fruit drinks at the discount store. Estimate the annual savings.

12. At a discount store, ground coffee averages $0.10 less per ounce than at a supermarket. Four college roommates purchase all their ground coffee at the discount store. Estimate the annual savings. (7.68 ounces of ground coffee make 1 gallon of coffee.)

13. At a supermarket, a 24-pack of 16.9-fluid-ounce bottles of water costs $4.99.

    a. At a discount store, a 35-pack of 16.9-fluid-ounce bottles of water costs $4.49. Suppose you buy all your bottled water at the discount store instead of at the supermarket. How much will you save per year?

    b. Your tap water rate is $2.62 per 1000 gallons of water. Suppose you use tap water instead of buying bottled water at the supermarket. How much will you save per year?

14. You use 0.5 fluid ounce of coffee creamer with each 6-fluid-ounce "cup" of coffee. The creamer that you use costs $2.19 for 16 fluid ounces. You can make homemade coffee creamer using the recipe shown.

Homemade Coffee Creamer (Makes 22 fluid ounces)

    * 1 14-oz can sweetened condensed milk   ($1.49)
    * 1½ cups milk   ($0.29)
    * 1 tbsp vanilla extract   ($0.70)

    Preparation: Pour the ingredients into a jar and shake well.

    a. Estimate how much you could save annually by making your own creamer.

    b. The vanilla extract is optional. Estimate how much you could save annually by making your own creamer without the vanilla instead of buying the creamer at the store.

▶ Extending Concepts

Product Downsizing Instead of increasing the price of a product, some manufacturers decrease the amount of the product in a package but charge the same price. In Exercises 15–18, analyze the effect of product downsizing on unit price.

15. What is the difference in unit price?

Old size: 7 oz   New size: 5 oz
Price: $0.89   Price: $0.89

16. What is the difference in unit price?

Old size: 64 fl oz   New size: 52 fl oz
Price: $2.59   Price: $2.59

17. A manufacturer decreases the amount of product in a package but keeps the price the same. What effect does this have on the unit price of the product? Explain your reasoning.

18. Is it possible to increase the unit price of a product by decreasing both the size and the price? Explain your reasoning.

Paying More for Less The weight or volume listed on a product may be greater than the actual amount of the product that you are receiving. In Exercises 19 and 20, analyze how you could end up paying more money for less product.

19. A 1-pound package of chicken costs $1.99.

   a. The label on the package says "up to 15% solution." This means that 15% of the weight of the package is a solution. What is the actual unit price per pound of chicken?

   b. Compare the unit price in part (a) to the unit price on the package.

20. Two brands of bleach and their prices are shown.

Brand A

96 fl oz
$1.79

Brand B

182 fl oz
$2.12

   a. Find the unit prices of the two brands in dollars per gallon. Based on your calculations, which brand is the better buy?

   b. Brand A is 6% bleach. Brand B is 3% bleach. You can make 3% bleach by mixing 1 fluid ounce of brand A with 1 fluid ounce of water. How many fluid ounces of 3% bleach can you make using brand A?

   c. Find the unit prices of the two brands in dollars per gallon of 3% bleach. Based on your calculations, which brand is the better buy?

2.2 Markup & Discount

▶ Find the markup on an item.

▶ Find the discount on an item.

▶ Find the final price after multiple discounts.

Finding the Markup on an Item

Markup is the difference between the retail price and the price the retailer pays for the item, or wholesale price.

Finding a Markup

The markup on an item is the difference between the retail price and the wholesale price.

$$\text{Markup} = \text{retail price} - \text{wholesale price}$$

To find the **markup percent,** divide the markup by the wholesale price.

$$\text{Markup percent} = \frac{\text{markup}}{\text{wholesale price}}$$

In fairness to retailers, you should remember that markup is the only way for a retailer to make a profit. Markup is not equal to profit. Out of the markup, a retailer has to pay for rent, utilities, taxes, salaries, benefits, and other business expenses.

EXAMPLE 1 Finding a Markup Percent

Markup percents vary greatly. Prescription drugs and jewelry tend to have high markup percents. Grocery items and everyday clothing tend to have low markup percents. Find the markup percent for each item.

a. Pair of earrings
Wholesale price: $89.00
Retail price: $459.00

b. Pair of athletic socks
Wholesale price: $1.13
Retail price: $1.69

SOLUTION

a. $\dfrac{459 - 89}{89} = \dfrac{370}{89}$

≈ 4.157

$\approx 416\%$

b. $\dfrac{1.69 - 1.13}{1.13} = \dfrac{0.56}{1.13}$

≈ 0.496

$\approx 50\%$

✓ **Checkpoint**   Help at *Math*.and**YOU**.com

Find the markup percent for each item.

c. Automobile
Wholesale price: $25,450
Retail price: $27,990

d. Leather chair
Wholesale price: $235
Retail price: $799

EXAMPLE 2   **Finding a Markup Percent**

You live near a designer handbag outlet store. The store sells discontinued and slightly damaged handbags at prices that are well below normal retail prices. You buy a handbag for $195. The handbag normally retails for $895. You then put the handbag on eBay® and sell it for $395.

a. What is your markup and markup percent?

b. Is your markup the same as your profit? Explain.

SOLUTION

Study Tip

"Wholesale price" is a relative term. In Example 2, you are the retailer, so the price you pay at the outlet store becomes your wholesale price.

a. Your markup is

$$\text{Markup} = 395 - 195 = \$200.$$

Your markup percent is

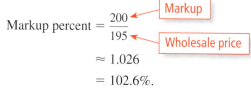

$$\text{Markup percent} = \frac{200}{195}$$

$$\approx 1.026$$

$$= 102.6\%.$$

b. Your markup in this transaction is *not* the same as your profit. To find your profit, you must also subtract your expenses.

eBay® insertion fee	$2.00	Fee charged for listing
Final value fee	$35.55	9% of $395
PayPal fee	$11.76	2.9% of $395 + $0.30
Sales tax	$11.70	6% paid at outlet store
Shipping	$22.50	Mail handbag to customer
Other expenses	$35.00	Supplies, transportation
Total	$118.51	

Your profit is

$$\text{Markup} - \text{Expenses} = 200.00 - 118.51$$

$$= \$81.49.$$

So, your markup is not the same as your profit. If you spend 5 hours in traveling, shopping, Internet use, correspondence, wrapping, and shipping, then your hourly wage is $16.30.

✓ Checkpoint

Help at *Math*.and**YOU**.com

A bookstore pays $140 for a textbook and sells it for $200. The bookstore manager determines that the retail price of the book is a 30% markup because $60 is 30% of $200. Is this a correct use of the term *markup*? Explain your reasoning.

Finding the Discount on an Item

Discount is the difference between the "regular" retail price of an item and the price that a consumer actually pays for the item.

Finding a Discount

The discount on an item is the difference between the regular retail price and the discounted, or sale, price.

$$\text{Discount} = \text{regular price} - \text{discounted price}$$

To find the **discount percent,** divide the discount by the regular price.

$$\text{Discount percent} = \frac{\text{discount}}{\text{regular price}}$$

EXAMPLE 3 Using Discounts and Markup

You work in a clothing store. You order 25 T-shirts in assorted sizes. The wholesale price of each shirt is $8.37. The regular retail price is $24.99. During the next 2 months, you repeatedly mark down the price until you finally sell all 25 shirts. Find the average markup percent for the 25 shirts.

- 13 shirts at $24.99
- 6 shirts at 50% off
- 4 shirts at 25% off
- 2 shirts at 75% off

SOLUTION

A spreadsheet works well to organize this problem.

	A	B	C	D	E	
	Wholesale Price	Regular Price	Discount Percent	Discount Price	Quantity Sold	Revenue
1						
2	$8.37	$24.99	0.0%	$24.99	13	$324.87
3	$8.37	$24.99	25.0%	$18.74	4	$74.96
4	$8.37	$24.99	50.0%	$12.50	6	$75.00
5	$8.37	$24.99	75.0%	$6.25	2	$12.50
6	Total					$487.33
7						

You paid $25(8.37) = \$209.25$ for the shirts. Your total markup was $487.33 - 209.25 = \$278.08$. So, your average markup percent was

$$\frac{\text{Total markup}}{\text{Total wholesale price}} = \frac{278.08}{209.25} \approx 1.329 = 132.9\%.$$

✓ Checkpoint

Find the average markup percent for the following sales record.

- 11 shirts at $24.99
- 3 shirts at 50% off
- 7 shirts at 25% off
- 4 shirts at 75% off

EXAMPLE 4   Buying Generic Prescription Drugs

In the United States, prescription drugs are often expensive. You can usually save money by asking your doctor to prescribe a generic version of a drug. Here are some sample savings.

Condition	Monthly Cost for Brand-Name Drug	Monthly Cost for Generic Substitute
High blood pressure	$128	$13
High cholesterol	$95	$37
Depression	$103	$37
Arthritis pain	$135	$30
Heartburn	$179	$24

Which of these represents the greatest discount percent?

SOLUTION

A spreadsheet works well to organize the information in this problem.

Brand − Generic

Discount Brand

DATA

	A	B	C	D	E
1	**Condition**	**Brand**	**Generic**	**Discount**	**Percent**
2	High blood pressure	$128.00	$13.00	$115.00	89.8%
3	High cholesterol	$95.00	$37.00	$58.00	61.1%
4	Depression	$103.00	$37.00	$66.00	64.1%
5	Arthritis pain	$135.00	$30.00	$105.00	77.8%
6	Heartburn	$179.00	$24.00	$155.00	86.6%
7					

The greatest discount percent is for the high blood pressure drug.

✓ Checkpoint

Help at *Math*.and**YOU**.com

Find the discount percents for the three drugs described in the article.

Crossing Borders for Prescription Drugs

Thousands of American seniors are going to Canada to stock up on something they can't seem to find at home—affordable prescription drugs.

Here are some examples. A stomach acid medication costs US$129 in the US, but only US$53 in Canada. An antihyperglycemic agent costs US$52 in the US, but only US$12 in Canada, Conjugated estrogens cost US$26 in the US, but only US$7 in Canada.

Finding the Final Price after Multiple Discounts

When you have two discounts on the same item, the final price can depend on the order in which you calculate the discounts.

EXAMPLE 5   **Calculating Multiple Discounts**

Each day in the United States, millions of coupons are distributed by mail, newspapers, and e-mails. You acquire a coupon for a pair of jeans.

$\$10$ off

Any pair of jeans.

Coupon must be presented
at the time of transaction.
Good through June 30, 2012.

A pair of jeans retails for $40 and is being sold at a store that is having a "25% off" sale on all clothing.

a. What is the final price when you first take 25% off, and then subtract $10?

b. What is the final price when you first subtract $10, and then take 25% off?

SOLUTION

a. Begin by taking 25% off.

Original price   25% discount

$$40 - 0.25(40) = 40 - 10 = \$30$$

Then use the coupon.

Coupon

$$30 - 10 = \$20$$

The final price is $20.

b. Begin by using the coupon.

$$40 - 10 = \$30$$

Then take 25% off.

$$30 - 0.25(30) = 30 - 7.50 = \$22.50$$

The final price is $22.50.

Jeans were invented in 1873 by
Levi Strauss and Jacob Davis.

✓ **Checkpoint**

Help at  Math.and**Y**U.com

Suppose the coupon in Example 5 is for 10% off, rather than $10 off. Would the order in which you apply the discounts make a difference in the final price? Explain your reasoning.

EXAMPLE 6   **Using Promotional Codes**

A car rental company offers 10% off the cost of any rental car. The company also sends you an email with a promotional code for 20% off. You decide to use both discounts to reserve a car online.

Promotional Code:

The regular price for renting a car is $70 per day. What is the final price after both discounts?

SOLUTION

Begin by taking 10% off.

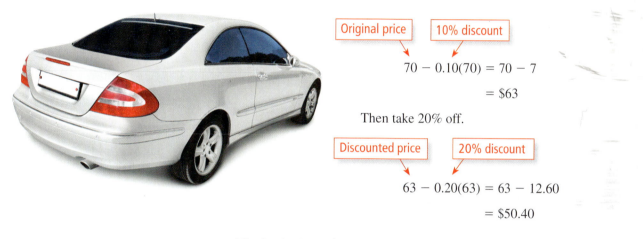

Original price   10% discount

$$70 - 0.10(70) = 70 - 7$$
$$= \$63$$

Then take 20% off.

Discounted price   20% discount

$$63 - 0.20(63) = 63 - 12.60$$
$$= \$50.40$$

The final price is $50.40.

✓ **Checkpoint**   Help at *Math*.and**YOU**.com

Use the information in the article to estimate a shopper's hourly wage for clipping grocery coupons from newspapers. Explain your reasoning.

Save Money by Using Coupons

In 2009, about 367 billion coupons were distributed in the United States. Of these, about 3.3 billion coupons were redeemed.

Studies have shown that shoppers who spend 20 minutes per week clipping coupons can save up to $1000 per year. That is a 20% savings for a family with a $5000 annual grocery bill.

2.2 Exercises

Apparel In Exercises 1–4, find the markup percent for the item. *(See Examples 1 and 2.)*

1.

Wholesale: $18
Retail: $40

2.

Wholesale: $36
Retail: $65.99

3.

Wholesale: $12.18
Retail: $19.99

4.

Wholesale: $12.89
Retail: $16.19

5. Necklace The wholesale price for a necklace is $10. The markup percent is 200%. What is the markup?

6. Watch The wholesale price for a watch is $25. The markup percent is 50%. What is the retail price?

DATA **Lawn and Garden** You work in the lawn and garden section of a local retail store. You order 26 garden statues. The wholesale price of each statue is $35. The regular retail price is $70. During the next 4 months, you repeatedly mark down the price until you finally sell all 26 statues. In Exercises 7–10, use the sales record shown. *(See Example 3.)*

- × 10 at $70
- × 5 at 50% off
- × 8 at 25% off
- × 3 at 75% off

7. Find the average markup percent for the 26 statues.

8. Is the revenue from the sale of the 26 garden statues equal to the profit? Explain your reasoning.

9. Your goal as a businessperson is to make a profit.

    a. How many garden statues did you discount below the wholesale price?

    b. How can you make a profit when you discount some of your inventory at less than the wholesale price? Explain your reasoning.

10. Create a spreadsheet similar to the one in Example 3 for the garden statue sales record. Experiment with the numbers in the "Quantity Sold" column. Remember that the total number of garden statues is 26.

	A	B	C	D	E	F
1	Wholesale Price	Regular Price	Discount Percent	Discount Price	Quantity Sold	Revenue
2	$35.00	$70.00	0.0%	$70.00	10	$700.00
3	$35.00	$70.00	25.0%	$52.50	8	$420.00
4	$35.00	$70.00	50.0%	$35.00	5	$175.00
5	$35.00	$70.00	75.0%	$17.50	3	$52.50
6	Total					$1,347.50
7						

    a. Find four values such that the revenue is greater than the cost. (Cost = number of garden statues purchased × wholesale price)

    b. Find four values such that the revenue is less than the cost.

    c. Explain why you need to know this information to make a profit.

Cereal In Exercises 11 and 12, find the discount percent for the generic cereal. *(See Example 4.)*

11.

 vs.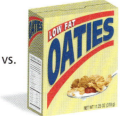

Generic $1.99   Brand Name $5.29

12.

 vs.

Generic $2.39   Brand Name $2.99

Television **You are purchasing a television from an electronics store. The television retails for $380, and the store is having a "15% off" sale on all items. In Exercises 13 and 14, use this information.** *(See Example 5.)*

13. You have the coupon shown.

    a. What is the final price when you first take 15% off, and then subtract $25?

    b. What is the final price when you first subtract $25, and then take 15% off?

14. You have the coupon shown.

    a. What is the final price when you first take 15% off, and then subtract $20?

    b. What is the final price when you first subtract $20, and then take 15% off?

    c. Suppose the coupon is for 20% off, rather than for $20 off. What is the final price when you use the coupon with the sale?

Footwear **A shoe store sends you a promotional code that you can use for a discount. In Exercises 15 and 16, use the information in the coupon.** *(See Example 6.)*

15.

    a. What is the percent discount?

    b. The store is having a "60% off" sale on all items. What is the final price when you use the promotional code with the sale?

16.

    a. What is the percent discount?

    b. The store is having a "50% off" sale on all items. What is the final price when you use the promotional code with the sale?

▶ Extending Concepts

Energy Drink In Exercises 17–20, use the display.

Cost Analysis of an Energy Drink

$3.50 — Retail price

$3.00 — Distributor price

$2.50 — Manufacturer price

$1.00 — Cost of materials

17. What is the markup percent from the manufacturer price to the retail price?

18. What is the markup percent from the distributor price to the retail price?

19. You buy one energy drink. You have a coupon for $0.75 off. What is the discount percent?

20. Suppose the store marks up the energy drink to $3.25 instead of $3.50. What is the markup percent from the distributor price to the retail price?

21. **Clearance** A store advertises that it is having a "30% off" sale on all items, with an additional 20% off clearance items. The additional 20% is taken off after the 30% discount is applied. You are buying a spice rack that is on clearance. The regular price of the spice rack is $30.

  a. What is the total discount?

  b. What is the total discount percent?

  c. Why do you think the store advertises its sale using two discount percents instead of one? Explain.

22. **Customer Service** You work at a store that is having a "50% off" sale. A customer has a coupon for 50% off any item and thinks that a $40 sweater should be free with the coupon. How would you explain why the sweater is not 100% off? Explain your reasoning.

2.1–2.2   Quiz

Liquid Hand Soap In Exercises 1–5, use the information below.

Brand A
7.5 fl oz
$1.99

Brand B
19.95 fl oz
$2.39

Brand C
32 fl oz
$2.99

1. Compare the unit prices of the three brands of liquid hand soap.

2. A family of 4 uses 128 fluid ounces of liquid hand soap annually. Suppose the family purchases brand C instead of brand A. Estimate the annual savings for the family.

3. The wholesale price of brand A is $0.99.

   a. Find the markup.

   b. Find the markup percent.

4. You have a coupon for 10% off brand B.

   a. How much can you expect to pay for brand B?

   b. How does the coupon affect the unit price of brand B? Explain your reasoning.

   c. Does brand B give you the most hand soap for your money after you use the coupon? Explain your reasoning.

5. Below is a recipe for homemade liquid hand soap.

   ### Homemade Liquid Hand Soap

   * 1 bar of soap   ($0.33)   * 1 tsp glycerin   ($0.14)
   * 1 tbsp honey   ($0.19)

   Grate the soap and put it in a blender with 1 cup of boiling water. Blend the soap and water. Then stir in the honey and glycerin. Let the mixture cool for 15 minutes, and then blend. Add cold water until you have 6 cups of mixture. Blend the mixture again and pour it into a container to let it cool. Do not put a lid on the container.

   a. What is the unit price of the homemade liquid hand soap?

   b. You make your own liquid hand soap instead of buying brand C. What is the discount percent?

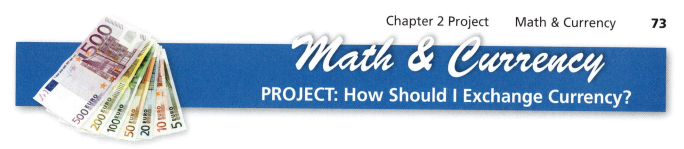

Math & Currency

PROJECT: How Should I Exchange Currency?

1. You are traveling from the United States to Europe and plan to take $2000 in cash. You are trying to decide the best way to do this. To start, you use the *Universal Currency Converter ®** at *Math.andYou.com*. Using the converter, how many euros will you receive for $2000 (USD)?

In Exercise 2, use the websites for the following to complete parts (a)–(f).

- **local bank**
- **airport currency exchange company**
- **credit card company**
- **foreign bank**

2. Which method of exchanging currency is the least costly?

   a. Exchange your dollars for euros at your local bank.

   b. Exchange your dollars for euros at the airport.

   c. Use your credit card to make purchases in Europe.

   d. Exchange your dollars for euros at a bank in the country you are visiting.

   e. Buy traveler's checks at your local bank. Then exchange the traveler's checks for euros at a bank in the country you are visiting.

   f. Use an ATM in the country you are visiting.

3. Using the methods in Exercise 2, will you receive as many euros as the *Universal Currency Converter®* indicates? If not, why?

*Provided by XE.com

2.3 Consumption Taxes

▶ Find the sales tax on an item.

▶ Find the excise tax on an item.

▶ Find the value-added tax on an item.

Finding the Sales Tax on an Item

A **consumption tax** is a tax on spending. There are three basic types of consumption taxes: sales tax, excise tax, and value-added tax.

Consumption taxes do not tax savings, income, or property. Although personal and corporate income taxes provide most of the revenue to the federal government, consumption taxes are a primary source of income for state and local governments. The United States implemented consumption taxes much earlier than income taxes. The paragraph at the left was written by Alexander Hamilton in the Federalist Papers in 1787.

EXAMPLE 1   Finding Sales Tax

In the United States, sales tax rates vary greatly. Several states have no sales tax. There are states in which cities and counties have sales taxes, which are added on to state sales taxes. Find the sales tax on each item.

a. Yacht: $214,000
Sales tax: 8%

b. Sports car: $65,900
Sales tax: 7%

SOLUTION

Retail price	8% tax rate

a. 214,000(0.08) = $17,120
The sales tax is $17,120.

Retail price	7% tax rate

b. 65,900(0.07) = $4613
The sales tax is $4613.

✓ Checkpoint

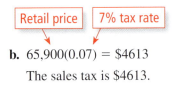

Help at *Math*.and**YOU**.com

Find the sales tax on each item.

c. High heel shoes: $250
Sales tax: 6%

d. Watch: $1350
Sales tax: 7.4%

The Federalist No. 21

"It is a signal advantage of taxes on articles of consumption, that they contain in their own nature a security against excess. They prescribe their own limit; which cannot be exceeded without defeating the end proposed, that is, an extension of the revenue. . . . If duties are too high, they lessen the consumption; the collection is eluded; and the product to the treasury is not so great as when they are confined within proper and moderate bounds."

Alexander Hamilton

Study Tip

In Example 1a, you can estimate the sales tax to be "a little less than 10%." Because 10% of $214,000 is $21,400, a sales tax of $17,120 is reasonable.

Sales tax has become a quagmire of rules, misconceptions, tax forms, legislation, and legal battles. The basic problem is that the United States has thousands of sales tax jurisdictions (counting states, counties, and cities). Most jurisdictions interpret sales tax law to mean that a resident of the jurisdiction must pay sales tax on each item the resident purchases, regardless of where the item is purchased.

For instance, if you live in Washington State (which has sales tax) and purchase an item in Oregon (which has no sales tax), Washington expects you to report your purchase and pay sales tax. Even more common is the issue of Internet sales. If you run an Internet business, then you might have to collect and report sales tax when someone purchases an item from you.

| EXAMPLE 2 | **Finding a Sales Tax Rate** |

You purchase a jacket in an airport mall. You are unfamiliar with the sales tax rates in the area. You are given the sales receipt shown at the left. Find the sales tax rates indicated by the receipt.

```
Jacket            $135.95
Subtotal:         $135.95

Sales tax city:   $  1.36
Sales tax county: $  4.35
Sales tax state:  $  9.25

TOTAL:            $150.91

- THANKS FOR SHOPPING -
```

SOLUTION

For sales tax, the retail price is always the **base**.

City sales tax → / Retail price →
$$\frac{1.36}{135.95} \approx 0.01 = 1\%$$

The city has a sales tax rate of about 1%.

County sales tax → / Retail price →
$$\frac{4.35}{135.95} \approx 0.032 = 3.2\%$$

The county has a sales tax rate of about 3.2%.

State sales tax → / Retail price →
$$\frac{9.25}{135.95} \approx 0.068 = 6.8\%$$

The state has a sales tax rate of about 6.8%.

✔ Checkpoint

Help at *Math*.and**YOU**.com

a. Suppose that Washington State loses the sales tax on 10,000 major appliances that are purchased in Oregon each year. The sales tax rate in Washington is 6.5%. Estimate the loss in tax revenue from these sales.

b. Suppose that each person in the United States avoids paying $500 in sales tax each year (by buying out of state, buying on the Internet, or buying at informal outlets such as garage sales). Estimate the total sales tax revenue lost each year.

Finding the Excise Tax on an Item

There are many types of **excise taxes.** The most common excise tax applies to a specific class of goods, typically alcohol, gasoline, cigarettes, and gambling. Rather than being a certain percent, excise taxes are often a fixed dollar amount. For instance, a state might charge $1 per gallon of gasoline or $2 per pack of cigarettes. The tax is reported to the consumer as part of the retail price. For instance, when gasoline is reported as $2.94 per gallon, this price includes any state or federal excise tax on the gasoline.

The first federal excise tax in the United States was levied in 1791 on distilled whiskey. The tax was unpopular with farmers on the western frontier who got their corn to market by distilling it into whiskey. The unpopularity of the tax grew into a revolt known as the Whiskey Rebellion.

EXAMPLE 3   **Estimating Excise Tax**

Estimate the amount of excise tax generated by cigarette sales each year in the United States.

SOLUTION

This is not an easy question, but you can still come up with a reasonable estimate. Assume the following.

- There are about 230 million people who are 18 years old or older.
- Of these, about 20% smoke cigarettes.
- A cigarette smoker smokes one pack per day.
- State and federal excise taxes total about $2.50 per pack.

With these assumptions, you can obtain the following estimate.

$$(230{,}000{,}000)(0.20)\left(\frac{1 \text{ pack}}{\text{day}}\right)\left(\frac{\$2.50}{\text{pack}}\right)\left(\frac{365 \text{ days}}{\text{year}}\right) = \frac{\$41{,}975{,}000{,}000}{\text{year}}$$

So, a reasonable estimate for the revenue generated by state and federal excise taxes on cigarettes is about $40 billion.

Excise taxes on tobacco, alcohol, and gambling are sometimes called "sin taxes."

✓ **Checkpoint**   Help at

The United States uses about 140 billion gallons of gasoline in a year. Estimate the amount of excise tax generated by gasoline sales each year in the United States.

EXAMPLE 4   **Finding Changes in Excise Taxes**

The graph shows the total excise tax revenue (actual and projected) that the U.S. federal government raises.

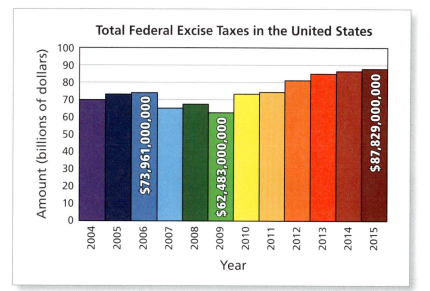

Total Federal Excise Taxes in the United States

$73,961,000,000

$62,483,000,000

$87,829,000,000

a. How much did the revenue decline from 2006 to 2009?

b. Why did the revenue decline during those years?

SOLUTION

a. From 2006 to 2009, the revenue declined by

$$73,961,000,000 - 62,483,000,000 = \$11,478,000,000.$$

This is a decline of about 15.5%.

b. One cause for the decline in revenue was an economic recession that began in 2007. There was, however, a less obvious cause. In 1898, the U.S. federal government implemented the first telephone excise tax in an effort to help fund the Spanish-American war. In 2005, a federal court declared it illegal to tax long-distance telephone services, a decision that cost the federal government billions of dollars in revenue.

The Telephone Excise Tax Refund was a onetime payment available on the 2006 federal income tax return. It refunded excise taxes collected on long-distance telephone services billed between February 28, 2003, and August 1, 2006. Most phone customers, including most cell phone users, qualified for the refund.

✓ **Checkpoint**

Help at *Math*.and**YOU**.com

Use the graph above to answer the questions.

c. How much is the revenue projected to increase from 2009 to 2015?

d. What is the percent increase, using 2009 as the base year?

e. Suppose you are asked to devise a plan to raise the total excise tax revenue to $250 billion by 2015. Describe such a plan and explain why you think it could raise the necessary tax revenue.

Value-Added Tax

A **value-added tax** (VAT) is a consumption tax. It is a tax on the market value that is added to a product or material at each stage of its manufacture or distribution. The tax is passed on to the consumer because it increases the ultimate retail price of the product. Value-added taxes differ from sales taxes because sales taxes are applied only at the point of purchase.

As of the writing of this text, the United States does not have a value-added tax. It has been considered by Congress but has always been rejected. Several other countries, including Canada, do have value-added taxes.

Proponents of value-added taxes argue that they inhibit consumer spending less than sales taxes, because they are hidden in the final retail price of the product.

EXAMPLE 5   **Comparing Sales Tax and Value-Added Tax**

During the production of a sheepskin coat, value is added by the following.

- A farmer: $10
- A leather tanning company: $25
- A coat making company: $150
- A retail clothing company: $210

Compare the effect of (a) a sales tax of 10% and (b) a valued-added tax of 10%.

SOLUTION

a. A sales tax of 10% is straightforward and easy to calculate. No consumption taxes are paid until the coat is sold to the consumer.

$$\text{Sales tax} = (0.10)(10 + 25 + 150 + 210) = \$39.50$$

The consumer pays $434.50, of which $39.50 is sales tax.

b. Following the trail of a value-added tax is more complicated. A spreadsheet helps keep track of the value-added tax at each stage.

The rules for collecting and paying a value-added tax vary. However, as you can see in Example 5, value-added taxes involve more paperwork.

	A	B	C	D
1	Current Value	Value Added	10% Value-Added Tax	New Value
2	$0.00	$10.00	$1.00	$11.00
3	$11.00	$25.00	$2.50	$38.50
4	$38.50	$150.00	$15.00	$203.50
5	$203.50	$210.00	$21.00	$434.50
6	Total	$395.00	$39.50	
7				

The total tax is the same: $39.50.

 **Checkpoint**    Help at *Math*.and**Y☺U**.com

In the example above, compare the effects of (a) a sales tax of 8% and (b) a value-added tax of 8%.

EXAMPLE 6   **Comparing Tax Revenue Sources**

The stacked area graph shows the sources of tax revenue for the U.S. federal government.

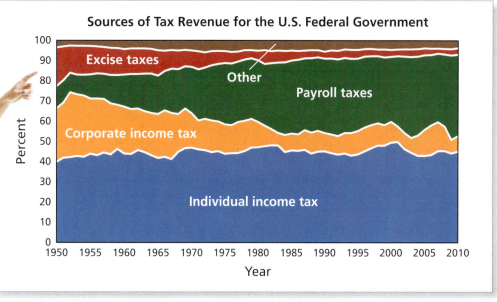

Sources of Tax Revenue for the U.S. Federal Government

Since 1950, excise taxes have become a smaller and smaller percent of the total tax revenue for the federal government.

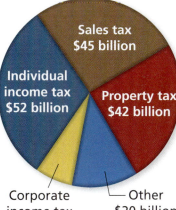

Describe the changes in the types of taxes from 1950 to 2010.

SOLUTION

Individual income tax   This source of tax revenue has held roughly constant during the 60-year period. It makes up about 45% of the total tax revenue.

Corporate income tax   This source of tax revenue declined from about 30% in 1950 to about 10% in 1985. From that time on, it has held fairly steady at about 10%.

Payroll taxes   This source of tax revenue increased from about 10% in 1950 to about 35% in 1985. From that time on, it has held fairly steady at about 35%.

Excise taxes   This source of tax revenue decreased from about 20% in 1950 to only about 3% in 2010.

Other   This source of tax revenue has held roughly constant during the 60-year period. It makes up about 5% of the total tax revenue.

✓ **Checkpoint**   Help at *Math*.and**YOU**.com

The circle graph shows the total tax revenue for state and local governments in California during a recent year.

a. Compare the percent of California's tax revenue raised from individual income taxes with the percent of federal tax revenue raised from individual income taxes.

b. Compare the percent of California's tax revenue raised from corporate income taxes with the percent of federal tax revenue raised from corporate income taxes.

State and Local Taxes in California

2.3 Exercises

Oil Painting In Exercises 1–6, use the table and the oil painting shown. *(See Examples 1 and 2.)*

U.S. State Sales Tax Rates (%) in 2010									
Alabama	4.0	Hawaii	nil	Massachusetts	6.25	New Mexico	5.0	South Dakota	4.0
Alaska	nil	Idaho	6.0	Michigan	6.0	New York	4.0	Tennessee	7.0
Arizona	5.6	Illinois	6.25	Minnesota	6.875	North Carolina	5.75	Texas	6.25
Arkansas	6.0	Indiana	7.0	Mississippi	7.0	North Dakota	5.0	Utah	4.7
California	8.25	Iowa	6.0	Missouri	4.225	Ohio	5.5	Vermont	6.0
Colorado	2.9	Kansas	5.3	Montana	nil	Oklahoma	4.5	Virginia	5.0
Connecticut	6.0	Kentucky	6.0	Nebraska	5.5	Oregon	nil	Washington	6.5
Delaware	nil	Louisiana	4.0	Nevada	6.85	Pennsylvania	6.0	West Virginia	6.0
Florida	6.0	Maine	5.0	New Hampshire	nil	Rhode Island	7.0	Wisconsin	5.0
Georgia	4.0	Maryland	6.0	New Jersey	7.0	South Carolina	6.0	Wyoming	4.0

1. You buy the oil painting in an art gallery in California. What is the state sales tax?

2. You buy the oil painting in an art gallery in Missouri. What is the state sales tax?

3. You buy the oil painting in Washington, D.C. The sales receipt shows that you paid $75.54 in sales tax. What is the sales tax rate in Washington, D.C.?

4. You buy the oil painting in New York. The sales receipt shows that you paid $88.13 in county sales tax. What is the county sales tax rate?

5. You can buy the oil painting in Mississippi or drive 20 miles to an art gallery in Alabama to make the purchase. Which option would you choose? Explain your reasoning.

Oil painting: $1259.00

6. Under what conditions is it beneficial to a consumer to drive to another state to make a purchase? Give specific examples.

T-shirt	$25.00
SUBTOTAL:	$25.00
Sales tax city:	$ 0.30
Sales tax state:	$ 1.50
TOTAL:	**$26.80**
- - THANK YOU - -	

7. **T-shirt** You are given the sales receipt shown after buying a T-shirt. Find the sales tax rates indicated by the receipt. *(See Example 2.)*

8. **Internet Sales** Some people avoid paying sales tax by purchasing items out of state. This is a big problem for many states. What would you do as a state legislator to fix the problem?

Beer Tax In Exercises 9–13, use the information in the map. *(See Examples 3 and 4.)*

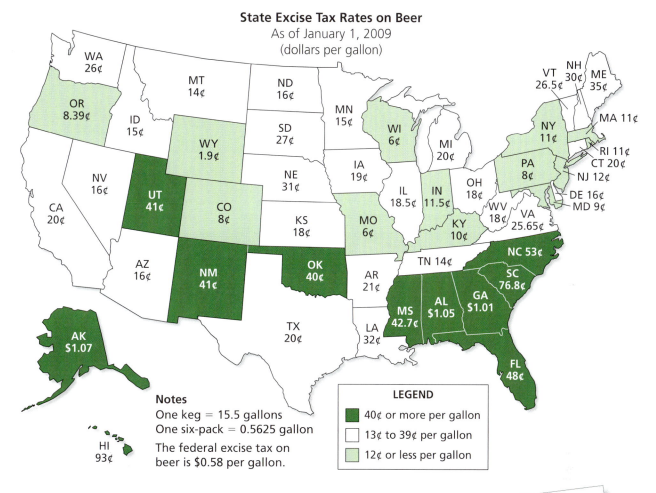

State Excise Tax Rates on Beer
As of January 1, 2009
(dollars per gallon)

WA 26¢
MT 14¢
ND 16¢
OR 8.39¢
ID 15¢
WY 1.9¢
SD 27¢
MN 15¢
WI 6¢
MI 20¢
NV 16¢
UT 41¢
CO 8¢
NE 31¢
IA 19¢
IL 18.5¢
IN 11.5¢
OH 18¢
CA 20¢
KS 18¢
MO 6¢
KY 10¢
WV 18¢
VA 25.65¢
AZ 16¢
NM 41¢
OK 40¢
AR 21¢
TN 14¢
NC 53¢
SC 76.8¢
TX 20¢
LA 32¢
MS 42.7¢
AL $1.05
GA $1.01
AK $1.07
FL 48¢
HI 93¢

VT 26.5¢
NH 30¢
ME 35¢
MA 11¢
NY 11¢
RI 11¢
CT 20¢
PA 8¢
NJ 12¢
DE 16¢
MD 9¢

Notes
One keg = 15.5 gallons
One six-pack = 0.5625 gallon

The federal excise tax on
beer is $0.58 per gallon.

LEGEND
- 40¢ or more per gallon
- 13¢ to 39¢ per gallon
- 12¢ or less per gallon

9. In which state is the excise tax rate on beer the least? the greatest?

10. What is the state excise tax on a keg of beer in Georgia? in Idaho?

11. What is the state excise tax on a 6-pack of beer in Alaska?

12. A customer in Florida buys a 12-pack of beer. How much does the customer pay in federal and state excise taxes?

13. A bar manager in South Carolina purchases a keg of beer for $110.00. What percent of the cost is allocated to federal and state excise taxes?

14. **Excise Tax vs. Sales Tax** How is an excise tax different from a sales tax? Explain your reasoning.

Car Manufacturing In Exercises 15–17, use the information below. *(See Example 5.)*

During the production of a car, value is added by the following.

- Raw materials manufacturers: $7000
- A car manufacturer: $11,000
- A car dealer: $5000

15. The sales tax rate is 6.0%. What is the sales tax on the car?

16. Use the spreadsheet to find the value-added tax of 6% at each stage. Compare the value-added tax of 6% with a sales tax of 6%.

	A	B	C	D	E
1		**Current Value**	**Value Added**	**6% Value-Added Tax**	**New Value**
2	Raw materials	$0.00	$7,000.00		$7,420.00
3	Manufacturer	$7,420.00			
4	Dealer				
5		**Total**			
6					

17. Using the value-added tax approach, what is the retail price of the car?

 Truck Manufacturing In Exercises 18–20, use the information below. *(See Example 5.)*

During the production of a truck, value is added by the following.

- Raw materials manufacturers: $11,000
- A truck manufacturer: $16,000
- A truck dealer: $7000

18. The sales tax rate is 8.25%. What is the sales tax on the truck?

19. Use a spreadsheet to find the value-added tax of 8.25% at each stage. Compare the value-added tax of 8.25% to a sales tax of 8.25%.

20. Using the value-added tax approach, what is the retail price of the truck?

Gasoline The graph shows the prices of gasoline in eight countries. In Exercises 21 and 22, use the graph. *(See Example 6.)*

21. Use percent to compare the taxes collected per gallon of gasoline in the United States with those collected in Germany.

22. Describe any patterns you see in the graph.

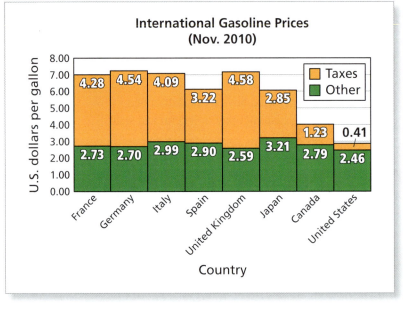

▶ # Extending Concepts

Sunglasses In Exercises 23 and 24, use the sales receipt.

Ray's Beachwear

ITEM	
Hotshot sunglasses	$120.00
SUBTOTAL	$120.00
Shipping	$0.00
Sales tax	$8.40
TOTAL	**128.40**

-THANK YOU-

23. While on vacation in Florida, you buy a pair of sunglasses for the beach. The state sales tax rate in Florida is 6%. What is the local sales tax rate?

24. Your friend loves the sunglasses you bought while on vacation in Florida. She decides to buy the same pair of sunglasses from the Ray's Beachwear website. What might be the total cost for her order? Explain.

25. Exempt Some items are not subject to sales tax in certain tax jurisdictions. The rules for exempt status can vary greatly from state to state. The state sales tax rate applied on the sales receipt shown at the right is 6%. Which item is exempt from sales tax?

26. Other Excise Taxes One of the main purposes of an excise tax is to discourage certain behavior. Name one item or service (not listed in this text) that you think should have an excise tax. Explain how you would define the tax structure, including the rate.

The CORNER **Convenience Store**

6 in. ham & cheese sub	$4.50
Bottle of cough syrup	$7.99
Pack of pens	$1.99
SUBTOTAL:	$14.48
STATE SALES TAX:	$0.39
TOTAL:	**$14.87**

THANK YOU

DATA **27. Dining Room Table Set** The value-added tax spreadsheet shows each stage of the production of a dining room table set. Complete the spreadsheet. What is the tax rate?

	A	B	C	D	E
1		Current Value	Value Added	Value-Added Tax	New Value
2	Raw materials	$0.00	$1,200.00		
3	Manufacturer	$1,290.00	$2,300.00		
4	Finisher	$3,762.50	$500.00		
5	Retailer	$4,300.00	$1,400.00		
6		Total	$5,400.00		

28. National Sales Tax There has been some discussion about whether the United States should implement a national sales tax in addition to existing state and local sales taxes. Suppose you are entering a debate about a national sales tax. Use the Internet to research the topic. Would you be *in favor of* or *opposed to* a national sales tax? Explain your reasoning.

2.4 Budgeting

▶ Create and balance a monthly budget.

▶ Write checks and balance a checkbook.

▶ Analyze a budget.

Creating and Balancing a Monthly Budget

A personal monthly **budget** is a plan that includes your income and your expenses. Here are some tips for creating a meaningful budget.

- If you have bills that are due quarterly, semiannually, or annually, include a monthly average in your monthly budget.

- Include a savings goal in your budget. Saving for a goal is not only smart financially, it is smart emotionally.

- Leave some leeway in your budget for unexpected expenses.

- Keep your budget in a paper or an electronic journal. Save your receipts. Do not fall behind on entering expenses in your budget.

| EXAMPLE 1 | **Comparing Budgeted and Actual Expenses** |

Find the difference between the actual amounts and the budgeted amounts.

CATEGORY	MONTHLY ACTUAL AMOUNT	−	MONTHLY BUDGETED AMOUNT	=	DIFFERENCE
Food					
Groceries	$287.60		$300.00		
Eating out, lunches, snacks	$234.86		$200.00		
Health and Medical					
Insurance (medical, dental, vision)	$165.00		$165.00		

SOLUTION

For each row, subtract the budgeted amount from the actual amount.

Groceries: $287.60 − 300.00 =   −$12.40$

Eating Out: $234.86 − 200.00 = 34.86

Insurance: $165.00 − 165.00 = 0.00

✓ **Checkpoint**   Help at *Math.andYOU.com*

Find the difference between the actual amounts and the budgeted amounts.

CATEGORY	MONTHLY ACTUAL AMOUNT	−	MONTHLY BUDGETED AMOUNT	=	DIFFERENCE
Utilities					
Electricity	$121.46		$125.00		
Water and sewer	$62.30		$58.00		

Study Tip

When the actual amount is less than the budgeted amount, the difference is a negative number. In accounting, negative numbers are often indicated by parentheses and shown in red. For instance, −$12.40 is written as ($12.40).

EXAMPLE 2   **Balancing a Monthly Budget**

Use the monthly budget form to find your total actual expenses. Compare your total actual expenses with your total budgeted expenses.

Study Tip

A negative number means you are under budget. So, it is "good" to have negative numbers for expense differences, but "bad" to have negative numbers for income differences.

DATA CATEGORY	MONTHLY ACTUAL AMOUNT ⊖	MONTHLY BUDGETED AMOUNT ═	DIFFERENCE
INCOME	$5500.00	$5500.00	$0.00
Income Taxes Withheld			
Federal income tax	$1100.00	$1100.00	$0.00
State and local income tax	$225.00	$225.00	$0.00
Social Security/Medicare tax	$412.50	$412.50	$0.00
Spendable Income	$3762.50	$3762.50	$0.00
EXPENSES			
Home			
Mortgage or rent	$455.00	$455.00	$0.00
Homeowners/renters insurance	$95.00	$95.00	$0.00
Property taxes	$234.00	$234.00	$0.00
Utilities			
Electricity	$121.46	$125.00	−$3.54
Water and sewer	$62.30	$58.00	$4.30
Natural gas	$158.16	$200.00	−$41.84
Telephone (landline, cell)	$138.92	$125.00	$13.92
Food			
Groceries	$287.60	$300.00	−$12.40
Eating out, lunches, snacks	$234.86	$200.00	$34.86
Health and Medical			
Insurance (medical, dental, vision)	$165.00	$165.00	$0.00
Medical expenses, co-pays	$0.00	$200.00	−$200.00
Transportation			
Car payments	$175.00	$175.00	$0.00
Gasoline/oil	$48.23	$60.00	−$11.77
Auto repairs/maintenance/fees	$0.00	$50.00	−$50.00
Auto insurance	$125.00	$125.00	$0.00
Debt Payments	$253.48	$253.48	$0.00
Entertainment/Recreation	$124.50	$150.00	−$25.50
Clothing	$0.00	$50.00	−$50.00
Investments and Savings	$125.00	$125.00	$0.00
Miscellaneous	$93.50	$200.00	−$106.50
Total Expenses		$3345.48	

SOLUTION

The total of your actual expenses is $2897.01. So, you spent $3345.48 − 2897.01 = 448.47 less than what you budgeted.

✓ **Checkpoint**   Help at

Complete the "difference" column of the monthly budget form. Explain how you can use the total in this column to check the total in the "actual" column.

Writing Checks and Balancing a Checkbook

A **check** is a written order directing a bank to pay money. Checks are written on preprinted forms as shown.

Write the dollar amount of the check. Use words for whole dollars and a fraction for cents, such as 45/100 for $0.45.

Write the name of the person or company who is receiving the check.

Write the date.

Write the dollar amount using numbers.

Use this part of the check as a reminder of what you are paying for.

Sign your name. Do not print.

Checkbooks come with a registry so that you can keep track of each transaction in your checking account.

EXAMPLE 3 Keeping a Checkbook Registry

Find the balance in your checking account as of 5/11/12.

Date	Check #	Transaction	Credit	Debit	Balance
		Balance Forward			100.00
5/11/12	996	Grocery Store		29.55	
5/11/12		Deposit Paycheck	482.75		
5/16/12		ATM Withdrawal		100.00	
5/17/12	997	Gym Membership		22.35	
5/17/12	998	Cell Phone Company		58.00	
5/17/12	999	Car Payment		82.66	
5/18/12		Deposit Paycheck	501.50		
5/18/12	1000	Birthday Gift		41.28	

SOLUTION

$100.00 - 29.55 = \$70.45$

$70.45 + 482.75 = \$553.20$

Date	Check #	Transaction	Credit	Debit	Balance
		Balance Forward			100.00
5/11/12	996	Grocery Store		29.55	70.45
5/11/12		Deposit Paycheck	482.75		553.20

The balance as of 5/11/12 is $553.20.

 **Checkpoint**

Help at *Math*.and**YOU**.com

Find the balance in your checking account as of 5/18/12.

A **bad check** is a check that is written for an amount that is greater than the balance in the checking account. A bad check is also called an *insufficient funds check* or a *bounced check*. Many banks charge penalties for writing bad checks. These penalties can be significant and can result in numerous problems.

In addition to bank and possible vendor penalties, writing a bad check is illegal. The penalty varies by state but can involve a fine and a prison term. Whether the state takes legal action usually depends on whether a person intentionally or accidentally writes a bad check.

EXAMPLE 4   Calculating Bad Check Penalties

You forget to record an ATM withdrawal of $100 (5/20/12) in your checkbook registry. This causes three checks to "bounce." Your bank charges $50 for each bounced check. You think the balance in your account as of 5/23/12 is $2.63. What is the actual balance after the bad check penalties are deducted?

Some common reasons overdrafts occur are errors in an account register, failure to enter ATM or debit card transactions in a register, and temporary holds on deposits.

Date	Check #	Transaction	Credit	Debit	Balance
		Balance Forward			332.85
5/20/12	406	Cell Phone Company		219.45	113.40
5/23/12	407	Pharmacy		23.56	89.84
5/23/12	408	Electric Company		48.67	41.17
5/23/12	409	Credit Card Payment		38.54	2.63

SOLUTION

After the penalties are deducted, your checkbook registry will look like this.

DATA

Date	Check #	Transaction	Credit	Debit	Balance
		Balance Forward			332.85
5/20/12		ATM Withdrawal		100.00	232.85
5/20/12	406	Cell Phone Company		219.45	13.40
5/23/12	407	Pharmacy		23.56	13.40
5/23/12		Insufficient Funds Penalty		50.00	−36.60
5/23/12	408	Electric Company		48.67	−36.60
5/23/12		Insufficient Funds Penalty		50.00	−86.60
5/23/12	409	Credit Card Payment		38.54	−86.60
5/23/12		Insufficient Funds Penalty		50.00	−136.60

Returned to vendor, not paid. (Pharmacy, 407)

Returned to vendor, not paid. (Electric Company, 408)

Returned to vendor, not paid. (Credit Card Payment, 409)

The actual balance in your account is −$136.60.

✓ Checkpoint

Help at *Math*.and**Y☺U**.com

Suppose that each vendor in Example 4 also charges $25 for a bounced check. How much does forgetting to record your ATM withdrawal cost you?

Analyzing a Budget

| EXAMPLE 5 | **Analyzing Annual Expenses** |

The circle graph shows the average annual expenses for a U.S. household, according to the Department of Labor. What percent of the expenses are spent on taxes and shelter?

SOLUTION

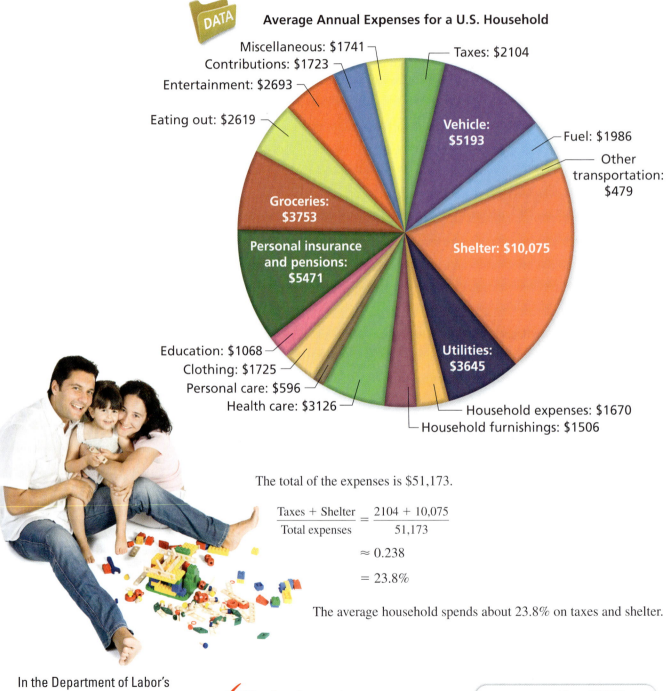

Average Annual Expenses for a U.S. Household

Miscellaneous: $1741
Contributions: $1723
Entertainment: $2693
Eating out: $2619
Groceries: $3753
Personal insurance and pensions: $5471
Education: $1068
Clothing: $1725
Personal care: $596
Health care: $3126
Taxes: $2104
Vehicle: $5193
Fuel: $1986
Other transportation: $479
Shelter: $10,075
Utilities: $3645
Household expenses: $1670
Household furnishings: $1506

The total of the expenses is $51,173.

$$\frac{\text{Taxes} + \text{Shelter}}{\text{Total expenses}} = \frac{2104 + 10{,}075}{51{,}173}$$

$$\approx 0.238$$

$$= 23.8\%$$

The average household spends about 23.8% on taxes and shelter.

In the Department of Labor's survey, the average household had 2.5 members and 1.3 wage earners.

 **Checkpoint**

Help at *Math*.and**YOU**.com

What percent of the expenses are spent on transportation? on food?

When you apply for a home mortgage, the lender will use your gross monthly income and expenses to calculate your **debt-to-income ratio.** There are often two qualifying measures, called the *28/36 rules*.

- **28% Rule:** The ratio of your monthly mortgage (including loan payment, property taxes, and insurance) to your gross monthly income is expected to not exceed 28%.

$$\frac{\text{Monthly mortgage}}{\text{Gross monthly income}} \leq 28\%$$

- **36% Rule:** The ratio of your total monthly debt payments (including mortgage, credit card minimum payments, loans, and all other debts) to your gross monthly income is expected to not exceed 36%.

$$\frac{\text{Monthly debt payments}}{\text{Gross monthly income}} \leq 36\%$$

EXAMPLE 6   Qualifying for a Home Mortgage

You are considering buying a home for $325,000. After your down payment, the monthly mortgage payment (including property taxes and insurance) would be $1450.00. Your gross annual income is $63,000, and you already have a monthly car payment and a monthly credit card payment totaling $250. According to the 28/36 rules, do you qualify for the home mortgage?

SOLUTION

Your gross monthly income is $63,000/12 = $5250.

28% Rule: $\quad\dfrac{\text{Monthly mortgage}}{\text{Gross monthly income}} = \dfrac{1450}{5250}$

$$\approx 0.276$$
$$= 27.6\%$$

36% Rule: $\quad\dfrac{\text{Monthly debt payments}}{\text{Gross monthly income}} = \dfrac{1450 + 250}{5250}$

$$\approx 0.324$$
$$= 32.4\%$$

Both of your debt-to-income ratios are in the acceptable range, so you do qualify for the home mortgage.

✓ **Checkpoint**

Help at *Math*.and**YOU**.com

You are considering buying a home for $425,000. After your down payment, the monthly mortgage payment (including property taxes and insurance) would be $1950.00. Your gross annual income is $73,000, and you already have a monthly car payment and a monthly credit card payment totaling $450. According to the 28/36 rules, do you qualify for the home mortgage?

2.4 Exercises

DATA

Monthly Budget In Exercises 1–7, balance the monthly budget by entering the correct amount in the green cells. *(See Examples 1 and 2.)*

	CATEGORY	MONTHLY ACTUAL AMOUNT	⊖	MONTHLY BUDGETED AMOUNT	⊜	DIFFERENCE
1.	**INCOME**	$4500.00		$4500.00		
	Payroll Deductions					
	Insurance (medical, dental, vision)	$145.00		$145.00		$0.00
2.	Retirement/401(k)	$225.00				$0.00
	Federal income tax	$560.00		$560.00		$0.00
	State and local income tax	$180.00		$180.00		$0.00
	Social Security/Medicare tax	$335.00		$335.00		$0.00
	Total Deductions	**$1445.00**		**$1445.00**		**$0.00**
3.	**Net Income**			**$3055.00**		**$0.00**
	EXPENSES					
	Home					
	Mortgage or rent	$760.00		$760.00		$0.00
	Homeowners/renters insurance	$15.00		$15.00		$0.00
	Property taxes	$0.00		$0.00		$0.00
	Utilities					
	Electricity	$107.26		$95.00		$12.26
	Water and sewer	$30.00		$30.00		$0.00
	Natural gas	$110.00		$110.00		$0.00
	Telephone (landline, cell)	$110.97		$115.00		($4.03)
	Food					
4.	Groceries	$268.34		$300.00		
	Eating out, lunches, snacks	$315.45		$250.00		$65.45
	Transportation					
	Car payments	$384.00		$384.00		$0.00
5.	Gasoline/oil			$75.00		($5.50)
	Auto insurance	$115.00		$115.00		$0.00
	Auto repairs/maintenance/fees	$0.00		$20.00		($20.00)
6.	**Credit Cards**			$45.00		$55.00
7.	**Entertainment/Recreation**	$205.75				($44.25)
	Clothing	$0.00		$50.00		($50.00)
	Investments and Savings	$200.00		$200.00		$0.00
	Miscellaneous	$142.54		$200.00		($57.46)
	Total Expenses	**$2933.81**		**$3014.00**		**($80.19)**

8. Expenses Interpret the difference between the actual and the budgeted total expenses. *(See Example 2.)*

Checkbook In Exercises 9–14, use your checkbook registry shown. *(See Examples 3 and 4.)*

Date	Check #	Transaction	Credit	Debit	Balance
		Balance Forward			520.25
6/24/12	214	Shoe Store		89.99	
6/25/12		ATM Withdrawal		50.00	
6/26/12		ATM Deposit	160.00		
6/27/12	215	Grocery Store		110.59	
6/29/12		Direct Deposit Paycheck	452.17		
6/30/12	216	Rent		450.00	
7/2/12	218	Car Payment		325.15	
7/2/12	219	Electric Company		62.38	
7/3/12		ATM Deposit	65.00		
7/3/12	220	Cell Phone Company		127.16	
7/6/12		Direct Deposit Paycheck	452.17		
7/8/12	221	Credit Card Payment		85.00	

9. Find the balance in your checking account as of 6/26/12.

10. Find the balance in your checking account as of 6/30/12.

11. Are there any bad checks shown in your checkbook registry? If so, which checks are bad?

12. You forget to record check #217 (shown below) in your checkbook registry. Explain the consequences of this omission.

13. The bad check policy at your bank is to return the bounced check to the vendor and charge your account $45 for each instance. Using the information in Exercise 12, find the actual balance in your checking account as of 7/8/12.

14. In addition to the bank charges in Exercise 13, each vendor shown in your checkbook registry charges a $35 penalty for an insufficient funds check. What is your total cost of bad check fees?

Where Does My Money Go? In Exercises 15–22, use the doughnut graph of your monthly budget. *(See Examples 5 and 6.)*

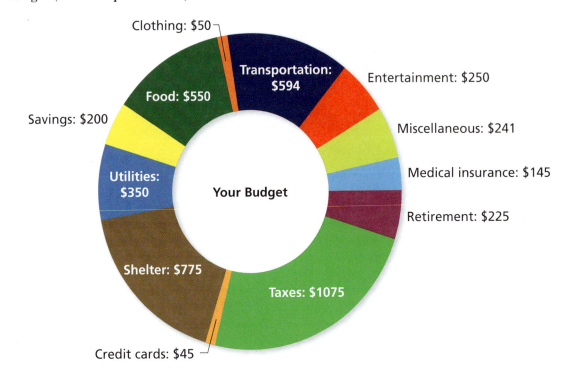

Clothing: $50

Transportation: $594

Food: $550

Entertainment: $250

Savings: $200

Miscellaneous: $241

Utilities: $350

Medical insurance: $145

Your Budget

Retirement: $225

Shelter: $775

Taxes: $1075

Credit cards: $45

15. Your entire monthly income is represented in the doughnut graph. Find your gross annual income.

16. What percent of your expenses are budgeted for food?

17. What percent of your expenses are budgeted for entertainment?

18. What percent of your expenses could you eliminate if needed?

19. You are interested in buying a house. Your realtor determines that the monthly mortgage (including property taxes and insurance) would be $1350.00. Of the $594 budgeted for transportation expenses, $384 is your car payment. According to the 28/36 rules, do you qualify for the home mortgage?

20. According to the 28/36 rules, how much do you think you will be able to spend on a monthly mortgage payment 10 years from now? Explain your reasoning.

21. You go to a financial adviser to get advice about household budgets. The financial adviser gives you the general percent guidelines shown. How does your budget compare with the guidelines?

22. Describe other ways you could analyze your budget.

Household Budget Guidelines

- 25-35% on housing and utilities
- 15-20% on transportation
- 12-15% on food
- 10% on retirement
- 4.5-6% on health care
- 5% on savings
- 5% on entertainment
- 0-5% on credit card/other debt
- 2-4% on miscellaneous

▶ # Extending Concepts

23. Gym Membership Monthly budgets should include expenses that occur quarterly, semiannually, or annually. When you divide these expenses to find a monthly average, you are *prorating* the expense. A gym membership is $150 each quarter. Prorate this expense for your monthly budget.

24. Prorated Expenses For some months, the actual amount spent on prorated expenses will be $0. As a result, the surplus in your monthly budget should increase by the budgeted amount for the expense. Why should you transfer this surplus to savings?

25. Mortgage You are planning to buy a new house, which will increase your shelter expenses by $400. What adjustments would you make to your budget to accommodate the increased expense?

26. Over Budget Many people make sincere attempts to follow a monthly budget. However, for some people, these attempts fail because they either spend more than they should on a certain category or they underestimate an expense in the budget. How can you handle an expense that seems to repeatedly go over budget?

Bank Fees **Many banks and credit unions charge their customers some type of monthly maintenance fee or various transaction fees such as an ATM withdrawal fee. In Exercises 27 and 28, use the portion of your bank statement shown.**

Banking Fees		Statement period		Account	
		2012 - 6 - 7 to 2012 - 7 - 7		0081-156487	
Date	Description	Ref.	Withdrawals	Deposits	Balance
6/16/12	ATM Withdrawal Fee		1.50		526.51
7/3/12	Balance Request Fee		0.50		60.59
7/3/12	ATM Withdrawal Fee		1.50		10.91
7/7/12	Monthly Maintenance Fee		8.00		2.91

27. Why should you account for the bank fees in your checkbook registry?

28. As an executive at a bank, you are in charge of setting up the fee structure, including monthly maintenance fees, bad check fees, and ATM transaction fees. Write a detailed policy explaining how and when these fees would be applied to a customer's account.

2.3–2.4   Quiz

Gas Station In Exercises 1–6, use the sales receipt.

1. Sales taxes are not applied to cigarettes and gasoline for this purchase.

   a. What is the state sales tax rate?

   b. What is the county sales tax rate?

   c. Use the table on page 80 to determine the state in which the gas station is located.

Gas Station

Magazine	$5.95
Sports drink	$1.89
Potato chips	$2.99
3 packs of cigarettes	$18.75
10 gallons of gas	$31.00
Subtotal:	**$60.58**
County sales tax:	$0.11
State sales tax:	$0.49
Total:	**$61.18**

*** THANK YOU ***

2. The state excise tax on a pack of cigarettes is $1.03.

   a. How much is paid in state excise tax for the cigarettes?

   b. What percent of the cost of the cigarettes is allocated to state excise tax?

3. The state excise tax on a gallon of gasoline is $0.17.

   a. How much is paid in state excise tax for the gasoline?

   b. What percent of the cost of the gasoline is allocated to state excise tax?

4. How would you enter this trip to the gas station in your monthly budget?

5. Your checkbook registry is shown. What is the balance in your checking account as of 9/9/12?

Date	Check #	Transaction	Credit	Debit	Balance
		Balance Forward			712.34
9/3/12	361	Rent		540.00	
9/3/12		ATM Withdrawal		80.00	
9/4/12		ATM Deposit	120.00		
9/6/12	363	Department Store		154.17	
9/7/12		Direct Deposit Paycheck	525.36		
9/9/12	364	Gas Station		61.18	

6. You forget to record check #362 in your checkbook registry.

   a. Explain the consequences of this omission.

   b. Your bank charges $40 for a bad check. Find the actual balance in your checking account as of 9/9/12.

362

DATE _Sept. 5, 2012_

PAY TO THE ORDER OF _Cable TV Company_ $ 64.00

Sixty-four and 00/100 - DOLLARS

FOR _Monthly bill_   _Your Signature_

⑈1234567891⑈   ⑈1234567⑈ 0362

Chapter 2   Summary

Section Objectives		How does it apply to you?
Section 1	Find the unit price of an item.	The unit price of an item is the amount you pay per unit of weight or volume. *(See Examples 1 and 2.)*
	Compare the unit prices of two or more items.	You can use unit prices to compare the costs of two or more items that have different sizes. *(See Examples 3 and 4.)*
	Find the annual cost of an item.	Small differences in unit prices can add up to large savings. *(See Examples 5 and 6.)*
Section 2	Find the markup of an item.	A markup is how much more you pay for an item than a retailer. *(See Examples 1 and 2.)*
	Find the discount of an item.	A discount is the amount you save when an item is on sale. *(See Examples 3 and 4.)*
	Find the final price after multiple discounts.	When you have two discounts on the same item, the final price can depend on the order in which you calculate the discounts. *(See Examples 5 and 6.)*
Section 3	Find the sales tax on an item.	Sales tax increases the amount you pay for an item. *(See Examples 1 and 2.)*
	Find the excise tax on an item.	The federal government charges excise taxes on goods such as alcohol, tobacco, gasoline, and gambling to generate revenue. *(See Examples 3 and 4.)*
	Find the value-added tax on an item.	In some countries, a value-added tax is included in the retail price. *(See Example 5.)*
Section 4	Create and balance a monthly budget.	A budget can help you manage your money. *(See Examples 1 and 2.)*
	Write checks and balance a checkbook.	Checks are a common form of payment. A balanced checkbook can help you keep track of your money and avoid overdrafts and bad checks. *(See Examples 3 and 4.)*
	Analyze a budget.	You can analyze your budget to determine if you are spending your money wisely. *(See Examples 5 and 6.)*

Chapter 2   Review Exercises

Section 2.1

Soda In Exercises 1–4, use the soda prices shown.

2-liter bottle

2 L
$0.89

6-pack

101.4 fl oz
$2.69

Case

144 fl oz
$3.79

1. Compare the unit prices of the products.

2. A 20-fluid-ounce bottle of soda sells for $1.50. Find the unit price of the 20-fluid-ounce bottle. Then compare it with the unit price of the 2-liter bottle.

3. A family of 4 drinks about 200 gallons of soda annually. All of the family's soda is purchased in 2-liter bottles instead of 6-packs. How much does the family save per year?

4. A store offers a deal of $10.00 for 4 cases of soda.

    a. Compare the unit price for four cases with the deal to the unit price for one case without the deal.

    b. How much do you save per case with the deal?

Soda The sizes of the bubbles in the bubble graph represent the unit prices of four sodas. The bigger the bubble, the greater the unit price. In Exercises 5–8, use the bubble graph.

5. Without changing the volume, how does increasing the price affect the size of bubble A?

6. Without changing the price, how does decreasing the volume affect the size of bubble C?

7. Which bubble represents the product with the most soda per dollar? Explain your reasoning.

8. Which bubble represents the product with the least soda per dollar? Explain your reasoning.

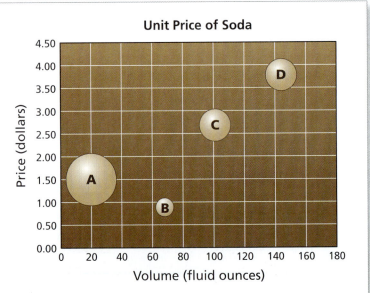

Section 2.2

Waffle Maker The waterfall graph shows the pricing history of waffle makers at an appliance outlet. In Exercises 9–16, use the waterfall graph.

Pricing History

9. Find the initial retail price.

10. Find the initial markup percent.

11. Find the total discount.

DATA 12. You work at the outlet. You order 50 waffle makers at the wholesale price of $35 each. During the next 6 months, you repeatedly mark down the price until you finally sell all 50 waffle makers. Use the sales record below to find the average markup percent.

- 20 sold at initial retail price
- 10 sold at the second discount
- 14 sold at the first discount
- 6 sold at the third discount

13. What is the discount percent on the initial retail price after the second discount?

14. When you apply the third discount, the outlet advertises a 50% discount.

   a. Did the outlet use the initial retail price or the price after the second discount to calculate the discount percent? Explain your reasoning.

   b. Suppose the outlet had used the other price in part (a). What is the discount percent?

15. You get an employee discount of 10% off any item. You buy a waffle maker after the first discount. How much do you pay?

16. You get an employee discount of 50% off any item.

   a. What is the final price when you apply the first discount and then apply the employee discount?

   b. What is the final price when you apply the employee discount and then apply the first discount?

Section 2.3

17. Nonexempt In many places, cigarettes are exempt from sales tax. However, some tax jurisdictions charge sales tax on cigarettes in addition to an excise tax. For instance, retailers in Colorado charge a 2.9% state sales tax on cigarettes. What is the sales tax on a $5.50 pack of cigarettes in Colorado?

18. Sales Tax Rate A convenience store charges $5.75 for a pack of cigarettes. The sales receipt shows that you paid a total of $5.98 including $0.23 in sales tax. Find the sales tax rate.

Cigarettes In Exercises 19–22, use the information in the map.

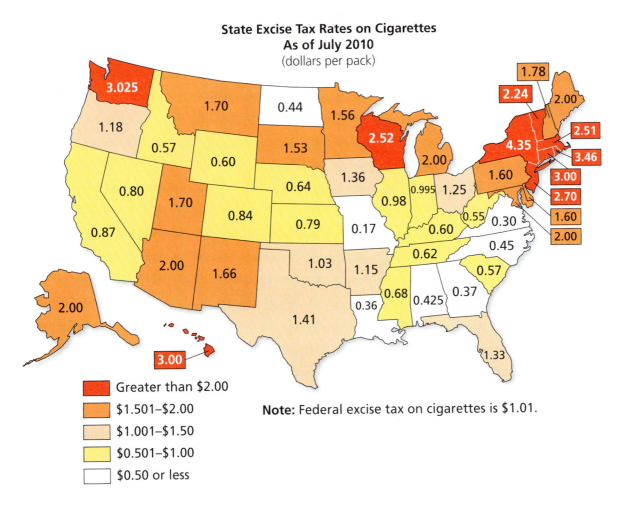

**State Excise Tax Rates on Cigarettes
As of July 2010**
(dollars per pack)

Greater than $2.00
$1.501–$2.00
$1.001–$1.50
$0.501–$1.00
$0.50 or less

Note: Federal excise tax on cigarettes is $1.01.

19. A customer buys four packs of cigarettes in Tennessee. How much does the customer pay in federal and state excise taxes?

20. In Connecticut, a customer purchases 2 packs of cigarettes for $11.50. What percent of the cost is allocated to federal and state excise taxes?

21. How is an excise tax different from a value-added tax?

22. In 1995, the federal excise tax on a pack of cigarettes was $0.24.

   a. Use a percent to describe the change in the federal excise tax from 1995 to 2010.

   b. How might the government justify the increase?

Section 2.4

Cell Phone **In Exercises 23–26, use the line graph.**

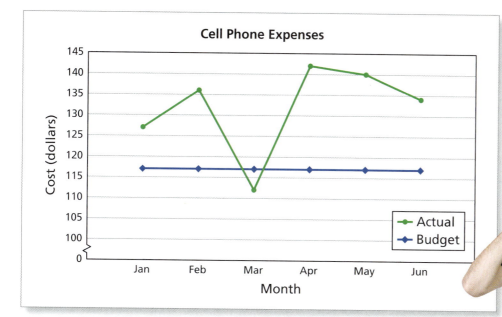

Cell Phone Expenses

23. Estimate the difference between the actual amount and the budgeted amount for each month.

24. What effect does exceeding the budgeted cell phone expense have on the budget surplus or shortage? Explain.

25. Based on the line graph, what adjustments should you make to your monthly budget?

DATA **26.** You pay April's cell phone bill with check #192. Your checkbook registry is shown. What is the balance in your checking account as of 4/5/12?

Date	Check #	Transaction	Credit	Debit	Balance
		Balance Forward			341.48
4/2/12		ATM Deposit	50.00		
4/5/12	191	Car Insurance		104.50	
4/5/12	192	Cell Phone Company		142.00	

27. **Home Mortgage** You are interested in buying a house. Your realtor determines that the monthly mortgage payment (including property taxes and insurance) would be $1075. Your gross annual income is $47,000, and you already have a monthly car payment and a monthly credit card payment totaling $340. According to the 28/36 rules, should you qualify for the home mortgage?

28. **Debt-to-Income Ratio** Use the Internet to research the 28/36 rules. How were the qualifying levels for a home mortgage determined to be 28% and 36%?

3 The Mathematics of Logic & the Media

3.1 Sets & Set Diagrams

▶ Use a union of two sets to represent *or*.

▶ Use an intersection of two sets to represent *and*.

▶ Use the complement of a set to represent *not*.

3.2 Statements & Negations

▶ Analyze statements that have the term *all*.

▶ Analyze statements that have the term *some* or *many*.

▶ Analyze negations of statements.

3.3 Deductive & Inductive Reasoning

▶ Use deductive reasoning with syllogisms.

▶ Know how a deductive reasoning system is created.

▶ Use inductive reasoning.

3.4 Fallacies in Logic

▶ Recognize deductive fallacies.

▶ Use set diagrams to detect fallacies.

▶ Recognize fallacies in advertisements.

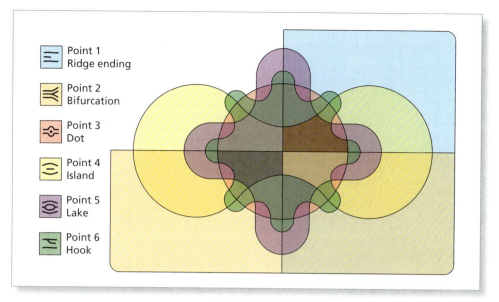

	Point 1 Ridge ending
	Point 2 Bifurcation
	Point 3 Dot
	Point 4 Island
	Point 5 Lake
	Point 6 Hook

Example 4 on page 105 explains why a 6-point match in fingerprinting is not sufficient for a positive identification. How many points do you need for a positive identification?

3.1 Sets & Set Diagrams

▶ Use a union of two sets to represent *or*.

▶ Use an intersection of two sets to represent *and*.

▶ Use the complement of a set to represent *not*.

The Union of Two Sets

Analyzing a statement is often easier when you can visualize the statement. In mathematics, you can use sets and **set diagrams** to visualize statements that deal with "belonging to a group" or "having a characteristic."

Finding the Union of Two Sets

The **union** of set A and set B is everything that is in set A *or* set B. A set diagram for the union of two sets is shown below.

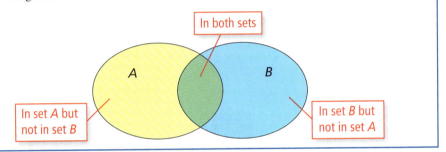

In both sets

In set A but not in set B

In set B but not in set A

EXAMPLE 1 **Drawing a Set Diagram**

There are about 6 million households in the United States that have 1 or more birds as pets. There are about 38 million households that have 1 or more cats as pets. Suppose about 2 million households have *both* birds and cats as pets. Use a set diagram to determine how many households have a bird *or* a cat as a pet.

SOLUTION

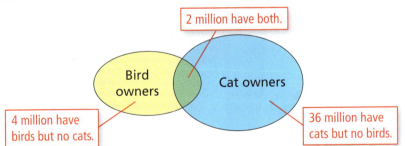

2 million have both.

Bird owners

Cat owners

4 million have birds but no cats.

36 million have cats but no birds.

So, 42 million households (4 + 2 + 36) have a bird or a cat as a pet.

Study Tip

Notice that a union of two sets is used as a visual for the word *or*. In logic, *or* is considered to be the "inclusive or," meaning that it includes one case, or the other case, or *both*.

✓ **Checkpoint**   Help at *Math*.and**YOU**.com

There are about 120 million women (18 years old or older) in the United States. Of these, about 5.7 million rode a motorcycle during the past year. In all, about 25 million Americans rode a motorcycle during the past year. Draw a set diagram that shows this information. Label each region.

EXAMPLE 2   **Describing Set Diagram Regions**

The set diagram has seven regions. Describe the characteristics of people who are in each region.

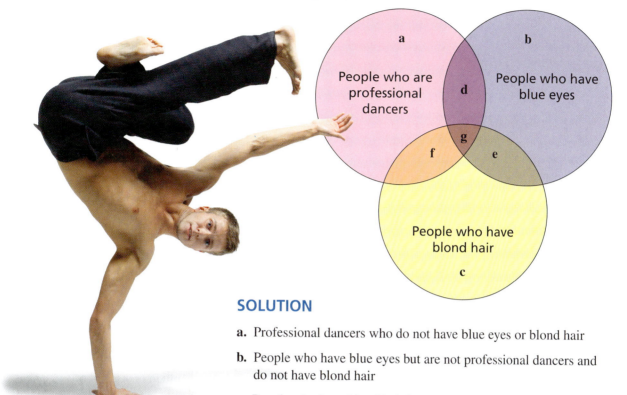

SOLUTION

a. Professional dancers who do not have blue eyes or blond hair

b. People who have blue eyes but are not professional dancers and do not have blond hair

c. People who have blond hair but are not professional dancers and do not have blue eyes

d. Professional dancers who have blue eyes but do not have blond hair

e. People who have blue eyes and blond hair but are not professional dancers

f. Professional dancers who have blond hair but do not have blue eyes

g. Professional dancers with blue eyes and blond hair

✓ **Checkpoint**   Help at *Math*.and**YOU**.com

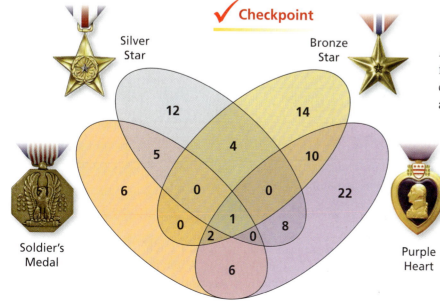

A group of military veterans gathers for a reunion dinner. Each veteran earned a Soldier's Medal, a Silver Star, a Bronze Star, or a Purple Heart.

h. How many earned a Purple Heart?

i. How many earned a Purple Heart *or* a Bronze Star?

j. How many earned a Purple Heart *and* a Bronze Star?

k. How many of the veterans earned all four medals? Explain.

The Intersection of Two Sets

Finding the Intersection of Two Sets

The **intersection** of set *A* and set *B* is everything that is in set *A* *and* set *B*. A set diagram for the intersection of two sets is shown below.

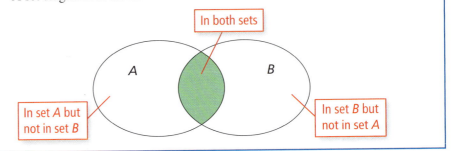

In both sets

In set *A* but not in set *B*

In set *B* but not in set *A*

There are two formulas in set theory that relate the number of items in the union and intersection of two sets.

Number in union	=	Number of items in *A*	+	Number of items in *B*	−	Number in intersection
Number in intersection	=	Number of items in *A*	+	Number of items in *B*	−	Number in union

EXAMPLE 3   **Using a Set Diagram**

You are reading a mystery novel about two crimes committed on a yacht. There are 11 passengers. Each passenger has at least 1 alibi, 8 have an alibi for 1 crime, and 7 have an alibi for the other crime. Assuming the alibis are valid, how many of the passengers are certain to have *not* committed either crime?

SOLUTION

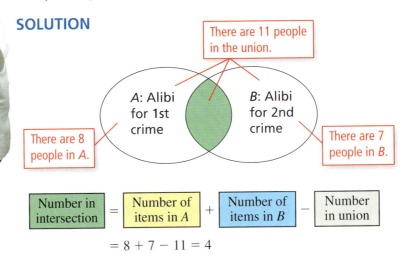

There are 11 people in the union.

A: Alibi for 1st crime

B: Alibi for 2nd crime

There are 8 people in *A*.

There are 7 people in *B*.

Number in intersection	=	Number of items in *A*	+	Number of items in *B*	−	Number in union

$$= 8 + 7 - 11 = 4$$

So, four of the passengers have alibis for both crimes. You can be certain that none of these four committed either crime.

✓ **Checkpoint**   Help at *Math*.and**Y☺U**.com

Suppose that of the 11 passengers, each has at least 1 alibi, 9 have an alibi for 1 crime, and 8 have an alibi for the other crime. Assuming the alibis are valid, how many of the passengers are certain to have *not* committed either crime?

EXAMPLE 4   Finding the Intersection of Sets

Suppose there are 1 billion people whose index fingerprint has the Point 1 characteristic. A similar number of people have each of the other five characteristics. Suppose that for any 2 of the 6 basic sets in the diagram, there is a 20% overlap. Explain why a 6-point match between 2 fingerprints is not a guarantee that the fingerprints came from the same person.

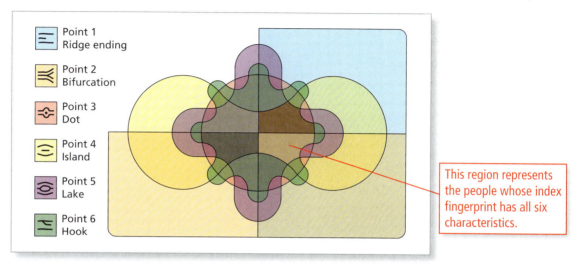

Point 1 Ridge ending
Point 2 Bifurcation
Point 3 Dot
Point 4 Island
Point 5 Lake
Point 6 Hook

This region represents the people whose index fingerprint has all six characteristics.

SOLUTION

Twenty percent (or 200 million) of the people have Points 1 and 2. Up to 40 million people have Points 1, 2, and 3. Continuing this pattern, up to 320,000 people have all 6 points. This gives you some idea why a 6-point match is not good enough to make a positive identification.

✓ **Checkpoint**

Help at *Math*.andY○U.com

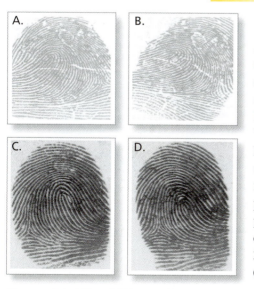

Fingerprints (A) and (B) come from the same finger but look different. Fingerprints (C) and (D) come from different fingers but look similar.

A 12-point match is often considered sufficient to conclude that 2 fingerprints come from the same person. In the research article "On the Individuality of Fingerprints," the authors state "the probability that a fingerprint with 36 minutiae points will share 12 minutiae points with another arbitrarily chosen fingerprint with 36 minutiae points is 6.10×10^{-8}."

Crossover
Bifurcation
Ridge ending
Core
Pore
Island
Delta

In Example 4, suppose that there are 12 different points. Continue the pattern described in the solution to find the number of people who are in the intersection of all 12 sets.

The Complement of a Set

Finding the Complement of a Set

The **complement** of set A consists of everything that is in some universal set U, but not in set A. A set diagram for the complement of a set is shown below.

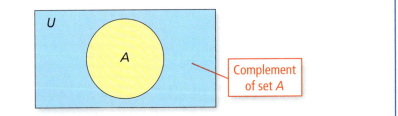

If you know how many items are in set A and in set U, then you can calculate the number of items in the complement of set A.

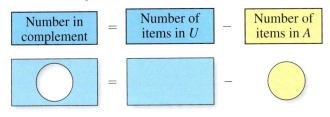

$$\text{Number in complement} = \text{Number of items in } U - \text{Number of items in } A$$

EXAMPLE 5 Finding the Complement of a Set

The numbers of unemployed people in the United States in a recent year are shown. The total number of employed people and unemployed people was 154,286,000. How many people were employed?

Age	Number Unemployed
16–19	1,285,000
20–24	1,545,000
25–44	3,553,000
45–64	2,276,000
65+	264,000

SOLUTION

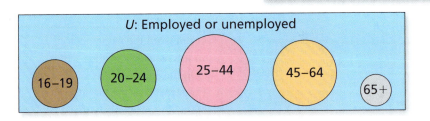

U: Employed or unemployed

$$\text{Unemployed} = 1{,}285{,}000 + 1{,}545{,}000 + 3{,}553{,}000 + 2{,}276{,}000 + 264{,}000$$
$$= 8{,}923{,}000$$
$$\text{Employed} = 154{,}286{,}000 - 8{,}923{,}000$$
$$= 145{,}363{,}000$$

About 145 million employed

✓ **Checkpoint** Help at *Math*.and**Y©U**.com

What was the unemployment rate for the year shown in Example 5?

EXAMPLE 6   **Finding the Complement of a Set**

There are about 62,300 species of vertebrates on Earth. They are classified as fish, amphibians, reptiles, mammals, and birds.

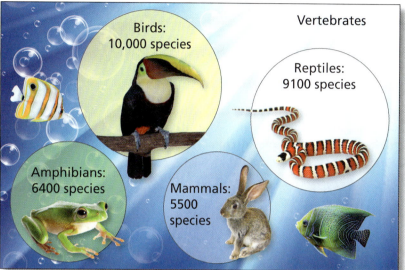

Vertebrates

Birds: 10,000 species

Reptiles: 9100 species

Amphibians: 6400 species

Mammals: 5500 species

a. How many species of fish are on Earth?

b. Would you say that there is more diversity among fish than among the other vertebrate classes? Explain your reasoning.

SOLUTION

Fish are found in nearly all aquatic environments, from high mountain streams to the depths of the deepest oceans.

a. Fish species $= 62{,}300 - (10{,}000 + 9100 + 6400 + 5500)$

$\qquad\qquad\quad\; = 62{,}300 - 31{,}000$

$\qquad\qquad\quad\; = 31{,}300$

There are about 31,300 species of fish on Earth.

b. There are more species of fish than in all of the other vertebrate classes combined. So, fish have a much greater species diversity than any of the other vertebrate classes.

✓ **Checkpoint**

Help at *Math*.and**YOU**.com

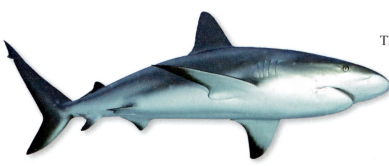

There are about 400 species of sharks.

Draw a set diagram that relates shark species with all other species of fish. Label the number of species in each region.

3.1 Exercises

College Testing **In Exercises 1 and 2, use the information below.** *(See Example 1.)*

In a high school graduating class, 128 students took the SAT, 100 students took the ACT, and 98 students took *both* the SAT and the ACT.

1. Draw a set diagram that shows the information.

2. How many students took the SAT *or* the ACT?

Colleges **The set diagram shows characteristics of colleges that a high school senior is considering. In Exercises 3–8, use the set diagram.** *(See Example 2.)*

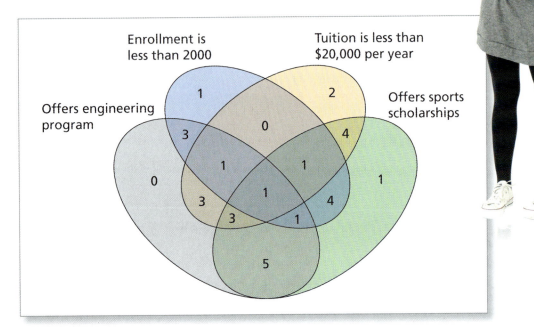

3. How many of the colleges offer an engineering program?

4. How many of the colleges offer sports scholarships *or* an engineering program?

5. How many of the colleges offer sports scholarships *and* have an enrollment less than 2000?

6. How many of the colleges have tuition less than $20,000 per year, offer an engineering program, *and* offer sports scholarships?

7. Use the set diagram to determine whether colleges with an enrollment less than 2000 are likely to have tuitions less than $20,000 per year. Explain why the answer is reasonable in real life.

8. List three characteristics of a college that are important to you. Research 20 colleges to determine whether each college meets your desired characteristics. Create a set diagram of the results.

Trivia **In Exercises 9 and 10, use the information below.** *(See Example 3.)*

A contestant on a game show must answer two trivia questions correctly to advance to the next qualifying round. Of the 10 contestants, 7 answer the first question correctly and 6 answer the second question correctly.

9. Draw a set diagram that shows the information.

10. How many of the contestants advance to the next qualifying round?

Game Show **In Exercises 11–16, use the information below.** *(See Example 4.)*

A game show contestant has a chance to win a car, a boat, and an all-terrain vehicle (ATV). There are eight keys that start the engine of each prize, two of which start the engine of all three prizes. There is a 50% overlap for any 2 of the sets in the diagram.

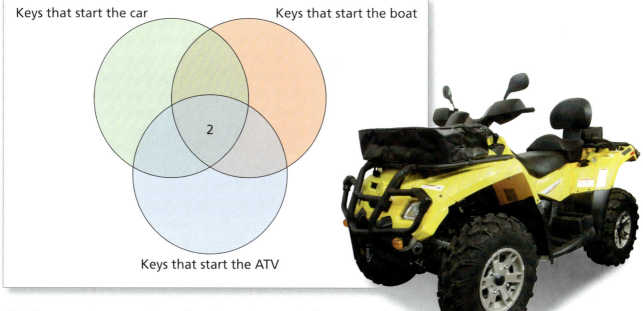

Keys that start the car

Keys that start the boat

2

Keys that start the ATV

11. How many keys start the engine of exactly one prize?

12. How many keys start the engines of exactly two prizes?

13. How many keys start the engine of the car *and* the boat?

14. How many keys start the engine of the car *or* the boat?

15. The contestant begins the game by choosing a key from a basket. How many keys are in the basket?

16. In the United States, game show contestants must pay taxes on items they win. As a result, contestants often refuse some or all of their prizes.

  a. You win the $40,000 car, the $15,000 boat, and the $5,000 ATV in December. Suppose you will owe taxes in April equal to 25% of the value of your winnings. How much will you owe? What will you do?

  b. You decide to keep all the prizes and sell the boat to cover the taxes. Can you be certain this will cover the taxes? What other problems might you encounter?

Nuclear Weapons **In Exercises 17 and 18, use the information below.** *(See Example 5.)*

A country's stockpile of nuclear weapons consists of operational warheads and reserve warheads.

U.S. stockpile: 5113 warheads

- Operational warheads:
  - 1968 strategic
  - 500 nonstrategic

17. Draw a set diagram that shows the information.

18. How many reserve warheads are in the U.S. stockpile?

Nuclear Weapons **In Exercises 19 and 20, use the information and the set diagram.** *(See Example 6.)*

Russia's 4600 operational warheads consist of nonstrategic warheads and the following strategic warheads.

- Intercontinental ballistic missiles (ICBMs)
- Submarine-launched ballistic missiles (SLBMs)
- Bombers/weapons

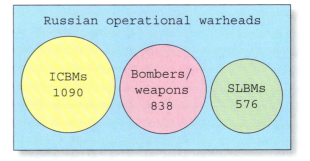

Russian operational warheads

ICBMs 1090   Bombers/weapons 838   SLBMs 576

19. How many Russian operational warheads are nonstrategic?

20. Can you use these data to determine whether the United States or Russia has a greater stockpile of nuclear weapons? Explain your reasoning.

21. **New START** In 2010, the United States ratified the New START (Strategic Arms Reduction Treaty) with Russia. This agreement reduces the number of strategic nuclear weapons in each country.

  a. How do you think New START will affect the set diagrams in Exercises 17–20?

  b. Discuss trends in the graph. How do you think New START will affect the graph?

U.S. Nuclear Weapons Stockpile (1945–2009)

Maximum warheads: 31,255

Cuban Missile Crisis

USSR disbands

y-axis: Warheads — 35,000 / 30,000 / 25,000 / 20,000 / 15,000 / 10,000 / 5,000 / 0

x-axis: 1945 1950 1955 1960 1965 1970 1975 1980 1985 1990 1995 2000 2005 2010

Fiscal years

22. **Doomsday Clock** The Bulletin of the Atomic Scientists assesses how close humanity is to doomsday. Midnight represents doomsday. Do you think the minute hand is moving away from midnight, toward midnight, or neither? Explain your reasoning.

▶ Extending Concepts

Bobbleheads In Exercises 23 and 24, a company is conducting a survey to determine which bobblehead would generate the most sales. Participants must choose at least one bobblehead from the following list.

A. Albert Pujols **B.** Drew Brees **C.** LeBron James **D.** Phil Mickelson

23. How many different sets of choices can result from the survey?

24. Which set diagram should the company use to display the results? Explain.

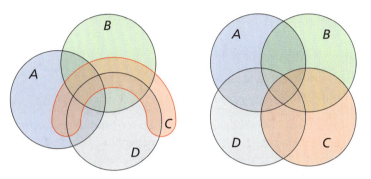

The first modern bobbleheads were introduced in the 1950s and were popularized with the help of Major League Baseball.

U.S. Banking Companies In Exercises 25–28, use the set diagram.

25. How many of the companies are *not* U.S. banking companies?

26. How many of the companies are *not* banking companies?

27. How many of the companies are neither U.S. companies nor banking companies?

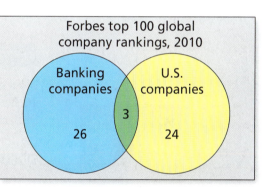

28. The graph shows the value of the KBW Bank Index (ticker: BKX) from 2008 to 2010. It is intended to reflect the evolving U.S. financial sector. How do you think the set diagram above will change when altered to show the list from early 2009 (during the U.S. financial crisis)? early 2008 (before the U.S. financial crisis)? Explain.

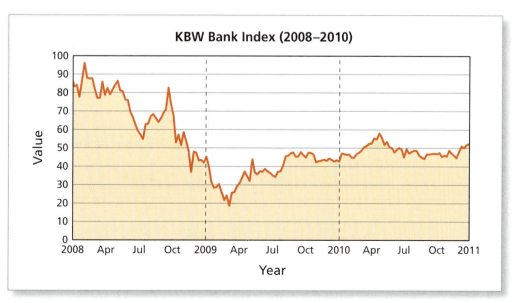

3.2 Statements & Negations

▶ Analyze statements that have the term *all*.

▶ Analyze statements that have the term *some* or *many*.

▶ Analyze negations of statements.

Statements Involving *All*

Logic is the study of the rules and terms used to make persuasive arguments. A **statement** is a declarative sentence that is either true or false. An **argument** is a string of two or more statements that imply another statement. Here is an example of an argument.

- Premise: All types of fruit are food.
- Premise: Tomatoes are a type of fruit.
- Conclusion: Therefore, tomatoes are food.

In this lesson, you will see how set diagrams can be used to analyze statements. In Section 3.3, you will learn rules for writing logical arguments. Finally, in Section 3.4, you will learn to recognize illogical arguments, which are called fallacies.

EXAMPLE 1 Analyzing a Statement Involving *All*

Use a set diagram to analyze the statement taken from the Declaration of Independence.

"We hold these truths to be self-evident, that all men are created equal, that they are endowed by their Creator with certain unalienable Rights, that among these are Life, Liberty and the pursuit of Happiness."

SOLUTION

There are several ways to use set diagrams to analyze this statement. Here are three of them.

Creatures who have the right to **Life** — All people

Creatures who have the right to **Liberty** — All people

Creatures who have the right to **Pursue Happiness** — All people

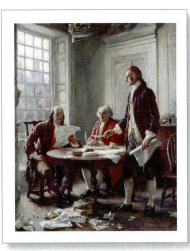

Virginia Historical Society, *www.vahistorical.org*

The Declaration of Independence was written primarily by Thomas Jefferson and adopted by the Continental Congress on July 4, 1776. It announced that the 13 American colonies were no longer part of the British Empire. The document uses logic to justify a people's right to abolish a government that denies them basic human rights.

Notice that the statement says nothing about animal rights. It does leave open the possibility that animals also have such rights.

✓ **Checkpoint**   Help at *Math*.and**Y⊕U**.com

Use a set diagram to analyze the 10th Amendment in the Bill of Rights:

"The powers not delegated to the United States by the Constitution, nor prohibited by it to the States, are reserved to the States respectively, or to the people."

EXAMPLE 2   **Analyzing a Statement Involving** *All*

Use a set diagram to analyze the pet policy in a renter's agreement.

> **PET POLICY** Resident agrees to pay a nonrefundable pet fee of $20 per month per pet. All pets found on the property, but not registered under this agreement, will be presumed to be strays and disposed of by the appropriate agency as prescribed by law. In the event a resident harbors an undisclosed pet, he or she agrees to pay a pet fee for the entire term of the agreement, regardless of when the pet was first introduced to the household.

SOLUTION

Here is one way to visualize the second sentence of the policy.

The home ownership rate in the United States is about 67%. That is, about 67% of all occupied housing units are occupied by the unit's owner. The other 33% of occupied housing units are occupied by renters.

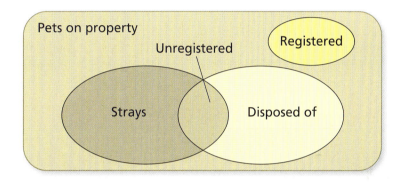

Using a set diagram helps you understand what is and is not being said by the policy. Here are some examples.

- It is possible that some strays are not disposed of.

- It is possible that some pets that are disposed of are not strays.

- All unregistered pets are considered strays and are disposed of.

✓ **Checkpoint**   Help at *Math*.and**YOU**.com

Use a set diagram to analyze the court cost clause in a renter's agreement.

> **COURT COSTS** Resident agrees to pay all court costs and attorneys' fees incurred by the owner in enforcing legal action or any of the owner's other rights under this agreement or any state law. In the event any portion of this agreement shall be found to be unsupportable under the law, the remaining provisions shall continue to be valid and subject to enforcement in the courts without exception.

Statements Involving *Some* or *Many*

EXAMPLE 3 **Analyzing a Statement Involving *Some***

Use a set diagram to analyze the statement, "In the United States, *some* presidential candidates have won the popular vote but still lost the election."

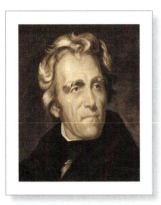

Andrew Jackson won the popular vote but lost the election in 1824.

Samuel Tilden won the popular vote but lost the election in 1876.

Grover Cleveland won the popular vote but lost the election in 1888.

Albert Gore won the popular vote but lost the election in 2000.

SOLUTION

Here is one way to use a set diagram to analyze the statement.

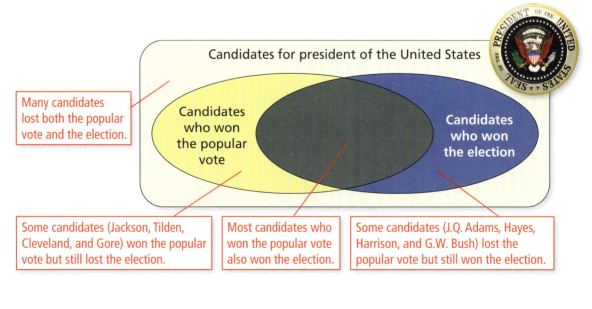

Candidates for president of the United States

Many candidates lost both the popular vote and the election.

Candidates who won the popular vote

Candidates who won the election

Some candidates (Jackson, Tilden, Cleveland, and Gore) won the popular vote but still lost the election.

Most candidates who won the popular vote also won the election.

Some candidates (J.Q. Adams, Hayes, Harrison, and G.W. Bush) lost the popular vote but still won the election.

✓ **Checkpoint**

Help at *Math*.and**Y⊕U**.com

Use a set diagram to represent the statement, "In the United States, 14 presidents served as vice presidents: J. Adams, Jefferson, Van Buren, Tyler, Fillmore, A. Johnson, Arthur, T. Roosevelt, Coolidge, Truman, Nixon, L. Johnson, Ford, and George H.W. Bush. Of these, Tyler, Fillmore, A. Johnson, Arthur, and Ford did not win a presidential election."

EXAMPLE 4   **Analyzing a Statement Involving *Many***

Use a set diagram to analyze the concepts in the paragraph.

"The late nineteenth and early twentieth centuries are often referred to as the time of the 'robber barons.' It is a staple of history books to attach this derogatory phrase to such figures as John D. Rockefeller, Cornelius Vanderbilt, and the great nineteenth-century railroad operators—Grenville Dodge, Leland Stanford, Henry Villard, James J. Hill, and others. To most historians writing on this period, these entrepreneurs committed thinly veiled acts of larceny to enrich themselves at the expense of their customers. Once again we see the image of the greedy, exploitative capitalist, but in many cases this is a distortion of the truth." *How Capitalism Saved America: The Untold History of Our Country, from the Pilgrims to the Present*, Thomas J. DiLorenzo

History Repeats Itself—The Robber Barons of the Middle Ages, and the Robber Barons of Today.

This 19th-century cartoon compares wealthy industrialists to the feudal lords of the Middle Ages who charged excessive fees to merchants who passed through their lands.

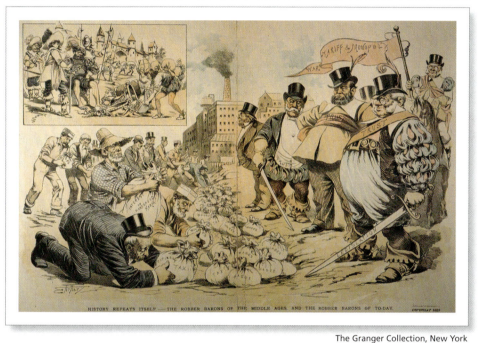

The Granger Collection, New York

SOLUTION

The paragraph implies that this region is not empty. There were many entrepreneurs in the late 19th and early 20th centuries who were not greedy and exploitative.

Entrepreneurs in the late 19th and early 20th centuries

Greedy, exploitative capitalists

✓ **Checkpoint**

Help at *Math*.and**YOU**.com

Use a set diagram to analyze the concepts in the paragraph.

Robber barons were often depicted as men wearing suits with top hats and walking sticks. Mr. Monopoly, originally known as Rich Uncle Pennybags, is an example of this characterization in popular culture.

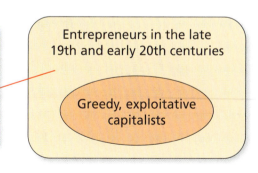

The Negation of a Statement

You might think that finding the negation of a statement is simple. Often, this is true. For instance, the negation of

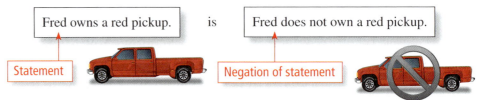

| Fred owns a red pickup. | is | Fred does not own a red pickup. |

Statement

Negation of statement

However, when a statement contains terms such as *and* or *or*, finding the negative can be a bit more challenging. In such cases, a set diagram can help describe the negation.

EXAMPLE 5 **Negating a Statement Involving *And***

Use a set diagram to visualize the negation of the "to love and to cherish" portion of this traditional wedding vow.

> I take you to be my wedded spouse. To have and to hold, from this day forward, for better, for worse, for richer, for poorer, in sickness or in health, to love and to cherish until death do us part.

SOLUTION

Start by drawing a set diagram. Identify the regions represented by the original statement. Then identify all the regions other than those represented by the original statement.

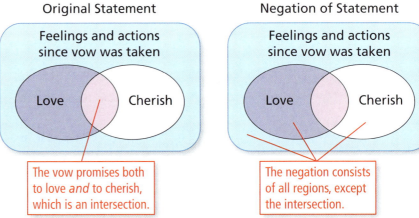

Original Statement

Feelings and actions since vow was taken

Love   Cherish

The vow promises both to love *and* to cherish, which is an intersection.

Negation of Statement

Feelings and actions since vow was taken

Love   Cherish

The negation consists of all regions, except the intersection.

So, the vow can be broken in three ways.

- Continue to love, but cease to cherish.
- Continue to cherish, but cease to love.
- Cease loving and cherishing.

✓ **Checkpoint**

Help at *Math*.and**YOU**.com

Use a set diagram to visualize the negation of this accusation in the game of CLUE: "It was Miss Scarlet with the wrench in the kitchen." If you are playing the game, how can you prove this accusation to be false?

EXAMPLE 6   **Negating a Statement Involving *Or***

Use a set diagram to visualize the negation of the statement. Then, describe how the original statement can be false.

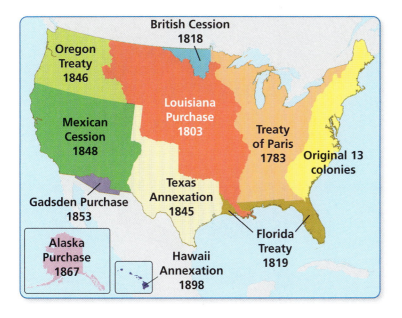

This summer I will visit the region acquired by the Oregon Treaty or the Louisiana Purchase.

The United States purchased or annexed its land (outside the original 13 colonies) in 10 stages.

1783:	Treaty of Paris
1803:	Louisiana Purchase
1818:	British Cession
1819:	Florida Treaty
1845:	Texas Annexation
1846:	Oregon Treaty
1848:	Mexican Cession
1853:	Gadsden Purchase
1867:	Alaska Purchase
1898:	Hawaii Annexation

SOLUTION

Start by drawing a set diagram. Identify the regions represented by the original statement. Then identify all the regions other than those represented by the original statement.

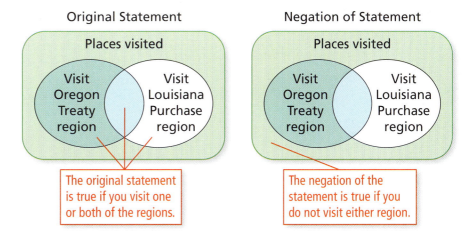

Notice that the original statement is still true if you visit both regions.

 **Checkpoint**   Help at *Math*.and**Y☺U**.com

Use a set diagram to visualize the negation of the statement. Then describe how the original statement can be false.

I am going to write about the Mexican Cession, the Gadsden Purchase, or the Texas Annexation for my term paper in American History.

3.2 Exercises

Social Networking In Exercises 1–7, use a set diagram to analyze the statement about a social networking website. *(See Examples 1 and 2.)*

1. You own all the content and information that you post.

2. All businesses can join the website and create fan clubs.

3. All individuals can join the website, post pictures, and manage their privacy settings.

4. You can receive a text message notification on your phone for every friend request and every email you receive on the website.

5. You will not conduct any contests on the website without prior authorization from the website administrators. All unauthorized contests found will be considered illegal activity and will be terminated. Your membership may also be terminated.

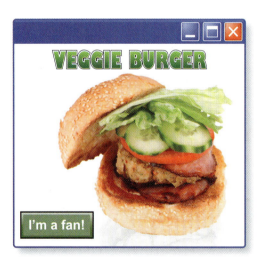

6. A fast-food restaurant has a fan page for its avocado burger and a separate fan page for its jalapeno burger. Every member who is a "fan" of both burgers will receive a coupon for a free burger of his or her choice.

7. All members who are fans of the avocado burger, jalapeno burger, and veggie burger are eligible for a grand prize. Add this information to your set diagram in Exercise 6.

8. **Privacy Settings** The recommended privacy settings on a social networking website are shown.

   a. Use a set diagram to analyze the different categories of members. Indicate what parts of your profile are accessible to each region in the set diagram.

   b. You have a circle of friends that you frequently spend time with as a group. Where in the set diagram are these friends most likely located?

	Friends	Friends of Friends	All Members
Name and main picture	✓	✓	✓
Family and relationships	✓	✓	✓
Photos and videos	✓	✓	
Political views	✓	✓	
Birthday	✓	✓	
Permission to write on your page	✓		
Contact information	✓		

Pearl Harbor In Exercises 9 and 10, use a set diagram to analyze the statement about the attacks at Pearl Harbor on December 7, 1941. *(See Example 3.)*

9. Some of the Japanese machinery destroyed in the attack were midget submarines.

10. Some of the U.S. ships heavily damaged in the attack were battleships.

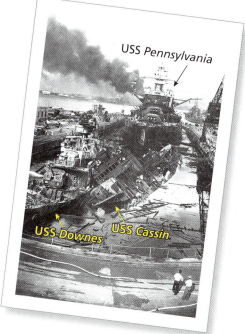

USS Pennsylvania

USS Downes   USS Cassin

The USS *Arizona* sank in Pearl Harbor, where it still lies today. A memorial was built above the battleship. You can see the image of the sunken ship in the picture above.

The USS *Pennsylvania* was dry-docked when the attacks occurred. The battleship received only minor damage. The destroyers USS *Downes* and USS *Cassin* were dry-docked with the USS *Pennsylvania*, but they were heavily damaged.

Winston Churchill In Exercises 11–14, use a set diagram to analyze the Winston Churchill quote. *(See Examples 3 and 4.)*

11. "Some see private enterprise as a predatory target to be shot, others as a cow to be milked, but few are those who see it as a sturdy horse pulling the wagon."

12. "Once in a while you will stumble upon the truth but most of us manage to pick ourselves up and hurry along as if nothing had happened."

13. "Though I have now stated the two great dangers which menace the homes of the people, war and tyranny, I have not yet spoken of poverty and privation which are in many cases the prevailing anxiety."
    "Sinews of Peace" speech

14. "From Stettin in the Baltic to Trieste in the Adriatic, an iron curtain has descended across the Continent. Behind that line lie all the capitals of the ancient states of Central and Eastern Europe. Warsaw, Berlin, Prague, Vienna, Budapest, Belgrade, Bucharest and Sofia, all these famous cities and the populations around them lie in what I must call the Soviet sphere, and all are subject in one form or another, not only to Soviet influence but to a very high and, in some cases, increasing measure of control from Moscow."   "Sinews of Peace" speech

Winston Churchill served as the British prime minister twice, with his first term during World War II. He was the first of seven foreigners to be named honorary citizens of the United States.

New Year's Resolutions In Exercises 15–18, use a set diagram to visualize the negation of the New Year's resolution. *(See Examples 5 and 6.)*

15. I will save $100 each month and go on a vacation overseas.

16. I will exercise 4 times each week and lose 10 pounds.

17. I will volunteer at a homeless shelter or a soup kitchen.

18. I will stop biting my fingernails or cracking my knuckles.

Holidays In Exercises 19–21, use a set diagram to visualize the negation of the statement about holidays in a company handbook. *(See Examples 5 and 6.)*

Holidays for Federal Employees

2013

January	February	March	April
S M T W T F S	S M T W T F S	S M T W T F S	S M T W T F S
(1) 2 3 4 5	1 2	1 2	1 2 3 4 5 6
6 7 8 9 10 11 12	3 4 5 6 7 8 9	3 4 5 6 7 8 9	7 8 9 10 11 12 13
13 14 15 16 17 18 19	10 11 12 13 14 15 16	10 11 12 13 14 15 16	14 15 16 17 18 19 20
20 (21) 22 23 24 25 26	17 (18) 19 20 21 22 23	17 18 19 20 21 22 23	21 22 23 24 25 26 27
27 28 29 30 31	24 25 26 27 28	24 25 26 27 28 29 30 31	28 29 30

May	June	July	August
S M T W T F S	S M T W T F S	S M T W T F S	S M T W T F S
1 2 3 4	1	1 2 3 (4) 5 6	1 2 3
5 6 7 8 9 10 11	2 3 4 5 6 7 8	7 8 9 10 11 12 13	4 5 6 7 8 9 10
12 13 14 15 16 17 18	9 10 11 12 13 14 15	14 15 16 17 18 19 20	11 12 13 14 15 16 17
19 20 21 22 23 24 25	16 17 18 19 20 21 22	21 22 23 24 25 26 27	18 19 20 21 22 23 24
26 (27) 28 29 30 31	23 24 25 26 27 28 29 30	28 29 30 31	25 26 27 28 29 30 31

September	October	November	December
S M T W T F S	S M T W T F S	S M T W T F S	S M T W T F S
1 (2) 3 4 5 6 7	1 2 3 4 5	1 2	1 2 3 4 5 6 7
8 9 10 11 12 13 14	6 7 8 9 10 11 12	3 4 5 6 7 8 9	8 9 10 11 12 13 14
15 16 17 18 19 20 21	13 (14) 15 16 17 18 19	10 (11) 12 13 14 15 16	15 16 17 18 19 20 21
22 23 24 25 26 27 28	20 21 22 23 24 25 26	17 18 19 20 21 22 23	22 23 24 (25) 26 27 28
29 30	27 28 29 30 31	24 25 26 27 (28) 29 30	29 30 31

19. Employees will not be paid for vacations, holidays, and sick leave.

20. Employees do not have to work the Friday before Easter, September 11, and holidays for federal employees.

21. Employees will be invited to a company picnic on Memorial Day, Independence Day, or Labor Day.

22. Compensation The following statement is in a company's employee handbook.

> Employees will receive a cost-of-living adjustment each year or a New Year's Day bonus.

a. Use a set diagram to visualize the negation of the statement. *(See Example 6.)*

b. The company changes the word *or* to *and*. How does this change the negation of the statement? *(See Example 5.)*

▶ **Extending Concepts**

ENERGY STAR In Exercises 23 and 24, use a set diagram to analyze the statement about **ENERGY STAR.**

23.

> The ENERGY STAR label was established to
>
> • reduce greenhouse gas emissions and other pollutants caused by the inefficient use of energy and
>
> • make it easy for consumers to identify and purchase energy-efficient products that offer savings on energy bills without sacrificing performance, features, and comfort.

24.

> **ENERGY STAR Rebates**
>
> Each state and territory will choose dollar amounts for the products selected. Most rebate amounts range from $50 to $500, depending upon the product being purchased, the purchase price, and other potential market factors. Some states give additional rebates for recycling.

25. Emissions Use the graph to write one statement involving "all" and "some" that you can make about the products in the graph. Use a set diagram to analyze the statement.

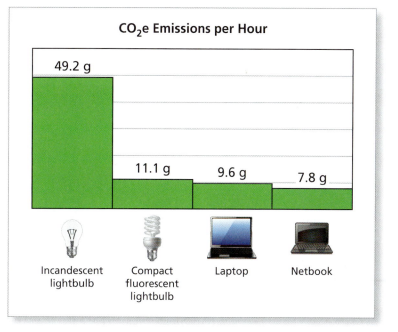

CO_2e Emissions per Hour

49.2 g

11.1 g   9.6 g   7.8 g

Incandescent lightbulb   Compact fluorescent lightbulb   Laptop   Netbook

Purchases In Exercises 26–28, use a set diagram to visualize the negation of the statement about planned purchases.

26. To help heat my apartment, I will buy a space heater and an electric fireplace, or I will buy an ENERGY STAR qualified woodstove.

27. To update my plumbing, I will buy a solar water heater or gas condensing water heater, and low-flow showerheads.

28. For my new home, I will buy a chest freezer or an upright freezer, and fluorescent lighting or ENERGY STAR LED lighting.

3.1–3.2 Quiz

Convictions The set diagram shows the offenses of the convicted inmates at a prison. In Exercises 1–3, use the set diagram.

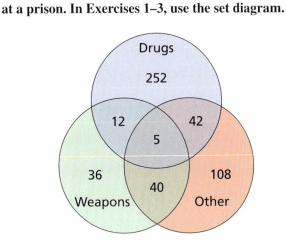

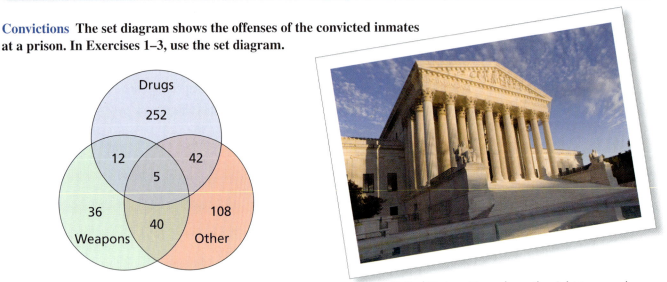

1. How many of the inmates were convicted of drug offenses?

2. How many of the inmates were convicted of weapons offenses *and* "other" offenses?

3. How many of the inmates were convicted of drug offenses *or* weapons offenses?

United States citizens have the right to appeal a conviction. Appeals are handled through appellate courts, but on rare occasions they can make it to the Supreme Court of the United States.

4. **Federal Inmates** There are 209,770 inmates in the U.S. federal prison system. The inmates are classified as citizens of the United States, Mexico, Colombia, Cuba, the Dominican Republic, or other. How many federal inmates are citizens of the United States?

Rights of Accused In Exercises 5–7, use a set diagram to analyze the statement about crime.

5. All defendants are presumed innocent until proven guilty.

6. Some people have been convicted of crimes they did not commit.

7. Habeas corpus is a safeguard of individual freedom stating that no person can be jailed without being charged with a crime.

8. **Sixth Amendment** Part of the Sixth Amendment states, "In all criminal prosecutions, the accused shall enjoy the right to a speedy and public trial." Use a set diagram to visualize the negation of this amendment.

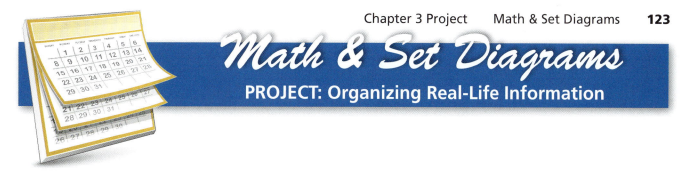

Math & Set Diagrams
PROJECT: Organizing Real-Life Information

1. Use the *Month Set Diagram* at *Math.andYou.com* to place the months in the correct locations.

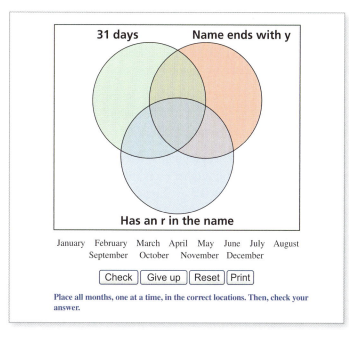

31 days   **Name ends with y**

Has an r in the name

January   February   March   April   May   June   July   August
September   October   November   December

[Check] [Give up] [Reset] [Print]

Place all months, one at a time, in the correct locations. Then, check your answer.

2. Use the *Alphabet Set Diagram* at *Math.andYou.com* to place the letters of the alphabet in the correct locations.

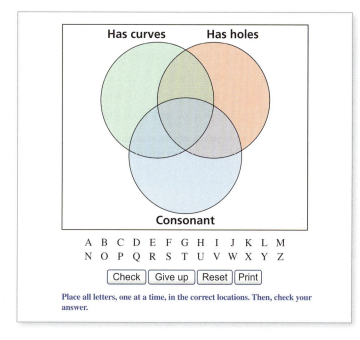

Has curves   **Has holes**

Consonant

A   B   C   D   E   F   G   H   I   J   K   L   M
N   O   P   Q   R   S   T   U   V   W   X   Y   Z

[Check] [Give up] [Reset] [Print]

Place all letters, one at a time, in the correct locations. Then, check your answer.

3. Give an example of a real-life situation you could model using set diagrams similar to those above.

3.3 Deductive & Inductive Reasoning

▶ Use deductive reasoning with syllogisms.

▶ Know how a deductive reasoning system is created.

▶ Use inductive reasoning.

Deductive Reasoning and Syllogisms

In this section, you will study two types of reasoning: **deductive reasoning** and **inductive reasoning.** In deductive reasoning, you start with two or more statements that you know or assume to be true. From these, you *deduce* or *infer* the truth of another statement. Here is an example.

- Premise: If this traffic doesn't clear up, I will be late for work.
- Premise: The traffic hasn't cleared up.
- Conclusion: I will be late for work.

This pattern for deductive reasoning is called a **syllogism.** The classical Greek philosopher Aristotle required that each premise be of the form "All A are B," "Some A are B," "No A are B," or "Some A are not B." Such syllogisms are called **categorical syllogisms.**

EXAMPLE 1 **Writing a Syllogism**

Suppose that aliens from Mars visit Earth. Also suppose that the physicist Stephen Hawking is correct on the outcome. What conclusion can you draw?

> "If aliens ever visit us, I think the outcome would be much as when Christopher Columbus first landed in America, which didn't turn out very well for the Native Americans."
> Stephen Hawking

SOLUTION

Here is one way to write this argument as a syllogism.

- Premise: All alien visits are bad for humans.
- Premise: All Martian visits are alien visits.
- Conclusion: All Martian visits are bad for humans.

You can use a set diagram to help see why the syllogism is valid.

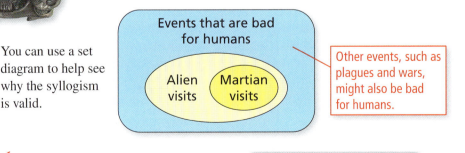

Events that are bad for humans

Alien visits

Martian visits

Other events, such as plagues and wars, might also be bad for humans.

✓ **Checkpoint**   Help at *Math*.and**Y♥U**.com

Any theory of gravity is a physical theory. Write a syllogism in which the first premise is the Stephen Hawking quote, "Any physical theory is always provisional, in the sense that it is only a hypothesis: you can never prove it."

When working with a syllogism, there are three main concerns.

1. Is the first premise true?
2. Is the second premise true?
3. Is the syllogism valid, having the correct conclusion?

For instance, in Example 1, regardless of whether the two premises are true, the syllogism is still valid. The point is, *if* Hawking's statement is true and *if* aliens visit Earth, then it does follow that things will be bad for humans.

EXAMPLE 2 Writing a Syllogism

Suppose that Daniel Defoe's premise in *The Education of Women* is true. Also suppose that women are given the advantages of an education. What conclusion can you draw?

> "I have often thought of it as one of the most barbarous customs in the world, considering us as a civilized and a Christian country, that we deny the advantages of learning to women. We reproach the sex every day with folly and impertinence; while I am confident, had they the advantages of education equal to us, they would be guilty of less than ourselves."
>
> *The Education of Women,* Daniel Defoe

Daniel Defoe (1660–1731) was an English writer. He was the author of what some consider the first English-language novel, *Robinson Crusoe.* He wrote more than 500 works, including books, pamphlets, and journals, on various topics including women's rights.

SOLUTION

Here is one way to write this argument as a syllogism.

- Premise: If women had the same opportunity for education as men, then they would be guilty of fewer offenses than men.
- Premise: Women have the same opportunity for education as men.
- Conclusion: Women are guilty of fewer offenses than men.

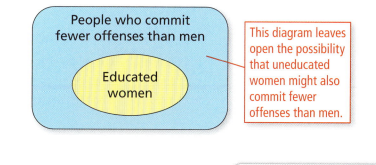

This diagram leaves open the possibility that uneducated women might also commit fewer offenses than men.

✓ Checkpoint

Help at *Math*.and**YOU**.com

Write a syllogism that involves the following Daniel Defoe quote. Then draw a set diagram to represent the syllogism.

> "But justice is always violence to the party offending, for every man is innocent in his own eyes."
>
> *The Shortest-Way with the Dissenters,* Daniel Defoe

Destruction of the Great Library of Alexandria

When Alexander the Great died in 323 B.C., his empire was divided into different regions. Ptolemy I, one of Alexander's generals, took control of Egypt. Euclid lived in the cosmopolitan city of Alexandria and had access to libraries and other mathematicians. The royal library in Alexandria is thought to have been the largest and most comprehensive library in the ancient world. Its destruction is considered to be one of the greatest losses of knowledge ever experienced by humanity.

Deductive Reasoning Systems

Unlike biology, chemistry, and physics, mathematical systems are not "discovered." Each of the logical systems within mathematics was invented and developed by humans. For instance, the logical system called Euclidean geometry was invented and developed by the Greek mathematician Euclid around 300 B.C.

It is meaningless to talk about whether a mathematical system is "true." No one can prove that Euclidean geometry is true. What you can do is decide whether the system is logically consistent, *assuming its premises are true.*

A logical system consists of four basic types of statements and concepts.

- Undefined terms
- Unproven postulates (statements)
- Defined terms
- Proven theorems (statements)

EXAMPLE 3 **Analyzing a Logical System**

Euclid understood that "you can't prove everything." At some point in developing a logical system, you have to use words that are not defined and you have to use premises that are not proven. For instance, Euclidean geometry does not define the concepts of "point" or "line." You just assume that you know what they are. Here are the five premises (or postulates) on which Euclidean geometry is based.

1. A unique straight line can be drawn between any two points.
2. A straight line segment can be extended to any finite length.
3. A circle can be described with any given point as its center and any distance as its radius.
4. All right angles are equal.
5. **Parallel Postulate:** At most, one line can be drawn through any point not on a given line parallel to the given line in a plane.

Write a syllogism that involves Euclid's first postulate and illustrate it.

SOLUTION

Here is one way to write a syllogism that involves Euclid's first postulate.

- Premise: Given any two points, there is a unique line that contains them.
- Premise: You are given points *A* and *B*.
- Conclusion: There exists a unique line containing *A* and *B*.

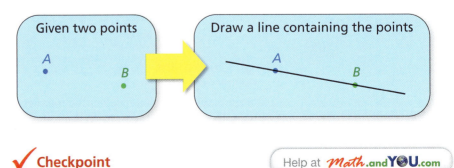

✓ **Checkpoint** Help at *Math*.and**YOU**.com

Write a syllogism that involves Euclid's fifth postulate and illustrate it.

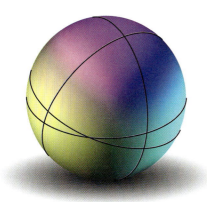

In spherical geometry, the undefined term *line* is imagined to be a great circle of a sphere. Longitude lines on Earth are great circles. Latitude lines, except for the equator, are not great circles.

EXAMPLE 4   **Comparing Logical Systems**

There are many possible "non-Euclidean" geometries. One is called spherical geometry. The difference between Euclidean and spherical geometry lies in what you assume the undefined term *line* to be and also in the Parallel Postulate. Here is one set of postulates for spherical geometry.

1. A unique straight line can be drawn between any two points, unless the points are antipodal, in which case they lie on many straight lines.

2. Any straight line segment can be extended indefinitely in a straight line. (At some point, it will connect with itself, but it can go around the sphere an infinite number of times.)

3. A circle can be described with any given point as its center and any distance as its radius, as long as the radius is less than half the circumference of the sphere.

4. All right angles are equal.

5. **Parallel Postulate:** No two lines are parallel. (Any two intersect.)

Explain why the Spherical Parallel Postulate makes sense.

SOLUTION

If you were to tell someone that "no two lines are parallel," that person might think that you are irrational. What needs to be understood is that your understanding of terms is different.

- In spherical geometry, you understand a "line" to be a great circle of a sphere. For you, a "plane" is the surface of the sphere.

- In Euclidean geometry, you understand a "line" to be straight, like a laser beam. For you, a "plane" is a flat surface.

- You both understand that two lines are "parallel" if they lie in the same plane and don't intersect.

To see why there are no parallel lines in spherical geometry, imagine putting a rubber band around a tennis ball. Make it a great circle so that it has a maximum circumference. Now try to put a second rubber band (great circle) on the ball. You can't do it without intersecting the first rubber band.

✓ **Checkpoint**

In Euclidean geometry, you know that the sides of a triangle are three straight line segments. You also know that the 3 angle measures total 180°. Sketch and describe a triangle in spherical geometry. What can you say about the sum of the angle measures of a triangle in spherical geometry?

Inductive Reasoning

Inductive reasoning occurs when you form conclusions based on repeated patterns. Although this leaves open the possibility of arriving at false conclusions, it is still the most common type of reasoning. Here is an example.

Conclusion Based on Pattern

- All of the fire I have felt is hot.
- Therefore, all fire is hot.

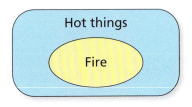

EXAMPLE 5 **Using Inductive Reasoning**

Explain how the Salk vaccine trials used inductive reasoning.

The Jonas Salk polio vaccine field trials of 1954 were the largest and most publicized clinical trials ever undertaken. Most of the reported polio cases occurred in children under 10 years of age. So, the trial targeted about 1.8 million children in the first 3 grades of elementary school at 211 test sites. In the experiment, about 440,000 children received the vaccine, about 200,000 received a placebo (a solution made to look like the vaccine, but containing no virus), and about 1,190,000 received neither. There were fewer cases of polio in children who received the vaccine than in children who received the placebo or nothing. The results, announced in 1955, were that the Salk vaccine was safe and effective in preventing polio.

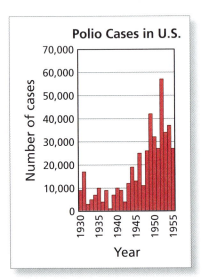

New polio cases dropped to under 6000 in 1957, the first year after the vaccine was widely available. In 1962, an oral vaccine became available. Today there are only a few polio cases in the United States.

SOLUTION

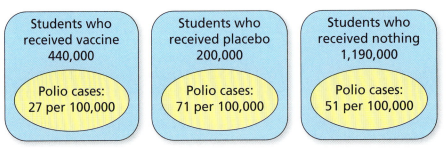

The announcement that the vaccine was both safe and effective was based on inductive reasoning. In other words, it was assumed that the 1.8 million children in the field trial were a good representation of all people.

✔ **Checkpoint** Help at *Math*.and**YOU**.com

Draw a set diagram that illustrates the inductive reasoning. Is the conclusion correct? Explain your reasoning.

- All the tigers I have seen are orange with black stripes.
- Therefore, all tigers are orange with black stripes.

| EXAMPLE 6 | **Using Inductive Reasoning** |

What comments can you make about the following article? Is the inductive reasoning valid? Explain your reasoning.

> **Mediterranean Bluefin Tuna Stocks Collapsing Now as Fishing Season Opens (April 2009), World Wildlife Fund**
>
> "As the Mediterranean's bloated fishing fleets ready themselves for the opening of the bluefin tuna fishery tomorrow, World Wildlife Fund (WWF) has released an analysis showing that the bluefin breeding population will disappear by 2012 under the current fishing regime.
>
> Global conservation organization WWF reveals that the population of breeding tunas has been declining steeply for the past decade—and will be wiped out completely in 3 years if fisheries managers and decision-makers keep ignoring the warnings from scientists that fishing must stop."

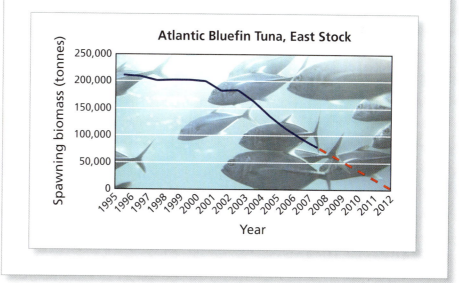

SOLUTION

Here is the basic argument.

- The Atlantic bluefin tuna breeding population has declined for 10 years.

- The decline is due to commercial fishing.

- Breeding Atlantic bluefin tuna will disappear unless commercial fishing stops.

This is a valid example of inductive reasoning. There are, of course, possible errors in the reasoning. Perhaps claiming that the breeding population will disappear is a bit alarmist. But common sense does indicate that the situation is dire.

 **Checkpoint**

 Help at *Math*.and**YOU**.com

Describe other examples of the use of inductive reasoning in environmental issues. In each case, clearly write the premises and the conclusion.

3.3 Exercises

Supreme Court Cases In Exercises 1 and 2, complete the syllogism. Then draw a set diagram to represent the syllogism. *(See Example 1.)*

1. *Brown v. Board of Education*

   - Premise: All unequal public schools are unconstitutional.

   - Premise: All segregated public schools are unequal.

   - Conclusion: _____

     _____

2. *Citizens United v. Federal Election Commission*

   - Premise: Any law that limits free speech is unconstitutional.

   - Premise: Any corporate spending on independent political broadcasts is free speech.

   - Conclusion: _____

     _____

Marbury v. Madison In Exercises 3 and 4, use the excerpt from the Supreme Court's majority opinion in the case *Marbury v. Madison.* *(See Examples 1 and 2.)*

> "If, then, the courts are to regard the Constitution, and the Constitution is superior to any ordinary act of the legislature, the Constitution, and not such ordinary act, must govern the case to which they both apply."
> *Marbury v. Madison*, 5 U.S. 137, 177, 178 (1803)

3. Suppose the Constitution is superior to any ordinary law. Also suppose that an ordinary law conflicts with the Constitution. What conclusion can you draw?

4. Explain why the Constitution must be considered superior to any ordinary law for the Supreme Court to conclude that any unconstitutional law is illegal.

Supreme Court Cases In Exercises 5–8, write a syllogism that involves the Supreme Court's decision. *(See Examples 1 and 2.)*

5. In *United States v. Virginia*, the Supreme Court held that the Virginia Military Institute's male-only admission policy was unconstitutional because it treated women unequally.

6. In *Miranda v. Arizona*, the Supreme Court held that statements made by Ernesto Miranda were inadmissible because Miranda had not been advised of his Fifth Amendment rights before he made the statements.

7. In *Georgia v. Randolph*, the Supreme Court held that it was unconstitutional for police to search a house without a warrant if one resident consents but another resident objects.

8. In *Roper v. Simmons*, the Supreme Court held that the execution of an offender who was under 18 years old at the time of the crime was "cruel and unusual punishment" and therefore unconstitutional.

Roads and Towns **In Exercises 9–16, consider the following postulates.**
(See Examples 3 and 4.)

> **Postulate 1:** Given any two towns, a road passes through them.
>
> **Postulate 2:** Given any road, there is at least one town that the road does not pass through.
>
> **Postulate 3:** There are at least two towns.

9. What are the undefined terms?

10. Does a road need to be a straight line? Explain your reasoning.

11. Write a syllogism that involves the first postulate and illustrate it.

12. Write a syllogism that involves the second postulate and illustrate it.

13. Use deductive reasoning to explain why there must be at least three towns.

14. Determine whether each model is valid. If a model is not valid, identify the postulate(s) that it violates. Explain your reasoning.

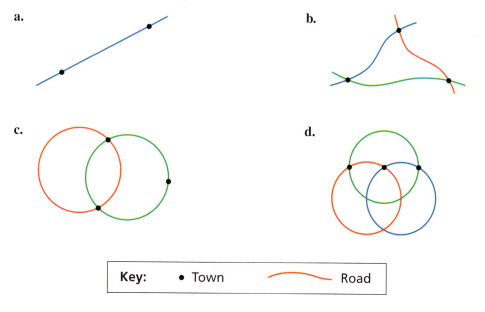

15. Do the postulates guarantee a town at every intersection of two roads? Explain your reasoning.

16. Consider the following replacement for Postulate 2.

    Postulate 2: Given any town, there is at least one road that does not pass through the town.

    a. Write a syllogism that involves the postulate and illustrate it.

    b. At least how many towns must exist? Explain your reasoning.

    c. At least how many roads must exist? Explain your reasoning.

Dinosaurs In Exercises 17–22, use the map. *(See Examples 5 and 6.)*

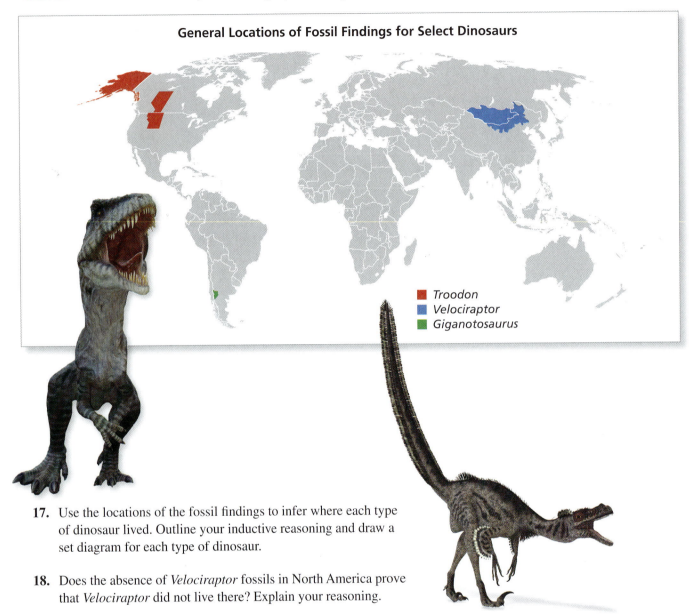

General Locations of Fossil Findings for Select Dinosaurs

■ *Troodon*
■ *Velociraptor*
■ *Giganotosaurus*

17. Use the locations of the fossil findings to infer where each type of dinosaur lived. Outline your inductive reasoning and draw a set diagram for each type of dinosaur.

18. Does the absence of *Velociraptor* fossils in North America prove that *Velociraptor* did not live there? Explain your reasoning.

19. *Giganotosaurus* fossils have been dated to about 100 to 95 million years ago. What can you infer about the time period in which all *Giganotosaurus* lived? Outline your inductive reasoning and draw a set diagram.

Contrary to the image popularized by the film *Jurassic Park*, *Velociraptor* actually had feathers.

20. *Troodon* fossils have been dated to about 75 to 65 million years ago. What can you infer about the time period in which all *Troodon* lived? Outline your inductive reasoning and draw a set diagram.

21. *Velociraptor* fossils have been dated to about 75 to 71 million years ago. Is it likely that *Velociraptor* encountered *Troodon* or *Giganotosaurus*? Explain your reasoning.

22. Research the locations in which *Tyrannosaurus rex* fossils have been found and the time period in which *Tyrannosaurus rex* lived. Is it likely that *Tyrannosaurus rex* encountered *Troodon, Velociraptor,* or *Giganotosaurus*? Explain your reasoning.

▶ Extending Concepts

Periodic Table **In Exercises 23–30, use the periodic table of elements.**

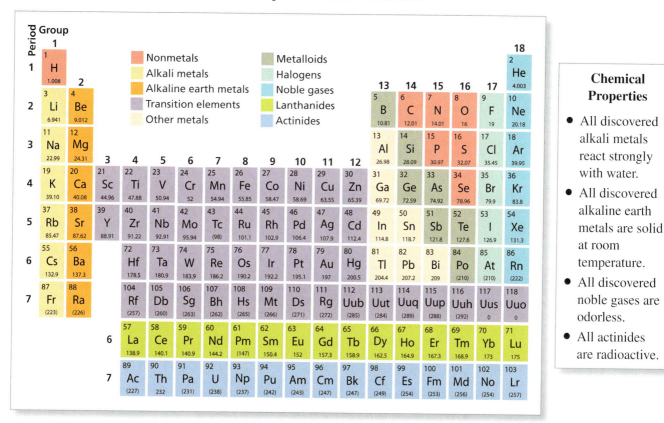

Chemical Properties

- All discovered alkali metals react strongly with water.
- All discovered alkaline earth metals are solid at room temperature.
- All discovered noble gases are odorless.
- All actinides are radioactive.

23. Write a syllogism that involves potassium (element 19). Then draw a set diagram.

24. Write a syllogism that involves xenon (element 54). Then draw a set diagram.

25. Write a syllogism that involves radium (element 88). Then draw a set diagram.

26. Write a syllogism that involves thorium (element 90). Then draw a set diagram.

27. There are hypothetical alkali metals that have not been discovered. Use inductive reasoning to draw a conclusion about all alkali metals, discovered or undiscovered.

28. There are hypothetical alkaline earth metals that have not been discovered. Use inductive reasoning to draw a conclusion about all alkaline earth metals, discovered or undiscovered.

29. Suppose ununennium (hypothesized element 119) is an undiscovered alkali metal. Write a syllogism that involves how ununennium reacts with water. Use your conclusion from Exercise 27 as your first premise.

30. Suppose unbinilium (hypothesized element 120) is an undiscovered alkaline earth metal. Write a syllogism that involves unbinilium's state at room temperature. Use your conclusion from Exercise 28 as your first premise.

Xenon is used in high-intensity discharge headlamp bulbs.

3.4 Fallacies in Logic

▶ Recognize deductive fallacies.
▶ Use set diagrams to detect fallacies.
▶ Recognize fallacies in advertisements.

Deductive Fallacies

A **fallacy** is an error in reasoning. This differs from a factual error, which is simply being wrong about a fact. A **deductive fallacy** is a deductive argument that is invalid. Here is an example.

- Premise: When it rains, the ground gets wet.
- Premise: The ground is wet.
- Conclusion: Therefore, it must have rained. ☹

Here is an example of valid reasoning.

- Premise: When it rains, the ground gets wet.
- Premise: It rained.
- Conclusion: Therefore, the ground got wet. ☺

An **inductive fallacy** occurs when the premises do not provide enough support for the conclusion.

EXAMPLE 1   **Detecting a Fallacy**

Is the logic in this description of lie detectors valid?

> To detect lies, a polygraph test evaluates a person's heart rate, breathing rate, blood pressure, and perspiration on fingertips. Sometimes a polygraph test also evaluates involuntary arm and leg movements and nervous tics, which often occur while being asked difficult questions. When people lie, their heart rates increase and they start sweating. So, when a polygraph test shows an increased heart rate and sweating, you can conclude that the person is lying.

SOLUTION

This argument is not valid. It is the same type of fallacy as shown above: *affirming the consequent.*

- Premise: When people lie, their heart rates increase and they start sweating.
- Premise: This man has an increased heart rate and is sweating.
- Conclusion: Therefore, he is lying. ☹

✓ **Checkpoint**   Help at *Math*.and**Y⊕U**.com

Is the logic in this description of craters valid? Explain.

> When meteors hit Earth, they form craters, some of which are over a mile in diameter. The diameter of Crater Lake in Oregon is about 5 miles. So, it must have been formed by a huge meteor.

There are many types of deductive fallacies. If you take a formal course in logic, you might encounter a dozen different types. Here is another example.

- Premise: When it rains, the ground gets wet.
- Premise: It isn't raining.
- Conclusion: Therefore, the ground is not wet. ☹

When examining arguments, keep reminding yourself, "It's not about whether the conclusion is true or false. It's about whether the conclusion was deduced in a logically valid way."

Oddly, the following syllogism *is* valid.

- Premise: When it rains, the ground gets wet.
- Premise: The ground is not wet.
- Conclusion: Therefore, it must not have rained. ☺

Study Tip

The type of fallacy at the right is called *denying the antecedent*.

- Premise: If *P*, then *Q*.
- Premise: *P* is not true.
- Conclusion: Therefore, *Q* is not true. ☹

EXAMPLE 2   Detecting a Fallacy

Is the logic in this Alan Turing quote valid?

> "If each man had a definite set of rules of conduct by which he regulated his life he would be no better than a machine. But there are no such rules, so men cannot be machines."
>
> "Computing Machinery and Intelligence," Alan Turing

SOLUTION

This argument is not valid. It is an example of *denying the antecedent*. In fact, in his article "Computing Machinery and Intelligence," Turing actually states that this is an example of an invalid argument.

- Premise: If each man had a definite set of rules of conduct by which he regulated his life, he would be no better than a machine.
- Premise: There is no set of rules of conduct.
- Conclusion: Therefore, men cannot be machines. ☹

Alan Turing was an English logician, cryptanalyst, and computer scientist. He is often considered the father of modern computer science. During World War II, Turing devised methods for breaking German codes, and because of this, some have called him "the man who saved the world."

✓ Checkpoint

Help at *Math*.and*YOU*.com

Outline the invalid syllogism described in the article.

> The buyer of a new vehicle brought claims against a manufacturer under Ohio's Lemon Law and for breaches of a warranty act. The trial court ruled in favor of the defendant on both claims. The court of appeals analyzed the trial court's logic. The trial court first addressed the plaintiff's Lemon Law claim and determined that it was invalid. Next, the trial court concluded that since the Lemon Law claim was not valid, the warranty act claim was not valid. The court of appeals rejected the trial court's reasoning, based on the fallacy of *denying the antecedent*.
>
> Summarized from "Conventional Logic: Using the Logical Fallacy of Denying the Antecedent as a Litigation Tool," Stephen Rice

Fallacies and Set Diagrams

A set diagram can be helpful in determining whether an argument is valid or invalid. Here is an example using a written argument.

> Taxes fund necessary services such as police, schools, and roads. Therefore, taxation is necessary.

First, outline the argument.

- **Premise:**   Police, schools, and roads are necessary.
- **Premise:**   Taxes fund police, schools, and roads.
- **Conclusion:**   Therefore, taxes are necessary. ☹

Then, organize the argument with a set diagram.

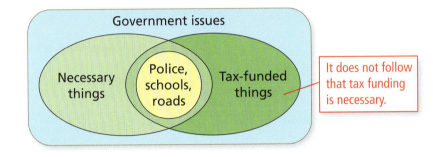

EXAMPLE 3   **Drawing a Set Diagram**

Use a set diagram to analyze the conclusion of the study.

> Teenagers who use a cell phone (for talking and texting) more than 15 times a day are more prone to disrupted sleep, restlessness, stress, and fatigue.
>
> Summarized from "Does Excessive Mobile Phone Use Affect Sleep in Teenagers?," Gaby Badre, MD, PhD

SOLUTION

Here is one way to analyze the conclusion of the study.

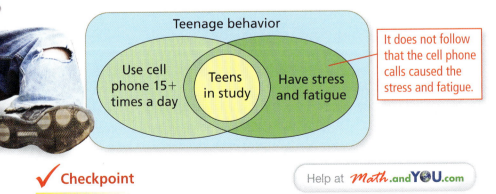

✓ **Checkpoint**   Help at *Math*.and**Y♥U**.com

Use a set diagram to analyze the conclusion of the argument.

> Our society is filled with violence, and there is a lot of violence on TV. It follows that the violence in society is caused by people watching TV.

Perhaps the most common type of fallacy is one in which the second premise does not refer to *all* the members of the category described in the first premise. This type of fallacy is called an *undistributed middle*.

Lee Grant was born Lyova Rosenthal in New York City in 1927. Grant established herself as an actress on Broadway in the show *Detective Story*. She made her film debut in the movie version of the same story.

EXAMPLE 4   **Drawing a Set Diagram**

Use a set diagram to analyze the argument for placing Lee Grant on the Hollywood blacklist.

> McCarthyism is the practice of making accusations of subversion or treason without proper evidence. It was originally coined to criticize the anti-communist activity of Senator Joseph McCarthy. A famous example of McCarthyism is the Hollywood blacklist, which was associated with hearings by the House Un-American Activities Committee. Actors who were put on the list had great difficulty finding acting roles.
>
> Lee Grant was one of the many actors who were blacklisted. She was blacklisted for refusing to testify against her husband, who was accused of being a communist. After several years, Grant was able to return to film and ended up winning an Oscar for her role in *Shampoo* (1975).

SOLUTION

Here is one way to analyze this event.

- Premise: All communists refused to testify against other communists.
- Premise: Lee Grant refused to testify against an accused communist.
- Conclusion: Therefore, Lee Grant must have been a communist. ☹

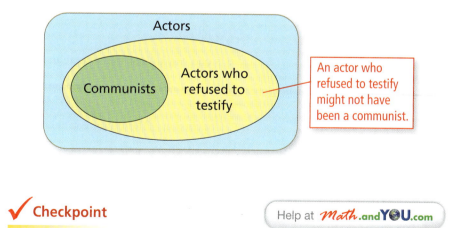

✔ **Checkpoint**   Help at *Math*.and**Y☺U**.com

In 1692, at the start of the Salem witch trials, several people were arrested and accused of witchcraft. Their accusers claimed that young girls were having fits whenever the accused were present. Analyze the argument against the accused. Describe how it is an example of a fallacy of the *undistributed middle*. Illustrate the argument with a set diagram.

Fallacies in Advertisements

Manufacturers have both the need and the right to advertise their products. Magazines, newspapers, television shows, Internet sites, and radio programs depend on the revenue from advertisements. And yet, as a field, advertising is filled with the use of logical fallacies. Moreover, it will most likely remain that way because fallacies can be effective ways to sell products.

EXAMPLE 5   **Analyzing an Advertisement**

In 1953, Marilyn Monroe was named "the Most Advertised Girl in the World" by the Advertising Association of the West. Is the advertisement a logical fallacy?

SOLUTION

Here is one way to look at the logic in the advertisement.

- Premise: If Marilyn Monroe uses a product, then you should use it.
- Premise: Marilyn Monroe uses Lustre-Creme shampoo.
- Conclusion: Therefore, you should use Lustre-Creme shampoo.

This is a valid syllogism. The problem is that the first premise is questionable. As such, this type of *appeal to celebrity* is considered a logical fallacy.

 **Checkpoint**   Help at

Do you agree with the following policy? Explain your reasoning.

China has passed laws banning celebrities from appearing in ads for drugs. A government spokesperson stated: "The move is to eradicate illegal drug advertisements that exaggerate drug's benefits and mislead customers."

Another type of logical fallacy that is often used in advertisements for products or political candidates is called *begging the question*. Arguments with this fallacy are circular. Here is a typical form.

- Premise: *A* is true because *B* is true.
- Conclusion: *B* is true because *A* is true.

EXAMPLE 6   **Analyzing Political Advertisements**

Analyze the following political advertisements.

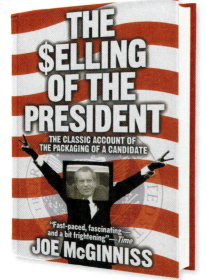

In *The Selling of the President*, Joe McGinniss describes how Richard Nixon used marketing tactics in his 1968 presidential campaign. Throughout the book, McGinniss claims that Nixon was sold like a product, which was a shocking idea at the time.

SOLUTION

If either of these ads makes you want to vote for the person, the ad is doing its job. Both ads are attractive, concise, and carry a message.

- Both ads *appeal to emotion* (a type of fallacy) by displaying the American flag and by using the colors of the flag.

- Both ads *beg the question*. The left ad assumes that America no longer represents the concept "of the people, by the people, and for the people." The right ad assumes there is corruption in D.C.

- The left ad also contains an *appeal to authority*, by using a quote from Lincoln's Gettysburg Address. The use of this quote implies that Lincoln endorses the candidate.

✓ **Checkpoint**   Help at *Math*.and**Y⊕U**.com

Analyze the following political advertisements.

3.4 Exercises

Iraq War In Exercises 1–8, outline the invalid syllogism and identify the logical fallacy.
(See Examples 1 and 2.)

1. A politician states, "If something is not there, you will not find it. We did not find weapons of mass destruction in Iraq. So, there are no such weapons in Iraq."

2. A representative states, "If Saddam had shipped his weapons out of Iraq, we would not have found them. We didn't find Saddam's weapons, so he must have shipped them out of Iraq."

3. A radio show host says, "If we found weapons of mass destruction in Iraq, it would prove that Iraq had such weapons. We haven't found weapons of mass destruction in Iraq, and this proves that Iraq did not have such weapons."

4. A political pundit states, "If the war in Iraq made America safer, there will have been no major terrorist attack since we invaded. There hasn't been any such attack. So, the war made America safer."

5. A talk show host says, "If you support the war in Iraq, you support America. The senator doesn't support the war, so he obviously doesn't support America."

6. An opinion columnist writes, "If a person is a great leader, then that person will do what he or she believes is right. George W. Bush did what he believed was right. He was a great leader."

7. A political science professor says, "If we had overthrown Saddam in the first Gulf War, we would not be fighting in Iraq today. It is clear that we are fighting in Iraq as a consequence of our decision not to end Saddam's regime in 1991."

8. A senator states, "The United States invaded Iraq on the premise that Saddam Hussein had weapons of mass destruction. Saddam did not have weapons of mass destruction, so the United States should not have invaded Iraq."

Iraq has a rich cultural heritage. It was home to the ancient city of Babylon. The lion shown above is from Babylon's Ishtar Gate, which was reconstructed at the Pergamon Museum in Berlin, Germany.

Everyday Fallacies In Exercises 9–16, draw a set diagram to analyze the conclusion.
(See Examples 3 and 4.)

9. When Jack is bluffing, he sets a chip on his cards. Jack set a chip on his cards, so I know he's bluffing.

10. When Rich wants a favor, he starts acting really friendly. Rich has been acting really friendly today, so I know he wants something.

11. Greg's friend Dylan is very conservative. Greg concludes that everyone in Dylan's family must be very conservative.

12. Erica awakens at midnight and discovers the power is out at her house. She looks out her window and sees that all of her neighbors' windows are dark. She concludes that they must have lost their power too.

13. Kyle is buying a dress for his wife. He knows she doesn't like the color blue. So, he picks out a yellow dress and concludes that she will like it because it isn't blue.

14. Kimberly's mother doesn't approve of boys with tattoos. Kimberly's new boyfriend doesn't have any tattoos, so she concludes that her mother will approve of him.

15. Tom is ordering pizza for his friends. He recalls that his friend David doesn't like pepperoni, so he orders a pizza with mushrooms and concludes that David will eat it because it doesn't have pepperoni on it.

16. I am feeling sick this morning. It must be because of the sushi I ate last night. Everyone who ate the sushi must be sick by now.

Commercial Advertising In Exercises 17–20, analyze the advertisement. *(See Example 5.)*

17.

SUPREME CLEAN
WHITENING TOOTHPASTE

9 out of 10 dentists can't be wrong!

18.

ALL-IN-ONE
ULTIMATE FITNESS SYSTEM

19.

...it's all in the shoes.

20.

We put safety first.

Campaign Posters In Exercises 21 and 22, analyze the political advertisement. *(See Example 6.)*

21.

I WANT YOU TO VOTE

N. Landon for Senate

22.

Declare your INDEPENDENCE from two-party politics

Vote Donna for Congress

23. **Weatherproof Ad** Weatherproof Garment Company controversially displayed an ad in Times Square showing President Obama wearing one of the company's jackets without his permission. The slogan on the ad was "A Leader In Style." Analyze the logic of the advertisement. *(See Example 5.)*

24. **Commercial Advertising** Find an example of a commercial advertisement and analyze its logic. *(See Example 5.)*

▶ Extending Concepts

Other Fallacies Six types of fallacies are given. In Exercises 25–30, determine which fallacy applies.

Ad hominem: An argument that attacks the character of a person rather than addressing the actual issue

Ad populum: An argument with the premise that if a majority believes that something is true, then it is true

Appeal to novelty: An argument that uses the premise that if something is new, it is automatically better

Composition: An argument that concludes that something is true for the whole because it is true for the parts of the whole

False dilemma: An argument that only gives two options when more than two are available

Self-refuting idea: A statement that is false as a consequence of being held true

25.
QUAYLE: ". . . I have as much experience in the Congress as Jack Kennedy did when he sought the presidency. . . ."

BENTSEN: "Senator, I served with Jack Kennedy, I knew Jack Kennedy, Jack Kennedy was a friend of mine. Senator, you are no Jack Kennedy."

1988 Vice-Presidential Debate

26.

NEW & IMPROVED

27.
"The only thing I know is that I know nothing."

Socrates

28.
"Either you are with us, or you are with the terrorists."

George W. Bush

29.

Everyone's watching...

TV's SHOW **#1**

30.
"We see that every city-state is a community of some sort, and that every community is established for the sake of some good (for everyone performs every action for the sake of what he takes to be good)."

Politics, Aristotle
(Translated by C.D.C. Reeve)

3.3–3.4   Quiz

Boiling Point **In Exercises 1–8, use the graph.**

1. Suppose you live near sea level (0 meters) and heat a pan of water until it boils.

   a. At what temperature will you observe the water boiling?

   b. What might you infer about the boiling point of all water?

   c. Draw a set diagram that represents the inference from part (b).

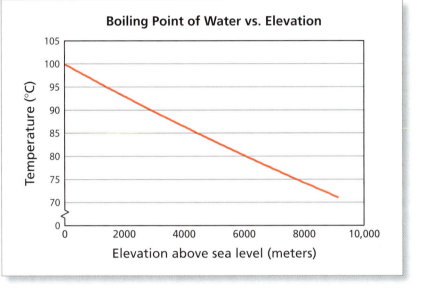

Boiling Point of Water vs. Elevation

Temperature (°C) vs. Elevation above sea level (meters)

2. The elevation of La Paz, Bolivia is about 3600 meters above sea level.

   a. At what temperature will a resident of La Paz observe water boiling?

   b. What might the resident infer about the boiling point of all water?

   c. Draw a set diagram that represents the inference from part (b).

3. Write a syllogism that involves the boiling point of water at sea level (0 meters).

4. Write a syllogism that involves the boiling point of water in La Paz (about 3600 meters above sea level).

5. Write a syllogism that involves the relationship between elevation and boiling point.

6. Is the logic in the statement valid? Draw a set diagram to analyze the argument.

   > If the elevation decreases, the boiling point increases. The boiling point increased, so the elevation must have decreased.

7. As elevation increases, atmospheric pressure decreases. Write a syllogism that involves the relationship between atmospheric pressure and elevation.

8. Suppose you perform the following experiment. You place water in a pressure cooker at an elevation of 5000 meters above sea level and adjust the pressure in the cooker to 1 atmosphere, which is the pressure at sea level (0 meters). When you heat the water, it boils at 100°C. What can you conclude?

Chapter 3   Summary

Section Objectives		*How does it apply to you?*
Section 1	Use a union of two sets to represent *or*.	⇨	Using a set diagram to visualize a statement can help you recognize what is in one set *or* in another set. *(See Example 1.)*
	Use an intersection of two sets to represent *and*.	⇨	Using a set diagram to visualize a statement can help you recognize what is in one set *and* in another set. *(See Example 3.)*
	Use the complement of a set to represent *not*.	⇨	Using a set diagram to visualize a statement can help you recognize what is not in a set. *(See Examples 5 and 6.)*
Section 2	Analyze statements that have the term *all*.	⇨	Using a set diagram to analyze a statement can help you recognize what is and is not being said in the statement. *(See Examples 1 and 2.)*
	Analyze statements that have the term *some* or *many*.	⇨	Using a set diagram to analyze a statement can help you make other statements. *(See Examples 3 and 4.)*
	Analyze negations of statements.	⇨	Using a set diagram to analyze a statement can help you negate the statement. *(See Examples 5 and 6.)*
Section 3	Use deductive reasoning with syllogisms.	⇨	Using a set diagram to visualize why syllogisms are valid can help you draw other conclusions. *(See Examples 1 and 2.)*
	Know how a deductive reasoning system is created.	⇨	Logical systems help you understand how different branches of mathematics were formed. *(See Examples 3 and 4.)*
	Use inductive reasoning.	⇨	You need to be able to make logical conclusions based on repeated patterns. *(See Examples 5 and 6.)*
Section 4	Recognize deductive fallacies.	⇨	You need to be able to distinguish valid logic from invalid logic. *(See Examples 1 and 2.)*
	Use set diagrams to detect fallacies.	⇨	Using a set diagram can help you determine whether statements are valid or invalid. *(See Examples 3 and 4.)*
	Recognize fallacies in advertisements.	⇨	You need to be able to distinguish between valid and invalid statements in advertising. *(See Examples 5 and 6.)*

Chapter 3 Review Exercises

Section 3.1

Oiled Birds **In Exercises 1 and 2, use the information below.**

Birds were collected in areas affected by the Deepwater Horizon oil spill in the Gulf of Mexico. Through November 2, 2010, a report indicates that of the birds collected, 4342 were visibly oiled, 6104 were dead, and 2263 were *both* visibly oiled and dead.

1. Draw a set diagram that shows the information.

2. How many birds were found visibly oiled *or* dead?

Kemp's Ridley Sea Turtles **The set diagram shows characteristics of sea turtles upon arrival at an oil spill rehabilitation center. In Exercises 3–6, use the set diagram.**

Kemp's ridley sea turtles are an endangered species. They were adversely affected by the 2010 Deepwater Horizon oil spill, as they lay eggs on Mexican beaches and forage from the Yucatan Peninsula to southern Florida.

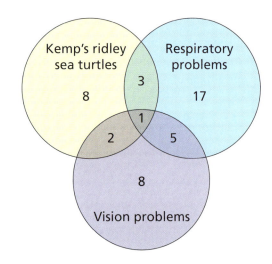

3. How many are Kemp's ridley sea turtles?

4. How many are not Kemp's ridley sea turtles?

5. How many sea turtles have respiratory problems *and* vision problems?

6. How many sea turtles have respiratory problems *or* vision problems?

U.S. Coast Guard In Exercises 7 and 8, use the information below.

The U.S. Coast Guard (USCG) sends British Petroleum bills that demand payment for equipment and personnel costs incurred due to the Deepwater Horizon oil spill. Information regarding bill #9 from January 11, 2011, is shown.

7. Draw a set diagram that shows the information.

8. How much did the USCG charge for the use of its personnel?

Nov/Dec 2010 **Costs:** $9,843,486.77
Equipment Costs:
• USCG Aircraft: $25,623.00
• USCG Vehicle: $43,484.92

Section 3.2

Ultraviolet Radiation In Exercises 9–14, use a set diagram to analyze the statement about tanning.

9. Suntanning is a result of exposure to ultraviolet (UV) radiation, which is a type of electromagnetic (EM) radiation. All EM radiation is made up of small particles that travel in a wave-like pattern at the speed of light.

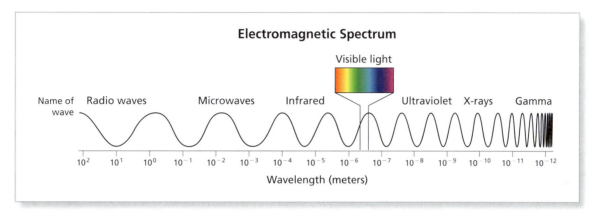

10. UV radiation from the Sun consists of three main types: UVA, UVB, and UVC. All UVC and some UVB are absorbed by Earth's ozone layer. Most tanning lamps emit both UVA and UVB rays, but some emit only UVA rays.

11. UV radiation damages DNA in cells. Your body responds by increasing the production of melanin, which causes tanning. UV radiation can also cause premature skin aging and skin cancer.

12. UV radiation can also damage your eyes. All UVB rays are absorbed by the cornea, but UVA rays pass through to the lens.

13. Sunscreens reduce the amount of UV radiation that penetrates the skin. Every broad-spectrum sunscreen blocks both UVA and UVB rays.

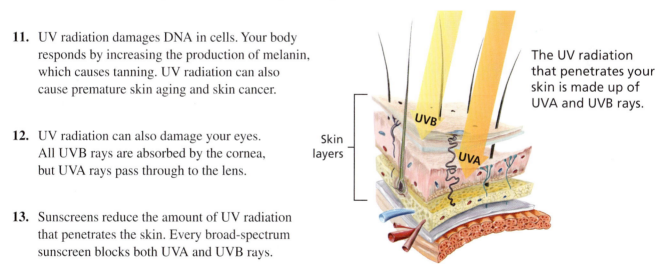

14. Some people tan easier than other people. Some people burn easier than other people.

Tanning In Exercises 15 and 16, use a set diagram to visualize the negation of the statement about tanning.

15. My best friend usually has a farmer's tan or a trucker's tan.

16. When I go to the beach, I will wear sunscreen with a sun protection factor (SPF) of at least 25 and I will use a beach umbrella to reduce my exposure to UV radiation.

Section 3.3

Universal Gravity **In Exercises 17 and 18, use the excerpt.**

"Why should that apple always descend perpendicularly
to the ground, thought [Newton] to himself; occasioned
by the fall of an apple, as he sat in a contemplative mood.
Why should it not go sideways, or upwards? But constantly
to the earth's center? Assuredly, the reason is, that the earth
draws it. There must be a drawing power in matter. . . . If
matter thus draws matter; it must be in proportion of its
quantity. Therefore, the apple draws the earth, as well as
the earth draws the apple."

Memoirs of Sir Isaac Newton's Life, William Stukeley

17. How does Newton conclude that there is a drawing power in matter?
What type of reasoning is this? Explain.

18. How does Newton conclude that, in addition to Earth pulling on the apple,
the apple is pulling on Earth? What type of reasoning is this?

Laws of Motion **In Exercises 19–23, use Newton's Laws of Motion.**

Law 1:   Every object in uniform motion will remain in uniform motion unless
an external force acts on it.

Law 2:   Force equals mass times acceleration.

Law 3:   For every action, there is an equal and opposite reaction.

19. What type of reasoning do you think Newton used to arrive at his three laws
of motion? Explain.

20. What type of reasoning do you think Newton used to
apply his three laws of motion? Explain.

21. Write a syllogism that involves
Newton's first law.

22. Write a syllogism that involves
Newton's third law.

23. In the twentieth century, it was discovered
that Newton's second law does not hold
at high velocities. Explain this revelation
in the context of inductive reasoning.

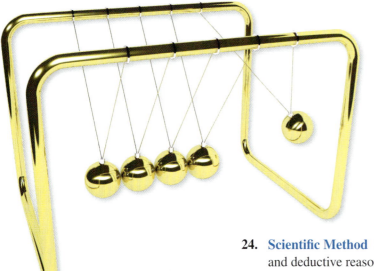

24. **Scientific Method** Explain the relationship between inductive
and deductive reasoning in science.

Section 3.4

Logical Fallacies In Exercises 25–28, draw a set diagram to analyze the conclusion.

25. When there is a lot of traffic, Laura is late for her appointments. Laura is late for her appointment, so there must have been a lot of traffic.

26. Anthony is from Texas. Texas is a "red" state. Therefore, Anthony is a Republican.

27. When Lynn is lying to me, she can't look me in the eye. Lynn's not looking me in the eye, so I know she's lying.

28. Marcus's girlfriend does not like horror movies. So, he rents an action movie and concludes that she will like it because it is not a horror movie.

Advertisements In Exercises 29–32, analyze the advertisement.

29.

30.

31.

32.

4 The Mathematics of Inflation & Depreciation

4.1 Exponential Growth

▶ Make a table showing exponential growth.

▶ Draw a graph showing exponential growth.

▶ Find an exponential growth rate.

4.2 Inflation & the Consumer Price Index

▶ Use a consumer price index.

▶ Use a graph to interpret a consumer price index.

▶ Compare inflation to the value of the dollar.

4.3 Exponential Decay

▶ Make a table and graph showing exponential decay.

▶ Calculate and use half-life.

▶ Find an exponential decay rate.

4.4 Depreciation

▶ Use straight-line depreciation.

▶ Use double declining-balance depreciation.

▶ Use sum of the years-digits depreciation.

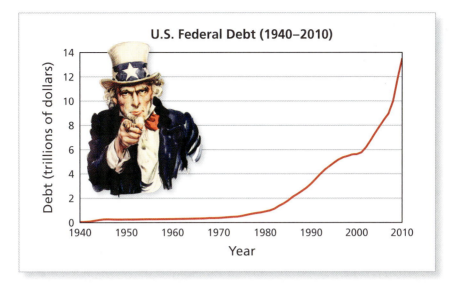

U.S. Federal Debt (1940–2010)

Debt (trillions of dollars) vs. Year

Example 6 on page 167 traces the federal debt from 1940 through 2010. Considering the consumer price index and inflation, was the government more in debt in 2010 than in 1983?

4.1 Exponential Growth

▶ Make a table showing exponential growth.
▶ Draw a graph showing exponential growth.
▶ Find an exponential growth rate.

Calculating Exponential Growth

A quantity has **exponential growth** when the quantity *increases* by the same percent from one time period to the next.

Formula for Exponential Growth

A quantity A that has exponential growth can be modeled by

$$A = P(1 + r)^n.$$

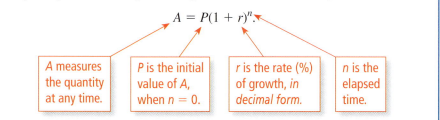

A measures the quantity at any time.

P is the initial value of A, when $n = 0$.

r is the rate (%) of growth, *in decimal form*.

n is the elapsed time.

EXAMPLE 1 **Making a Table**

The growth rate of a bacteria culture is 52% each hour. Initially, there are two bacteria. Make a table showing how many bacteria there are each hour for up to 12 hours.

SOLUTION

The formula for this exponential growth is

$$A = P(1 + r)^n = 2(1 + 0.52)^n. \qquad P = 2, r = 52\% = 0.52$$

Hours, n	Formula	Number of Bacteria
0	$A = 2(1.52)^0$	$A = 2$
1	$A = 2(1.52)^1$	$A = 3$
2	$A = 2(1.52)^2$	$A = 4$
⋮	⋮	⋮
10	$A = 2(1.52)^{10}$	$A = 131$
11	$A = 2(1.52)^{11}$	$A = 200$
12	$A = 2(1.52)^{12}$	$A = 304$

A nonzero number raised to an exponent of 0 is defined to be 1. Try it on your calculator.

✔ **Checkpoint** 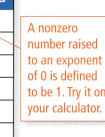 Help at *Math*.and**Y☺U**.com

The bacteria culture grows exponentially for a full day. How many bacteria are in the culture?

EXAMPLE 2   **Using a Spreadsheet**

The following excerpt is from *The Ring of Truth: An Inquiry into How We Know What We Know* by Philip and Phylis Morrison.

> "Chef Mark kneaded high-gluten white flour carefully along with the other ingredients of noodle dough in correct proportion: three cups of flour, half as much water, one-quarter teaspoon each of salt and baking soda. He vigorously swung and stretched the lump of dough out into a heavy single strand the length of his full two-arm span. Then he folded that long thick strand in half, and pulled the dough out again into its original length, so that two thinner strands now passed from one hand to the other. Repeat, repeat, repeat . . .
>
> CHEF MARK: Hello, everybody. I am the chef of the Dragon House in Wildwood, New Jersey. Today I will make the kind of noodles called *so*. Make the dough strong and smooth, keep the dough smooth and strong, and you will have noodles on the table.
>
> Fold one time: the dough becomes two noodles. Two times, it becomes four noodles. Three, four times . . . ten, eleven, now twelve doublings, or four thousand and ninety-six noodles."

Philip Morrison (1915–2005) was a professor of physics at MIT and a member of the Manhattan Project. His works include the film *Powers of Ten* (1977) and the 1987 PBS television series *The Ring of Truth*.

Use a spreadsheet to illustrate the number of dragon's beard noodles in each folding.

SOLUTION

The number of noodles doubles with each folding. This means that the rate of growth is 100%.

$$A = 1(1 + 1)^n$$

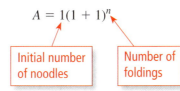

Initial number of noodles

Number of foldings

Entering this formula into a spreadsheet produces the table shown.

	A	B
1	**Foldings, *n***	**Number of Noodles**
2	0	1
3	1	2
4	2	4
5	3	8
6	4	16
7	5	32
8	6	64
9	7	128
10	8	256
11	9	512
12	10	1024
13	11	2048
14	12	4096
15		

✓ **Checkpoint**   Help at

Analyze the following statement, taken from the Morrisons' book. Is the claim true? Use a spreadsheet to justify your answer.

> "The tantalizing nature of the doubling process is that the subdivision is so rapid. Some forty-six doublings would make noodles of true atomic fineness, in principle. But note that such an incredible feat would produce not a mere few miles of dragon's beard, but noodles long enough to stretch to Pluto and beyond!"

Graphing Exponential Growth

While exponential growth often has a low growth rate (as in savings accounts), it can also have a dramatic growth rate. One example is the world flu pandemic of 1918–1920 in which 3–5% of Earth's population died. Another is the chain reaction that occurs during nuclear fission. Example 3 describes yet another example—that in which a species multiplies rapidly.

In the Australian grain belt, mouse population levels are normally low. Favorable seasonal conditions, however, can trigger extensive breeding. Mouse plagues erupt about every three years there.

EXAMPLE 3   Graphing Exponential Growth

In Australia, mice breed from August to May, which is about 42 weeks. For reasons that are not entirely known, every 3 or 4 years, the mouse population explodes and produces a plague of millions of mice. One breeding pair of mice and their offspring can produce 500 mice in just 21 weeks, which is a rate of 30% per week. At this rate, how many mice can one breeding pair produce in 42 weeks? Graph the results.

SOLUTION

Use a spreadsheet to evaluate the formula

$$A = 2(1 + 0.3)^n$$

from $n = 0$ to $n = 42$. The population after 42 weeks is about 122,082. Then use the spreadsheet to graph the results as shown.

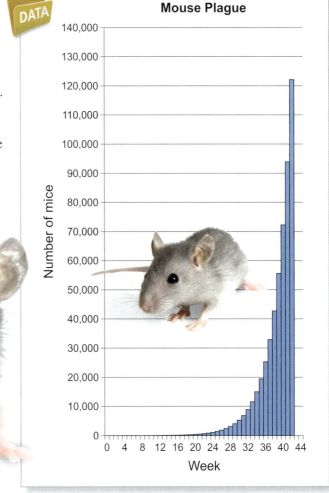

Checkpoint   Help at *Math*.and**Y☺U**.com

How many mice can be produced by 1000 breeding pairs in 42 weeks?

EXAMPLE 4   **Graphing Exponential Growth**

In 1975, Gordon Moore said that the number of transistors that can be placed inexpensively on an integrated circuit will double approximately every 2 years. This trend is known as Moore's Law and has continued for more than 35 years. One explanation for the accuracy of Moore's prediction is that the law is used in the semiconductor industry to guide long-term planning and to set goals for research and development. In this sense, the law has been a self-fulfilling prophecy.

In 1978, the Intel® 8086 held 29,000 transistors on an integrated circuit. According to Moore's Law, how many transistors could be placed on an integrated circuit in 2010? Graph the results.

SOLUTION

From 1978 to 2010, the number of transistors doubled 16 times.

Gordon Moore is a cofounder of © Intel Corporation. In 1954, Moore received a Ph.D. in chemistry and physics from Caltech. After his success in the semiconductor industry, Moore and his wife donated $600 million to Caltech, the largest gift ever to an institution of higher education.

	A	B
DATA		
1	**Year**	**Number of Transistors**
2	1978	29,000
3	1980	58,000
4	1982	116,000
5	1984	232,000
6	1986	464,000
7	1988	928,000
8	1990	1,856,000
9	1992	3,712,000
10	1994	7,424,000
11	1996	14,848,000
12	1998	29,696,000
13	2000	59,392,000
14	2002	118,784,000
15	2004	237,568,000
16	2006	475,136,000
17	2008	950,272,000
18	2010	1,900,544,000

Transistors on an Integrated Circuit

According to Moore's Law, about 2 billion transistors could be placed on an integrated circuit in 2010.

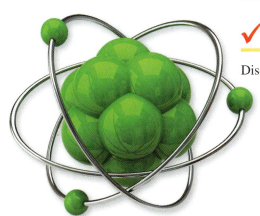

✓ **Checkpoint**

Help at Math.andYOU.com

Discuss the following claim that Moore's Law will reach its limit.

"Moore's Law, the central driver of our age, is based on the idea that circuit lines can be drawn ever-closer together. But there has to be a limit. The atomic scale. You can't make a circuit smaller than an atom."

"Moore's Law Reaches its Limit with Quantum Dot Amplifier,"
Dana Blankenhorn

Finding an Exponential Growth Rate

Exponential Growth Rate

If A_0 and A_1 are the quantities for any two times, then the growth rate between those times, r, is given by

$$\frac{A_1}{A_0} = 1 + r.$$

EXAMPLE 5 **Finding an Exponential Growth Rate**

You purchase 100 shares of a stock for $4.35 per share. One month later, the value of the stock is $4.55 per share.

a. Linear Growth: The value of the stock continues to increase by the *same dollar amount* each month. How much will your investment be worth in 2 years?

b. Exponential Growth: The value of the stock continues to increase by the *same percent* each month. How much will your investment be worth in 2 years?

$$\frac{4.55}{4.35} \approx 1.046$$

SOLUTION

a. Linear Growth: If the stock continues to increase by $0.20 per month, each share will be worth $4.35 + 24(0.2) = \$9.15$. So, your investment will be worth $100(9.15) = \$915.00$.

b. Exponential Growth: The rate of growth from $4.35 to $4.55 is about 4.6%. If the stock continues to grow at this rate, in 2 years each share will be worth $4.35(1.046)^{24} = \$12.80$. So, your investment will be worth about $100(12.8) = \$1280.00$.

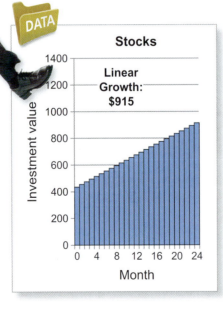

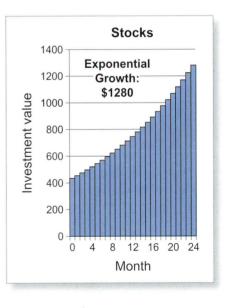

✓ Checkpoint Help at *Math*.and**YOU**.com

Using each type of growth, how much will the stock be worth in 4 years? Illustrate each type with a graph.

EXAMPLE 6   **Finding an Exponential Growth Rate**

Moore's Law states that the number of transistors that can be placed inexpensively on an integrated circuit will double every *two years* (see Example 4 on page 155). What is the *annual* rate of growth represented by Moore's Law?

SOLUTION

As it turns out, the answer to this question has a fascinating history in mathematics. The question comes down to this: *Can you find a number whose square is 2?*

Square Root of 2

The answer is denoted by $\sqrt{2}$, which is approximately 1.414213562. So, the annual rate of growth for Moore's Law is about 41.42%.

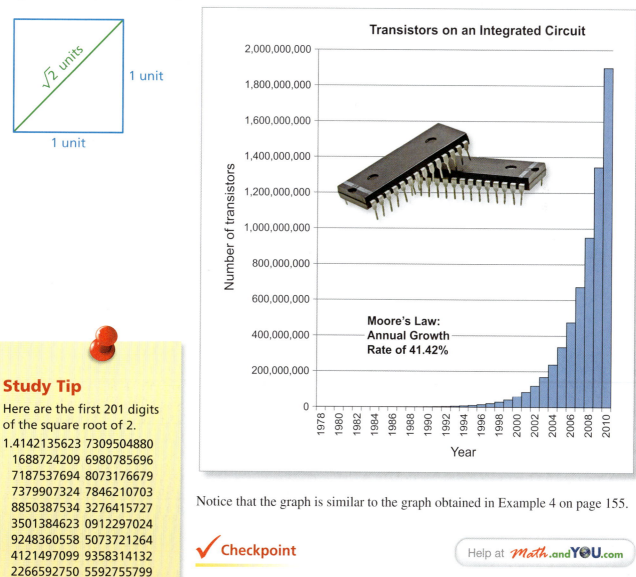

The ancient Greeks were able to prove that the square root of 2 cannot be represented by a fraction like 707/500. They were also able to construct a line segment with a length of exactly $\sqrt{2}$ units. To do this, draw a square with side lengths of 1 unit. Using the Pythagorean Theorem, it follows that the length of each diagonal of the square is $\sqrt{2}$ units.

$\sqrt{2}$ units

1 unit

1 unit

Study Tip

Here are the first 201 digits of the square root of 2.

1.4142135623 7309504880
 1688724209 6980785696
 7187537694 8073176679
 7379907324 7846210703
 8850387534 3276415727
 3501384623 0912297024
 9248360558 5073721264
 4121497099 9358314132
 2266592750 5592755799
 9505011527 8206057147

Notice that the graph is similar to the graph obtained in Example 4 on page 155.

✓ **Checkpoint**   Help at *Math*.and**YOU**.com

Using an initial value of 29,000 in 1978 and an annual growth rate of 41.42%, how many transistors could be placed on an integrated circuit in 2010?

4.1 Exercises

DATA **Gossip** Gossip is informal conversation, usually about the personal details of other people's lives. Gossip tends to spread very quickly and can be distorted and exaggerated from person to person. **In Exercises 1–4, use the following information.** *(See Example 1.)*

You win the lottery and share the news with three family members. You want to keep it a secret until you cash in your ticket and do all the paperwork. Somehow your secret gets out, and the number of people who hear about your good fortune grows by 25% each hour.

1. Write a formula that can be used to model the exponential growth of the gossip about you winning the lottery.

2. Use the formula from Exercise 1 to complete the table.

Time, n	Informed People
0 hr	3
2 hr	
4 hr	
6 hr	
8 hr	

3. The gossip about you winning the lottery continues to grow exponentially. How many people will know about you winning the lottery after 24 hours?

4. Suppose the gossip growth rate is changed from 25% each hour to 50% each hour. Does the second column of the table double? Explain your reasoning.

	A	B
1	**Minutes, n**	**Informed People**
2	0	100
3	3	
4	6	
5	9	
6	12	
7	15	
8	18	
9	21	
10	24	
11	27	
12	30	

DATA **5. Celebrity Wedding** As a result of the Internet and modern technology, gossip spreads much faster today than it did a few decades ago. This is especially true with news from Hollywood. Suppose a celebrity has a private, unannounced wedding with 100 guests. Once the news breaks, the number of people who know about the wedding grows by 45% each minute. Use a spreadsheet to illustrate the number of informed people every 3 minutes for up to 30 minutes. *(See Examples 1 and 2.)*

6. Social Network Do you think social networking websites and blogs are effective ways to communicate or get information? Explain your reasoning.

DATA   **Pandemic** A pandemic is an outbreak of an infectious disease or a condition that spreads through a large part of the human population. There have been numerous pandemics throughout history such as smallpox and tuberculosis. Current pandemics include HIV and certain strains of influenza. Once a pandemic reaches Phase 6, the number of infected people can grow exponentially. In Exercises 7–10, use exponential growth models. *(See Examples 3 and 4.)*

Stages of a Pandemic

Interpandemic period

Phase 1: No new influenza virus subtypes have been detected in humans. An influenza virus subtype that has caused human infection may be present in animals. If present in animals, the risk of human infection or disease is considered to be low.

Phase 2: No new influenza virus subtypes have been detected in humans. However, a circulating animal influenza virus subtype poses a substantial risk of human disease.

Pandemic alert period

Phase 3: Human infection(s) with a new subtype but no human-to-human spread, or at most rare instances of spread to a close contact.

Phase 4: Small cluster(s) with limited human-to-human transmission but spread is highly localized, suggesting that the virus is not well adapted to humans.

Phase 5: Larger cluster(s) but human-to-human spread still localized, suggesting that the virus is becoming increasingly better adapted to humans but may not yet be fully transmissible (substantial pandemic risk).

Pandemic period

Phase 6: Pandemic: increased and sustained transmission in general population.

Source: Centers for Disease Control and Prevention

---

7. From June 2009 through August 2010, the swine flu (H1N1) was considered a pandemic by the World Health Organization. Ten people in a community are infected with the swine flu. The next day, 26 people are infected. The growth rate is 160% per day. At this rate, how many people will be infected in 1 week? Graph the results.

8. Use the rate in Exercise 7 to find the number of people who will be infected after 1 week when the initial number of people infected is 50.

9. The 1918–1920 flu pandemic, commonly known as the Spanish flu, killed about 40 million people worldwide. That is more than twice the number of lives claimed by World War I. Estimates for the growth rate vary. Suppose 2 people are infected and the number of infected people triples every 3 days. How many people will be infected in 21 days? Graph the results.

10. Use the rate in Exercise 9. How many people will be infected in 42 days?

11. **Graph** Describe the graph of an exponential growth model.

12. **Growth Rate** Many factors affect the spread of a disease. Describe some factors that would influence the growth rate of an infectious disease. Would these factors increase or decrease the growth rate?

To help prevent the spread of germs, you should cough or sneeze into your elbow.

Investing **Investments often grow exponentially. In Exercises 13–18, use exponential growth rates.** *(See Examples 5 and 6.)*

13. You open a savings account and deposit $1250.00. One year later, the balance in the account is $1262.50. No other transactions were posted to the account. What is the annual rate of growth for the savings account?

14. You open a savings account and deposit $4200.00. One month later, the balance in the account is $4208.40. No other transactions were posted to the account. What is the monthly rate of growth for the savings account?

15. You buy 50 shares of stock for $18.25 per share. One month later, the value of the stock is $18.98 per share.

   a. The value of the stock continues to increase by the same dollar amount each month. How much will your investment be worth in 1 year?

   b. The value of the stock continues to increase by the same percent each month. How much will your investment be worth in 1 year?

16. You buy 65 shares of stock for $12.00 per share. One month later, the value of the stock is $12.78 per share.

   a. The value of the stock continues to increase by the same dollar amount each month. How much will your investment be worth in 3 years?

   b. The value of the stock continues to increase by the same percent each month. How much will your investment be worth in 3 years?

17. During the first year of operation, a company makes a profit of $40,000. During the third year of operation, the company makes a profit of $120,000. Estimate its annual rate of growth.

18. You buy a coin collection for $10,000. Two years later, you sell the coin collection for $12,500. Estimate its annual rate of growth.

DATA 19. **Mathematical Models** Mathematical models rarely match real-life data perfectly. In general, a model will give an approximate value for each input. The table shows the balances of an investment account at the end of six continuous years. Would the data in the table be better represented by a linear growth model or an exponential growth model? Explain your reasoning.

Year	2006	2007	2008	2009	2010	2011
Balance	$11,503	$12,348	$13,335	$14,293	$15,024	$15,841

20. **Risk Aversion** Investing in the stock market can be nerve-racking due to the uncertainty of how a stock will perform. A risk-averse investor might put money in bonds or a savings account with a low but guaranteed interest rate. How would you invest a $10,000 windfall? Explain your reasoning.

▶ Extending Concepts

Logistic Growth In general, the size of a population (people, bacteria, cancer cells, etc.) follows a logistic growth pattern. The graph illustrates the growth patterns of a logistic curve. In Exercises 21–23, use the graph.

21. Compare an exponential curve to a logistic curve.

22. Why is the size of a population over time better represented by a logistic growth model than an exponential growth model?

23. Based on the information in the table, what part of the graph represents the world population? Explain.

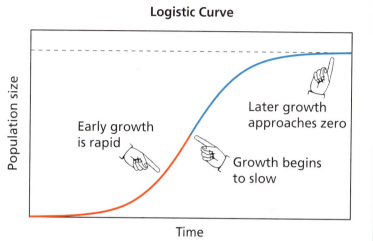

Logistic Curve

Early growth is rapid

Later growth approaches zero

Growth begins to slow

Population size

Time

Year	World Population (in billions)	Annual Growth Rate
1980	4.5	1.8%
1985	4.9	1.7%
1990	5.3	1.6%
1995	5.7	1.4%
2010	6.9	1.1%

24. **Example 6** To find the annual growth rate represented by Moore's Law, use the formula for exponential growth, $A = P(1 + r)^n$. Because the number of transistors doubles every 2 years, you can substitute $2P$ for A and 2 for n in the formula. Then solve for r as shown. Suppose the number of transistors triples every two years. What is the annual growth rate?

$A = P(1 + r)^n$	Formula for exponential growth
$2P = P(1 + r)^2$	Substitute $2P$ for A and 2 for n.
$\dfrac{2P}{P} = \dfrac{P(1 + r)^2}{P}$	Divide both sides by P.
$2 = (1 + r)^2$	Simplify.
$\sqrt{2} = 1 + r$	Take the square root of each side.
$\sqrt{2} - 1 = r$	Subtract 1 from each side.

Year	Tuition
1	$6200.00
3	$8252.20

Tuition In Exercises 25 and 26, use the table.

25. What is the annual growth rate of tuition?

26. At this rate, what will tuition be in three more years?

4.2 Inflation & the Consumer Price Index

▶ Use a consumer price index.

▶ Use a graph to interpret a consumer price index.

▶ Compare inflation to the value of the dollar.

Reading a Consumer Price Index

Inflation is a rise in the general level of prices for goods and services. A **consumer price index** (CPI) is used as a measure of inflation.

The U.S. Bureau of Labor Statistics defines a CPI as "a measure of the average change over time in the prices paid by urban consumers for a market basket of consumer goods and services." The Bureau of Labor Statistics lists many different CPIs. Here is one for *all* goods and services. (*Note:* The base year is 1983.)

UNITED STATES BUREAU OF LABOR STATISTICS							
Year	CPI	Year	CPI	Year	CPI	Year	CPI
1931	15.2	1951	26.0	1971	40.5	1991	136.2
1932	13.7	1952	26.5	1972	41.8	1992	140.3
1933	13.0	1953	26.7	1973	44.4	1993	144.5
1934	13.4	1954	26.9	1974	49.3	1994	148.2
1935	13.7	1955	26.8	1975	53.8	1995	152.4
1936	13.9	1956	27.2	1976	56.9	1996	156.9
1937	14.4	1957	28.1	1977	60.6	1997	160.5
1938	14.1	1958	28.9	1978	65.2	1998	163.0
1939	13.9	1959	29.1	1979	72.6	1999	166.6
1940	14.0	1960	29.6	1980	82.4	2000	172.2
1941	14.7	1961	29.9	1981	90.9	2001	177.1
1942	16.3	1962	30.2	1982	96.5	2002	179.9
1943	17.3	1963	30.6	1983	99.6	2003	184.0
1944	17.6	1964	31.0	1984	103.9	2004	188.9
1945	18.0	1965	31.5	1985	107.6	2005	195.3
1946	19.5	1966	32.4	1986	109.6	2006	201.6
1947	22.3	1967	33.4	1987	113.6	2007	207.3
1948	24.1	1968	34.8	1988	118.3	2008	215.3
1949	23.8	1969	36.7	1989	124.0	2009	214.5
1950	24.1	1970	38.8	1990	130.7	2010	218.1

This famous photograph came to represent the hardship of the Great Depression in America. It was taken by Dorothea Lange in 1936. The photograph is of a migrant worker, Florence Thompson, and three of her seven children. Florence was 32 years old at the time the photo was taken.

EXAMPLE 1 **Reading a Consumer Price Index**

Was the Great Depression (1930s) a time of inflation?

SOLUTION

Using the CPI, you can see that the prices of consumer goods and services did not rise during the 1930s. In fact, from 1931 through 1939, prices fell slightly. So, the Great Depression was not a time of inflation.

✓ **Checkpoint**   Help at *Math*.and**YOU**.com

During which of the eight decades shown in the CPI did prices increase by the greatest percent? Explain your reasoning.

The primary use of a CPI is to compare prices for goods and services between two different years. For instance, you can use a CPI to determine the hourly rate in 2010 that was equivalent to $15 per hour in 1980.

Calculating Prices

If you know the price of an item or service in year A, then the price of that same item in year B is

$$\text{Price in year B} = \frac{\text{CPI in year B}}{\text{CPI in year A}}(\text{price in year A}).$$

Math.and*YOU*.com

You can access an inflation calculator at *Math.andYou.com*.

EXAMPLE 2   **Calculating an Inflated Price**

Your grandfather bought a section (640 acres) of land in Montana in 1942 for $5000. You inherited the property in 2010. The property was evaluated to be worth $2,400,000. Did the value of the property "keep up with inflation" or did it exceed inflation? Explain your reasoning.

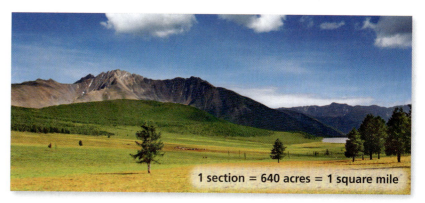

1 section = 640 acres = 1 square mile

SOLUTION

The CPI values for 1942 and 2010 were 16.3 and 218.1, respectively. If the value of the property had kept up with inflation, it would be worth

$$\text{Value in 2010} = \frac{\text{CPI in 2010}}{\text{CPI in 1942}}(\text{value in 1942})$$

$$= \frac{218.1}{16.3}(5000)$$

$$\approx \$66,902.$$

So, the value of the property far exceeded inflation.

✔ **Checkpoint**   Help at *Math*.and*YOU*.com

In 1942, a Bendix automatic washing machine cost about $150. Suppose the cost of the washing machine kept up with inflation. What would it have cost in 2010?

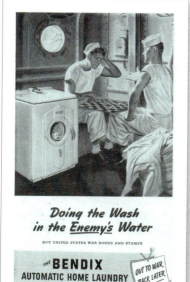

1942 ad for Bendix washing machines

Graphing a Consumer Price Index

EXAMPLE 3 **Graphing a Consumer Price Index**

Graph the CPI from 1931 through 2010. Use the graph to describe the rates of inflation during these years.

SOLUTION

One way to do this is to enter the CPI values into a spreadsheet. Then use the spreadsheet to graph the values.

In early 1980, inflation and interest rates rose to about 18%, which was much higher than during any other time of peace. Business people and bankers panicked and began talking about financial collapse, bankruptcies, and a drop in the American standard of living. Jimmy Carter gave a speech not only to announce a new program to fight inflation, but also to calm the panicking public.

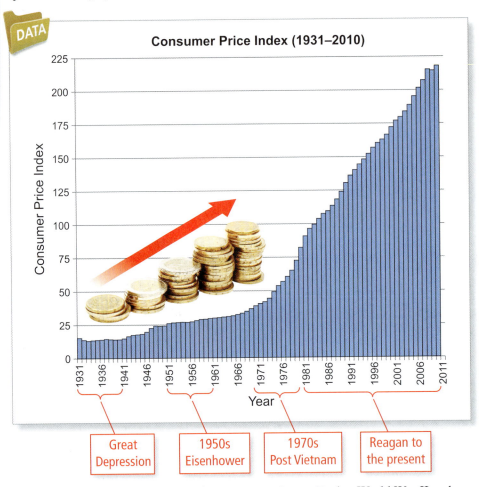

During the Great Depression, inflation was nonexistent. During World War II and the 1940s, inflation started to increase, reaching an annual rate of 14.4% in 1947. During the 1950s, America entered a period of stability, with an annual inflation rate of about 1.9%. With the Vietnam War in the 1960s, inflation started to increase again. By the 1970s, ending with the Carter years, inflation was rampant. The annual rate of inflation during this decade was about 6.5%. From 1981 through 2010, the CPI followed a linear pattern, which implies that the rate of inflation was gradually decreasing during that 30-year period.

✓ **Checkpoint** Help at *Math*.and*YOU*.com

From the graph in Example 3, does it appear that either political party (Democratic or Republican) is more associated with higher inflation rates? Explain your reasoning.

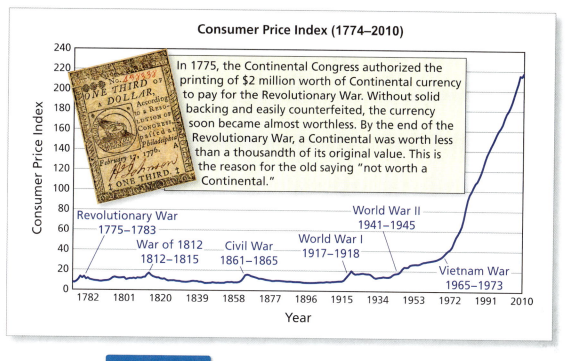

Consumer Price Index (1774–2010)

In 1775, the Continental Congress authorized the printing of $2 million worth of Continental currency to pay for the Revolutionary War. Without solid backing and easily counterfeited, the currency soon became almost worthless. By the end of the Revolutionary War, a Continental was worth less than a thousandth of its original value. This is the reason for the old saying "not worth a Continental."

Revolutionary War
1775–1783

War of 1812
1812–1815

Civil War
1861–1865

World War I
1917–1918

World War II
1941–1945

Vietnam War
1965–1973

Year

EXAMPLE 4   **Analyzing a Graph**

Use the graph of the CPI from 1774 through 2010 to discuss the statement from *InflationData.com*.

> "The very nature of War results in the destruction of goods. But in normal times money is spent to produce goods which makes the world a richer place. During a war, however, things are produced but . . . they are not productive things but destructive. The money is spent to destroy things. Often this is combined with an increase in the money supply in order to pay for the destruction. This increase in the money supply combined with a decrease in goods is classic inflation."

SOLUTION

Up through the Vietnam War, it does appear that war corresponded to inflation. Since that time, however, inflation appears to have been a constant part of U.S. economics, regardless of whether the country was at war.

✓ **Checkpoint**   Help at *Math*.and**Y⊙U**.com

Discuss the paragraph from *Blogvesting.com*. Do you agree with the argument? Explain.

> "Inflation occurs when the monetary base grows faster than the amount of goods and services in the economy. Deficit spending, applied to boost consumption, will result in inflation because the amount of goods and services in the economy is unchanged while the monetary base is increased. On the other hand, deficit spending applied to boost production will increase the amount of goods and services, and is not necessarily inflationary."

Inflation and the Value of the Dollar

There are two related ways to think about inflation.

- With inflation, the cost of a particular item or service *increases*.
- With inflation, the value of the dollar *decreases*.

The second concept is called **devaluation** of the dollar.

The Value of the Dollar

The value V of the dollar in year A is related to the value of the CPI in year A by the following formula.

$$V \text{ in year A} = \frac{100}{\text{CPI in year A}}$$

This depends on the base year, 1983, in which a dollar was worth $1.

EXAMPLE 5 Graphing the Value of the Dollar

Graph the value of the dollar from 1931 through 2010. Compare your graph with the graph in Example 3 on page 164.

SOLUTION

You can use the same data you used in Example 3. However, in your spreadsheet, create a new column that divides 100 by the CPI column.

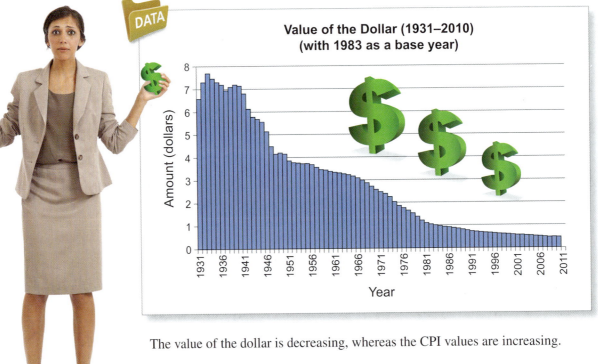

Value of the Dollar (1931–2010)
(with 1983 as a base year)

The value of the dollar is decreasing, whereas the CPI values are increasing.

✓ Checkpoint

Help at *Math*.and**YOU**.com

Why might the U.S. government encourage devaluation of the dollar during times of deficit spending? Explain your reasoning.

EXAMPLE 6   **Comparing Debt Between Two Years**

Was the federal government more in debt in 2010 than in 1983? Explain
your reasoning.

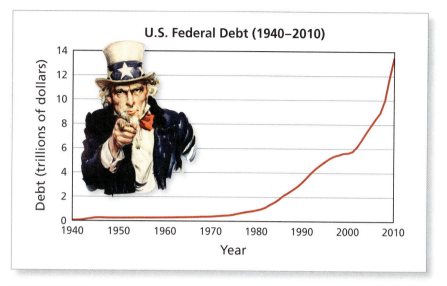

U.S. Federal Debt (1940–2010)

SOLUTION

The simple answer is yes. The government owed about $1.4 trillion in 1983 and
about $14 trillion in 2010. This simple answer, however, does not account for the
fact that the dollar was worth less in 2010 than it was in 1983. You can account
for this difference as follows.

The value of the dollar in 2010 (compared to 1983 dollars) was

$$V \text{ in } 2010 = \frac{100}{\text{CPI in } 2010}$$

$$= \frac{100}{218.1}$$

$$\approx \$0.459.$$

So, in terms of 1983 dollars, the 2010 federal debt was about 45.9% of
$14 trillion, or about $6 trillion. In other words, it was about 4 times greater than
the federal debt in 1983.

✓ **Checkpoint**   Help at *Math*.and**YOU**.com

Discuss the following loan plan. Is it valid? Explain.

This scenario happened many
times during the early 2000s.

1. Take out a mortgage for $250,000 in 2005.
   Down payment: $30,000, Rate: 6%, Term: 30 years

2. Make a monthly payment of $1320.

3. Depend on an annual inflation rate of 4%.

4. Depend on housing prices increasing by 10% annually.

5. Sell the house in 10 years for $650,000.
   The balance of your mortgage will be $184,000.

4.2 Exercises

Minimum Wage The table shows the federal minimum hourly wage in the years that the wage changed since 1978. In Exercises 1–4, use the table and the CPI. *(See Example 1.)*

Effective Year	Minimum Hourly Wage
1978	$2.65
1979	$2.90
1980	$3.10
1981	$3.35
1990	$3.80
1991	$4.25
1996	$4.75
1997	$5.15
2007	$5.85
2008	$6.55
2009	$7.25

The first federal minimum wage law, the Fair Labor Standards Act, was passed in 1938. This law set the minimum wage at $0.25 per hour, limited a workweek to a maximum of 44 hours, and banned child labor.

1. What is the percent increase in the CPI for the following years?

    a. From 1978 to 1981

    b. From 1978 to 1997

    c. From 1990 to 2009

    d. From 1978 to 2009

2. What is the percent increase in the federal minimum hourly wage for the following years?

    a. From 1978 to 1981

    b. From 1978 to 1997

    c. From 1990 to 2009

    d. From 1978 to 2009

3. During which of the time periods in Exercises 1 and 2 did the buying power of the federal minimum hourly wage outpace inflation? Explain your reasoning.

4. Do changes in the federal minimum hourly wage depend on the CPI? Explain your reasoning.

5. **Starting Wage** The starting wage at a company in 1992 was $14.50 per hour. In 2010, the starting wage was $21.25 per hour. Did the starting wage at the company keep up with inflation? Explain your reasoning. *(See Example 2.)*

6. **CEO Compensation** In 1980, the compensation for a CEO at a company was $2.5 million. By 2010, the compensation for the CEO was $15 million. Did the compensation for the CEO keep up with inflation? Explain your reasoning. *(See Example 2.)*

HICP In Exercises 7–12, use the information below.
(See Examples 3 and 4.)

The European Central Bank uses the harmonized index of consumer prices (HICP) to measure the average change over time in prices paid by consumers for normal goods and services across countries in the European Union. Unlike the CPI, the HICP excludes owner-occupied housing costs and incorporates rural consumers along with urban consumers in the calculations. The HICP is an internationally comparable measure of inflation.

Year	Austria	Germany	Italy	Netherlands	Spain	Sweden
HICP by Country						
1996	87.2	88.6	81.8	80.4	77.9	87.5
1997	88.2	90.0	83.3	81.9	79.4	89.1
1998	89.0	90.5	85.0	83.4	80.8	90.0
1999	89.4	91.1	86.4	85.1	82.6	90.5
2000	91.2	92.4	88.6	87.1	85.5	91.7
2001	93.3	94.1	90.7	91.5	87.9	94.1
2002	94.8	95.4	93.1	95.1	91.0	95.9
2003	96.1	96.4	95.7	97.2	93.9	98.2
2004	97.9	98.1	97.8	98.5	96.7	99.2
2005	100.0	100.0	100.0	100.0	100.0	100.0
2006	101.7	101.8	102.2	101.7	103.6	101.5
2007	103.9	104.1	104.3	103.3	106.5	103.2
2008	107.3	107.0	108.0	105.5	110.9	106.7
2009	107.7	107.2	108.8	106.6	110.6	108.7

7. Which year is the base year for the HICP?

8. Graph the HICP for Italy from 1996 through 2009. The vertical axis should have a range of 80 to 110. Use the graph to describe the rate of inflation during these years.

9. Graph the HICP for Sweden from 1996 through 2009. The vertical axis should have a range of 80 to 110. Use the graph to describe the rate of inflation during these years.

10. For the years shown, was the pricing for consumer goods and services more stable in Italy or Sweden? Explain your reasoning.

11. Did all the countries shown in the table experience inflation from 1996 to 2009? Explain your reasoning.

12. Discuss the statement. Do you agree with the argument? Explain.

> The HICP can be used to compare developments in inflation between different countries in the European Union, and to calculate inflation for the eurozone as a whole. The national CPIs of various other countries are not suitable for this because of differences in composition.

Average Pay The table shows the average hourly wage of production and nonsupervisory employees in private industries across the United States from 1966 to 2010. In Exercises 13–18, use the table. *(See Examples 5 and 6.)*

Average Hourly Wage of Production and Nonsupervisory Employees					
Year	Hourly Wage	Year	Hourly Wage	Year	Hourly Wage
1966	$2.73	1981	$7.44	1996	$12.04
1967	$2.85	1982	$7.87	1997	$12.51
1968	$3.02	1983	$8.20	1998	$13.01
1969	$3.22	1984	$8.49	1999	$13.49
1970	$3.40	1985	$8.74	2000	$14.02
1971	$3.63	1986	$8.93	2001	$14.54
1972	$3.90	1987	$9.14	2002	$14.97
1973	$4.14	1988	$9.44	2003	$15.37
1974	$4.43	1989	$9.80	2004	$15.69
1975	$4.73	1990	$10.20	2005	$16.13
1976	$5.06	1991	$10.52	2006	$16.76
1977	$5.44	1992	$10.77	2007	$17.43
1978	$5.88	1993	$11.05	2008	$18.08
1979	$6.34	1994	$11.34	2009	$18.62
1980	$6.85	1995	$11.65	2010	$19.04

13. Graph the average hourly wage of production and nonsupervisory employees from 1966 through 2010. Use the graph to describe the average hourly wage during these years.

14. Compare your graph from Exercise 13 to the graph of the CPI on page 164. Do you think the average hourly wage of production and nonsupervisory employees depends on the CPI? Explain your reasoning.

15. Did the average hourly wage of production and nonsupervisory employees have more buying power in 1990 than in 1970? Explain your reasoning.

16. Did the average hourly wage of production and nonsupervisory employees have more buying power in 2010 than in 1990? Explain your reasoning.

17. Would you be better off financially to earn $50,000 in 1990 or $75,000 in 2010? Explain your reasoning.

18. There is a discrepancy in buying power between the federal minimum hourly wage and the average hourly wage of production and nonsupervisory employees. Since 1978, do you think this discrepancy has increased, decreased, or remained unchanged? Explain your reasoning.

▶ Extending Concepts

DATA   **Creating a Price Index** In Exercises 19–22, use the information below and the table.

To create a price index, first choose a base year. The index in that year is 100. The price index reflects the ratio of the price from a given year to the price in the base year. Use the formula

$$\text{Index in year A} = 100 \times \frac{\text{Price in year A}}{\text{Price in base year}}$$

to find the index for a given year A.

19. Complete the table.

20. Make a double line graph of the gasoline index and the diesel index. Compare the rate of inflation for gasoline to the rate of inflation for diesel fuel.

Year	Unleaded Regular Gasoline		Diesel Fuel	
	Average Price Per Gallon	Gasoline Index	Average Price Per Gallon	Diesel Index
1995	$1.15	100.0	$1.11	100.0
1996	$1.23	107.0	$1.24	
1997	$1.23		$1.20	
1998	$1.06		$1.04	
1999	$1.17		$1.12	
2000	$1.51		$1.49	
2001	$1.46		$1.40	
2002	$1.36		$1.32	
2003	$1.59		$1.51	
2004	$1.88		$1.81	
2005	$2.30		$2.40	
2006	$2.59		$2.71	
2007	$2.80		$2.89	
2008	$3.27		$3.80	
2009	$2.35		$2.47	

21. For the years shown, use a spreadsheet to compare the rates of inflation for gasoline and diesel fuel to the rate of inflation indicated by the CPI.

22. Diesel engines are 30% more fuel efficient than similar-sized gasoline engines. From 2005 through 2009, would the annual fuel cost be more for a diesel engine or a similar-sized gasoline engine? Explain your reasoning.

23. **Inflation Rate** Why might the inflation rate published by the Bureau of Labor Statistics not match an individual's inflation experience?

24. **Economy** Discuss the effects of inflation on an economy. Explain your reasoning.

4.1–4.2 Quiz

Average Inflation Rate In Exercises 1–6, use the information below.

From 1983 to 2010, the average annual inflation rate in the United States was about 3%.

1. Write a formula that can be used to model the exponential growth of the CPI from 1983 to 2010.

2. Use the formula from Exercise 1 and a spreadsheet to project the CPI from 1983 to 2020.

Year	CPI
1983	100.0
1984	
1985	
1986	
1987	
1988	
1989	
1990	
1991	
1992	

Year	CPI
1993	
1994	
1995	
1996	
1997	
1998	
1999	
2000	
2001	
2002	

Year	CPI
2003	
2004	
2005	
2006	
2007	
2008	
2009	
2010	
2011	
2012	

Year	CPI
2013	
2014	
2015	
2016	
2017	
2018	
2019	
2020	

3. Use a spreadsheet to create a double bar graph to compare the projected CPI to the actual CPI from 1983 to 2010.

4. You buy a new car for $25,000 in 2010. Using the actual CPI for 2010, what would you expect to pay for a new car with similar features in 2020?

5. On "Black Tuesday," October 29, 1929, the stock market lost over $14 billion. Estimate this loss in terms of 2020 dollars. The CPI for 1929 is 17.1.

6. You invest $10,000 in 2005. Your investment earns an average of 2% annually for 15 years. In 2020, is the buying power of your investment higher than in 2005? Explain your reasoning.

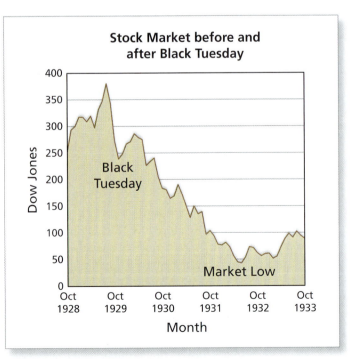

Stock Market before and after Black Tuesday

Math & Accepting a Job
PROJECT: Comparing Costs of Living

1. Use the *Cost of Living Calculator** at *Math.andYou.com*.

   Choose your home city, a job title that you are interested in, and a salary that you would expect this job to pay in your home city. Then compare the cost of living and salary in your home city to the cost of living and salary in each of the following cities. Organize your results graphically, as though you were presenting to a committee.

   - New York, New York
   - Boston, Massachusetts
   - Anchorage, Alaska
   - San Diego, California
   - Philadelphia, Pennsylvania
   - Grand Rapids, Michigan
   - Seattle, Washington

   - Minot, North Dakota
   - Arlington, Texas
   - Topeka, Kansas
   - Columbus, Ohio
   - Tulsa, Oklahoma
   - Missoula, Montana
   - Fort Worth, Texas

Cost of Living Calculator

Calculate the difference from city to city

Moving from
Georgia - Atlanta

Moving to
California - San Francisco

Job Title
Human Resources (HR) Specia

$ Salary
45000

Calculate

*Supplied by *PayScale.com*

> Complete all four fields. For your job title, choose a title that best describes the job you have or want.

> After pressing "Calculate," the calculator will display this type of result.

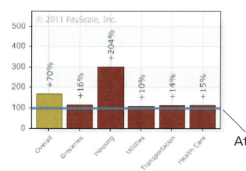

Cost of living in San Francisco, California by Expense Category

© 2011 PayScale, Inc.

+70% +16% +204% +10% +14% +15%

Overall · Groceries · Housing · Utilities · Transportation · Health Care

Atlanta

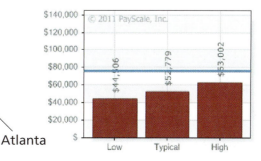

Salary Range for a Human Resources (HR) Specialist in San Francisco, California

© 2011 PayScale, Inc.

$44,906 $52,779 $53,002

Low   Typical   High

— Equivalent Salary: $76,534

You are currently earning **$45,000** in Atlanta, Georgia as a Human Resources (HR) Specialist.

You need to earn **$76,534** to maintain the same standard of living in San Francisco, California.

4.3 Exponential Decay

▶ Make a table and graph showing exponential decay.

▶ Calculate and use half-life.

▶ Find an exponential decay rate.

Calculating Exponential Decay

A quantity has **exponential decay** when the quantity *decreases* by the same percent from one time period to the next.

Study Tip

As with exponential growth, be sure you see that the rate of decay is written in decimal form.

Formula for Exponential Decay

A quantity A that has exponential decay can be modeled by

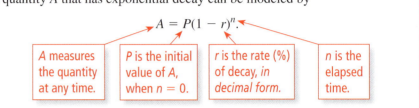

$$A = P(1 - r)^n.$$

A measures the quantity at any time.	*P* is the initial value of *A*, when $n = 0$.	*r* is the rate (%) of decay, *in decimal form*.	*n* is the elapsed time.

EXAMPLE 1   **Making a Table**

As young salmon pass through a turbine on a hydroelectric dam (on their way to the ocean), about 15% are killed. Make a table showing how many of 100,000 young salmon survive after passing through 6 turbines.

SOLUTION

The formula for this exponential decay is

$$A = P(1 - r)^n = 100,000(1 - 0.15)^n. \quad \text{\textcolor{red}{$P = 100{,}000, r = 15\% = 0.15$}}$$

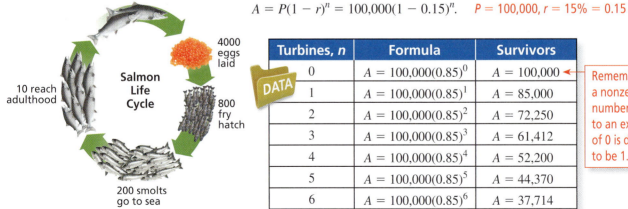

Turbines, *n*	Formula	Survivors
0	$A = 100,000(0.85)^0$	$A = 100,000$
1	$A = 100,000(0.85)^1$	$A = 85,000$
2	$A = 100,000(0.85)^2$	$A = 72,250$
3	$A = 100,000(0.85)^3$	$A = 61,412$
4	$A = 100,000(0.85)^4$	$A = 52,200$
5	$A = 100,000(0.85)^5$	$A = 44,370$
6	$A = 100,000(0.85)^6$	$A = 37,714$

Remember that a nonzero number raised to an exponent of 0 is defined to be 1.

2 return to spawn

4000 eggs laid

Salmon Life Cycle

10 reach adulthood

800 fry hatch

200 smolts go to sea

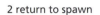 **Checkpoint**

Help at *Math*.and*YOU*.com

In the Columbia and Snake River system, some young salmon pass through 12 turbines. Of 100,000 young salmon, how many will survive?

EXAMPLE 2   **Graphing Exponential Decay**

As the height above sea level increases, atmospheric pressure decreases exponentially at a rate of about 3.8% per 1000 feet. At sea level, atmospheric pressure is denoted by 1 atmosphere (atm). Commercial jets typically fly at about 35,000 feet. Sketch a graph showing the decrease in atmospheric pressure as a jet climbs from sea level to 35,000 feet. Use the graph to estimate the atmospheric pressure at 35,000 feet.

SOLUTION

The formula for this exponential decay is

$$A = P(1 - r)^n = 1(1 - 0.038)^n. \qquad P = 1, r = 3.8\% = 0.038$$

Enter this formula into a spreadsheet and graph the results as shown.

In the 17th century, scientists discovered that air actually has weight. Evangelista Torricelli, one of the first to discover atmospheric pressure, said, "We live submerged at the bottom of an ocean of the element air." Earth's gravitational field pulls on air, and this pull is called atmospheric pressure. Torricelli went on to develop the mercury barometer to measure atmospheric pressure.

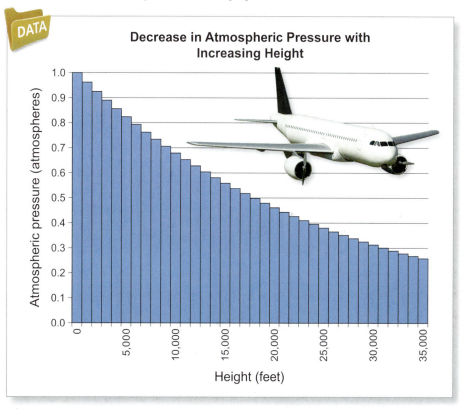

From the graph, the atmospheric pressure at 35,000 feet appears to be about 0.25 atmosphere. You can check this using the formula.

$$A = P(1 - r)^n = 1(1 - 0.038)^{35} = 1(0.962)^{35} \approx 0.258 \text{ atm}$$

So, at 35,000 feet above sea level, the atmospheric pressure is about a quarter of what it is at sea level.

✓ **Checkpoint**   Help at *Math*.and**Y⊙U**.com

There are many mountains in the United States that have heights of 14,000 feet or greater. They are all in Alaska, California, Colorado, and Washington. Mount McKinley in Alaska has a height of about 20,320 feet. How much more does the atmospheric pressure decrease as you climb from 14,000 feet to 20,320 feet?

Calculating and Using Half-Life

The **half-life** of a substance is the time it takes for the substance to lose half of its pharmacological, physiological, or radiological characteristic.

Using Half-Life to Calculate Exponential Decay

Consider a substance that has a half-life of T. The remaining quantity A that exists after an elapsed time of t is

$$A = P\left(\frac{1}{2}\right)^{t/T}.$$

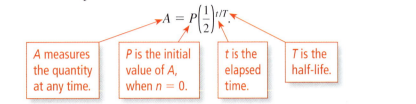

A measures the quantity at any time.

P is the initial value of A, when $n = 0$.

t is the elapsed time.

T is the half-life.

EXAMPLE 3 **Making a Table**

The half-life of a drug is the amount of time it takes for 50% of the drug to be removed from a person's body. Suppose you are injected with 500 milligrams of a drug that has a half-life of 3 hours. How much of the drug will be in your bloodstream after 24 hours?

SOLUTION

The formula for the amount left in your bloodstream is

$$A = 500\left(\frac{1}{2}\right)^{t/3}. \qquad P = 500,\ T = 3$$

Enter this formula into a spreadsheet and graph the results as shown.

500(0.5)^(3/3)

	A	B
1	Time, t	Amount, A
2	0	500.00
3	3	250.00
4	6	125.00
5	9	62.50
6	12	31.25
7	15	15.63
8	18	7.81
9	21	3.91
10	24	1.95

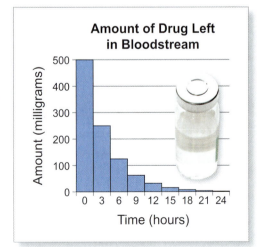

Amount of Drug Left in Bloodstream

About 2 milligrams will be in your bloodstream after 24 hours.

✓ **Checkpoint**

Help at *Math*.and**YOU**.com

How much longer will it take for there to be less than 1 milligram of the drug in your bloodstream?

Carbon dating uses the radioisotope carbon-14 to determine the age of organic materials up to about 50,000 years. Carbon-14 has a half-life of 5730 years. Atmospheric carbon dioxide has a concentration of about 1 atom of carbon-14 per every 10^{12} atoms of carbon-12. Living organisms take in carbon dioxide from the environment and have the same ratio of carbon-14 to carbon-12 as the atmosphere. When an organism dies, it stops taking in carbon. Then the carbon-14 in the organism starts to decay, which changes the ratio of carbon-14 to carbon-12. By measuring how much the ratio is lowered, it is possible to estimate how much time has passed since the organism lived.

EXAMPLE 4   Carbon Dating a Skeleton

You discover the skeleton of a mastodon. By having it tested, you find that the ratio of carbon-14 to carbon-12 is about one-quarter of that occurring in the atmosphere. How long ago did the mastodon live?

SOLUTION

The half-life of carbon-14 is 5730 years. Because the amount of carbon-14 in the sample is about one-quarter of the amount in the skeleton when the mastodon was alive, you can conclude that the mastodon lived about 2 half-lives ago, or about $2(5730) = 11,460$ years ago.

Fossils of the American mastodon have been found from Alaska to Florida. Its main habitat was cold, spruce woodlands, and it is believed to have traveled in herds. It is thought to have disappeared from North America about 10,000 years ago.

✓ Checkpoint

 Help at *Math*.and**Y❂U**.com

In the article, soil samples were estimated to be 3000 years old. Estimate the ratio of carbon-14 to carbon-12 in the soil samples. Explain your reasoning.

"Whale remains found at a 3,000-year-old site in northwestern Alaska called Old Whaling, for instance, were once considered evidence of early hunting. But a re-examination of the site in recent years has suggested that people there were simply scavenging dead whales that had washed ashore. There are some dramatic rock carvings in southeastern Korea that show bands of hunters going after whales. But these are nearly impossible to pin down with an exact date, says Odess. In contrast, the newfound ivory carving was pegged as being 3,000 years old by nearly a dozen radiocarbon dates on the soil in which it was embedded. The previous eldest solid evidence for whaling is some 2,000 years old."

"Whaling Scene Found in 3,000-Year-Old Picture," Alexandra Witze

Finding an Exponential Decay Rate

Exponential Decay Rate

If A_0 and A_1 are the quantities for any two times, then the decay rate between those times, r, is given by

$$\frac{A_1}{A_0} = 1 - r.$$

EXAMPLE 5 **Finding an Exponential Decay Rate**

Your *basal metabolic rate* (BMR) is the number of calories you expend per day while in a state of rest. Informally, it is the amount of calories your body uses to stay alive. Studies show that your BMR decreases in adulthood.

A man's BMR was 1800 calories when he was 20 years old. It decreased to 1710 calories when he was 30 years old. Assuming his BMR decays exponentially, estimate the man's age when his BMR will be about 1390 calories.

SOLUTION

Begin by dividing the BMR at age 30 by the BMR at age 20.

$$\frac{A_1}{A_0} = \frac{1710}{1800} = 0.95 = 1 - r$$

This implies that $r = 0.05$ and the rate of exponential decay is 5% every decade. Use a spreadsheet to calculate a 5% per decade decrease from the age of 20 for several decades.

	A	B
1	**Decades**	**BMR (calories)**
2	0	1800.0
3	1	1710.0
4	2	1624.5
5	3	1543.3
6	4	1466.1
7	5	1392.8
8	6	1323.2

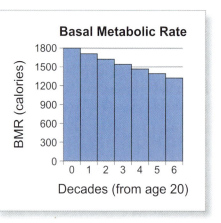

From the spreadsheet, you can see that his BMR will be about 1390 calories after 5 decades. So, he will be 70 years old.

As you get older and your BMR decreases, you may need to exercise more or eat less to keep from gaining weight.

✓ **Checkpoint**

Help at *Math*.and**YOU**.com

In the above example, suppose the man's BMR decreased to 1764 calories when he was 30 years old. Estimate the man's age when his BMR will be about 1660 calories assuming it decreases exponentially. Explain the results.

There are many different forms of radioactive waste. Here are a few examples.

- Used nuclear fuel
- Discarded parts from nuclear reactors
- Filters used to separate radioactive materials from water
- Protective clothing of workers in contaminated areas
- Medical supplies used in connection with radioactive materials
- Remains of lab animals injected with radioactive materials for research

EXAMPLE 6   Finding an Exponential Decay Rate

Ten grams of plutonium-239 are placed into a hazardous materials waste dump. Use the spreadsheet to determine how long it will take the 10 grams of plutonium-239 to decay to about 1 gram.

	A	B
1	Years	Grams Remaining
2	0	10.0
3	10,000	7.5
4	20,000	5.6
5	30,000	4.2

SOLUTION

The rate of exponential decay is given by

$$\frac{A_1}{A_0} = \frac{7.5}{10} = 0.75 = 1 - r.$$

This implies that $r = 0.25$ and the rate of decay is 25% every 10,000 years. Use a spreadsheet to calculate the amount remaining.

DATA	A	B
1	Years	Grams Remaining
2	0	10.0
3	10,000	7.5
4	20,000	5.6
5	30,000	4.2
6	40,000	3.2
7	50,000	2.4
8	60,000	1.8
9	70,000	1.3
10	80,000	1.0

It will take about 80,000 years for the 10 grams of plutonium-239 to decay to about 1 gram.

In the United States, there are three low-level radioactive waste disposal facilities. They are located in Barnwell, South Carolina; Clive, Utah; and Richland, Washington.

 **Checkpoint**   Help at *Math*.and**Y○U**.com

How much longer will it take the plutonium-239 to decay to about one-tenth of a gram? Explain your reasoning.

4.3 Exercises

DATA   Glaciers   In Exercises 1–6, use the information below. *(See Examples 1 and 2.)*

The size of a glacier is measured in the same way as land, in acres or hectares. In 1900, the size of a glacier was about 320 hectares. Between 1900 and 2000, its size decreased by about 14% per decade.

1. Write a formula that represents the size of the glacier at the beginning of each decade from 1900 to 2000.

2. Make a table showing the size of the glacier at the beginning of each decade from 1900 to 2000.

3. Sketch a graph showing the decrease in the area of the glacier from 1900 to 2000.

4. Use the graph in Exercise 3 to estimate the area of the glacier in 2000.

5. A general rule is that a moving piece of ice and snow is called a glacier when its size is at least 10 hectares. Assuming the trend continues, when will the glacier be too small to be considered a glacier?

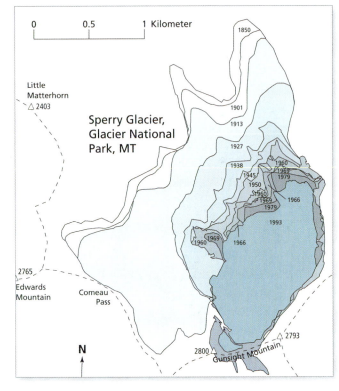

Sperry Glacier, Glacier National Park, MT

6. Suppose that a recent study predicts that the glacier will be completely melted by 2040.

   a. Does this prediction agree with your answer in Exercise 5? What does the research suggest about the melting rate of the glacier after 2000?

   b. When do you think an exponential model would be appropriate to model glacial melting? When do you think a linear model would be appropriate? Explain.

Sperry Glacier in 1913 (top) and 2008 (bottom). The more recent image shows how the glacier has diminished and separated into smaller pieces. Recent studies indicate that there may be no glaciers left in Glacier National Park by 2030.

Snake Venom In Exercises 7–12, use the information below. *(See Examples 3 and 4.)*

The half-life of a snake's venom is the amount of time it takes for 50% of the venom to be removed from an animal's body.

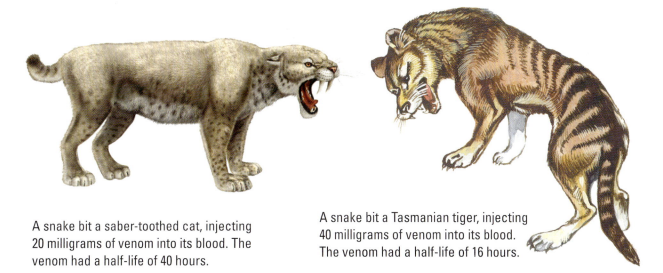

A snake bit a saber-toothed cat, injecting 20 milligrams of venom into its blood. The venom had a half-life of 40 hours.

A snake bit a Tasmanian tiger, injecting 40 milligrams of venom into its blood. The venom had a half-life of 16 hours.

7. Assume the saber-toothed cat was still alive after 80 hours. How much of the venom remained in its bloodstream?

8. Assume the Tasmanian tiger was still alive after 2 days. How much of the venom remained in its bloodstream?

DATA 9. How long did it take the amount of venom in each animal's bloodstream to drop to 10 milligrams?

10. The venom eventually killed the saber-toothed cat. An archaeologist discovers the remains and calculates that the ratio of carbon-14 to carbon-12 is about one-eighth of that occurring in the atmosphere. How long ago did the saber-toothed cat live?

11. The venom eventually killed the Tasmanian tiger. An archaeologist discovers the remains and calculates that the Tasmanian tiger lived 500 years ago. Estimate the ratio of carbon-14 to carbon-12 in the remains.

12. Suppose the half-life of the venom injected into the saber-toothed cat was only 20 hours. Would this increase or decrease the answer to Exercise 7? Explain.

13. Prostate Cancer A hospital experienced a large increase in the number of prostate cancer cases in the late 1980s after the implementation of the PSA blood test. Starting in 1992, when there were 200 new cases, the number of new cases began to decay exponentially. In 1993, there were 176 new cases. Estimate the year when there were about 136 new cases. *(See Example 5.)*

14. Lung Cancer The number of lung cancer deaths at a hospital decreased from 50 in 2005 to 45 in 2006. Assuming the number of yearly deaths decays exponentially, estimate the year when there were about 30 deaths. *(See Example 5.)*

DATA **Uranium Decay** **In Exercises 15–17, use the information below.** *(See Example 6.)*

Uranium-238 decays into thorium-234, which decays into protactinium-234m. This chain of decaying continues as shown in the diagram, which also includes radium-226 and radon-222.

15. A sample of uranium ore originally contained about 100 grams of uranium-238. Use the spreadsheet to determine how long it will take the sample to decay to about 61 grams.

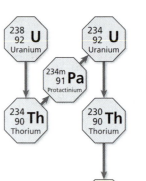

Uranium is a very common element on Earth and is present in many forms. The mineral above, autunite, contains uranium.

Exercise 15

	A	B
1	Years	Grams Remaining
2	0	100.0
3	400,000,000	94.0
4	800,000,000	88.4
5	1,200,000,000	83.1
6	1,600,000,000	78.1

Exercise 16

	A	B
1	Years	Grams Remaining
2	0	10.0
3	400	8.4
4	800	7.1
5	1200	5.9
6	1600	5.0

16. A scientist separates 10 grams of radium-226 from uranium ore and places it in a storage container. Use the spreadsheet to determine how long it will take the 10 grams of radium-226 to decay to about 3 grams.

17. Consider all the radon gas produced by the decaying of radium in the storage container after 1200 years.

 a. Would all this radon gas still be in the storage container? Explain.

 b. Suppose a jar contains 50 grams of radon-222. About 34.7 grams of radon-222 remain after 2 days. Approximate the half-life of radon-222. Explain your reasoning.

18. Radon Testing Radon itself is inert, so it is typically unreactive. However, the U.S. Environmental Protection Agency recommends that all homes be tested for radon. Why do you think elevated levels of radon gas in your home are a health hazard?

▶ Extending Concepts

Yosemitebear Mountain Giant Double Rainbow 1-8-10

Social Media In Exercises 19–23, use the information below.

A video posted on a social media website has 1024 views on day 1. The number of views increases by 50% each day through day 10, and then decreases by 10% each day through day 30.

19. Write a formula that represents the number of daily views for the first 10 days.

20. How many views does the video have on day 10?

21. Write a formula that represents the number of daily views for day 10 through day 30.

22. Sketch a graph showing the number of daily views for the 30 days.

23. Does the video have more views on day 1 or day 30?

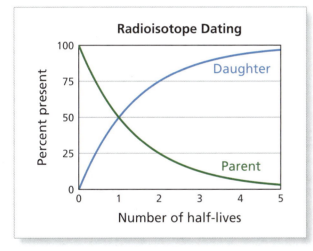

Radioisotope Dating

24. **Radioisotope Dating** There are several other methods that use radioactive decay to date objects. When a radioactive substance decays into another substance, the original substance is called the "parent" and the new substance is called the "daughter."

   a. When is the number of parents the same as the number of daughters?

   b. After two half-lives, what is the ratio of the number of parents to the number of daughters?

   c. Does the graph of the parents exhibit exponential decay? Does the graph of the daughters exhibit exponential growth? Explain.

   d. Several methods of dating objects are shown in the diagram. Why do you think uranium-lead dating can be used to date older objects that carbon-14 dating cannot date?

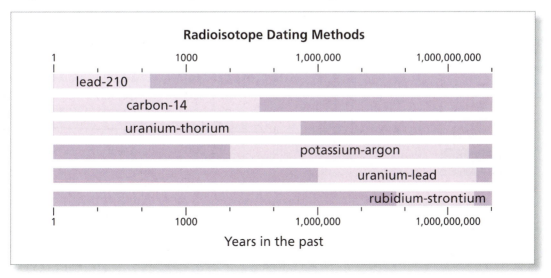

Radioisotope Dating Methods

4.4 Depreciation

▶ Use straight-line depreciation.
▶ Use double declining-balance depreciation.
▶ Use sum of the years-digits depreciation.

Straight-Line Depreciation

Most assets lose their value over time. In other words, they **depreciate** and must be replaced once their useful life is reached. Several accounting methods are used to determine an asset's depreciation expense over the period of its useful life. The simplest method of depreciation is the straight-line method, in which the same amount is expensed each year.

Study Tip

When you purchase an asset for a business, you are not allowed to expense the cost of the asset in the first year. You must determine the useful life in years and the salvage value, and then deduct only a portion of the expense each year during the useful life.

Straight-Line Depreciation

Straight-line depreciation is calculated by dividing the difference of the purchase price and the salvage value by the years of useful life.

$$\text{Annual depreciation} = \frac{(\text{purchase price}) - (\text{salvage value})}{\text{years of useful life}}$$

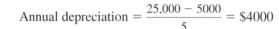

You can access depreciation schedule calculators at *Math.andYou.com*.

EXAMPLE 1 **Making a Depreciation Schedule**

You start a small business and rent an office. You furnish the office with $25,000 worth of office equipment. The useful life of the equipment is 5 years, and the salvage value is $5000. Make a straight-line depreciation schedule showing the depreciation you are allowed to expense each year.

SOLUTION

$$\text{Annual depreciation} = \frac{25{,}000 - 5000}{5} = \$4000$$

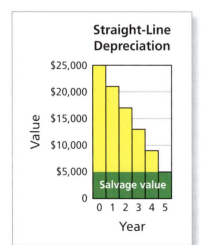

Straight-Line Depreciation

	A	B	C	D
1	Year	Value before Depreciation	Depreciation	Value after Depreciation
2	1	$25,000	$4,000	$21,000
3	2	$21,000	$4,000	$17,000
4	3	$17,000	$4,000	$13,000
5	4	$13,000	$4,000	$9,000
6	5	$9,000	$4,000	$5,000

✓ **Checkpoint**   Help at *Math.andYou.com*

Make a straight-line depreciation schedule for the office equipment using a useful life of 7 years and a salvage value of $4000.

Wheat farmers in the United States harvest about 2 billion bushels of wheat each year.

EXAMPLE 2   **Making a Depreciation Schedule**

A wheat farmer buys a new grain harvester for $300,000. Make a straight-line depreciation schedule using a useful life of 10 years and a salvage value of $50,000.

A combine harvester is a machine that harvests grain. It combines reaping, binding, and threshing into one operation. The straw left behind is either chopped and spread on the field or baled for feed and bedding for livestock.

SOLUTION

$$\text{Annual depreciation} = \frac{300,000 - 50,000}{10} = \$25,000$$

Year	Value before Depreciation	Depreciation	Value after Depreciation
1	$300,000	$25,000	$275,000
2	$275,000	$25,000	$250,000
3	$250,000	$25,000	$225,000
4	$225,000	$25,000	$200,000
5	$200,000	$25,000	$175,000
6	$175,000	$25,000	$150,000
7	$150,000	$25,000	$125,000
8	$125,000	$25,000	$100,000
9	$100,000	$25,000	$75,000
10	$75,000	$25,000	$50,000

In straight-line depreciation, the same amount is expensed each year. Graphically, this creates a linear pattern, hence the name "straight-line depreciation."

✔ **Checkpoint**   Help at *Math*.and**Y☺U**.com

Depreciation schedules do not necessarily represent the actual value that an asset could be sold for after a certain number of years of use.

Suppose that after claiming 5 years of depreciation, the farmer sells the combine for $180,000. How should he account for the income on his tax return? Explain your reasoning.

Double Declining-Balance Depreciation

Whereas straight-line depreciation uses the same amount of depreciation each year, double declining-balance depreciation uses the same rate of depreciation each year.

Double Declining-Balance Depreciation

To find the rate for **double declining-balance depreciation,** divide 2 by the years of useful life.

$$\text{Annual rate of depreciation} = \frac{2}{\text{years of useful life}}$$

To find the depreciation, multiply this rate by the current value.

EXAMPLE 3 **Making a Depreciation Schedule**

Make a double declining-balance depreciation schedule for the office equipment in Example 1.

SOLUTION

Annual rate of depreciation $= \frac{2}{5} = 40\%$

	A	B	C	D
	Year	Value before Depreciation	Depreciation	Value after Depreciation
1				
2	1	$25,000	$10,000	$15,000
3	2	$15,000	$6,000	$9,000
4	3	$9,000	$3,600	$5,400
5	4	$5,400	$400	$5,000
6	5	$5,000	$0	$5,000

0.4(25,000)

This value must be adjusted so that the value after depreciation does not go below the salvage value.

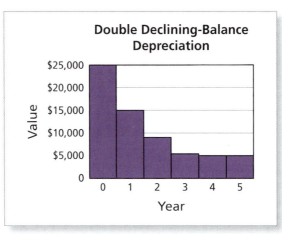

Double Declining-Balance Depreciation

✓ **Checkpoint** Help at *Math*.and**Y☺U**.com

Make a double declining-balance depreciation schedule for the office equipment in Example 1 using a useful life of 7 years and a salvage value of $4000.

EXAMPLE 4   **Making a Depreciation Schedule**

You own an Internet business. You purchase $200,000 worth of servers that you will depreciate as business expenses. Make a double declining-balance depreciation schedule using a useful life of 10 years and a salvage value of $25,000.

SOLUTION

Annual rate of depreciation $= \dfrac{2}{10} = 20\%$

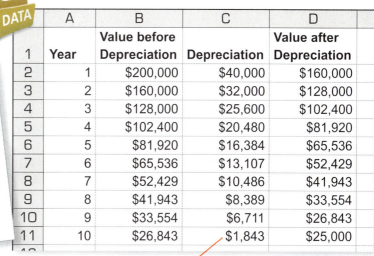

	A	B	C	D	
1	Year	Value before Depreciation	Depreciation	Value after Depreciation	
2	1	$200,000	$40,000	$160,000	
3	2	$160,000	$32,000	$128,000	
4	3	$128,000	$25,600	$102,400	
5	4	$102,400	$20,480	$81,920	
6	5	$81,920	$16,384	$65,536	
7	6	$65,536	$13,107	$52,429	
8	7	$52,429	$10,486	$41,943	
9	8	$41,943	$8,389	$33,554	
10	9	$33,554	$6,711	$26,843	
11	10	$26,843	$1,843	$25,000	

This value must be adjusted so that the value after depreciation does not go below the salvage value.

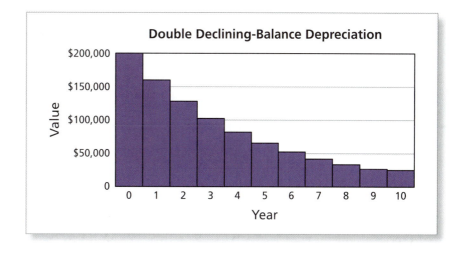

Double Declining-Balance Depreciation

✓ **Checkpoint**   Help at *Math*.and**YOU**.com

Using double declining-balance depreciation, how much more of the value did you expense during the first 4 years compared to straight-line depreciation?

Study Tip

In Examples 3 and 4, you saw that double declining-balance depreciation often requires an adjustment in the last year or last few years. This is not true of straight-line depreciation or of sum of the years-digits depreciation.

Sum of the Years-Digits Depreciation

A third commonly used depreciation method is the sum of the years-digits method. Like double declining-balance depreciation, this method expenses more of the purchase price in the early years.

Sum of the Years-Digits Depreciation

For **sum of the years-digits depreciation,** the depreciation rate for year k using a useful life of n years is given by dividing $(n + 1 - k)$ by the sum of the years of useful life digits.

$$\text{Depreciation rate for year } k = \frac{n + 1 - k}{\text{sum of the years of useful life digits}}$$

To find the depreciation, multiply this rate by the difference between the purchase price and the salvage value.

EXAMPLE 5   Making a Depreciation Schedule

You open a pizza shop and buy 2 delivery vans for a total of $60,000. Make a sum of the years-digits depreciation schedule using a useful life of 5 years and a total salvage value of $15,000.

SOLUTION

The sum of the years digits is $1 + 2 + 3 + 4 + 5 = 15$.

1st Year Rate	2nd Year Rate	3rd Year Rate	4th Year Rate	5th Year Rate
$\frac{5}{15}$	$\frac{4}{15}$	$\frac{3}{15}$	$\frac{2}{15}$	$\frac{1}{15}$

These are always in reverse order.

The difference between the purchase price and the salvage value is $60,000 - 15,000 = \$45,000$.

	A	B	C	D
1	Year	Value before Depreciation	Depreciation	Value after Depreciation
2	1	$60,000	$15,000	$45,000
3	2	$45,000	$12,000	$33,000
4	3	$33,000	$9,000	$24,000
5	4	$24,000	$6,000	$18,000
6	5	$18,000	$3,000	$15,000
7				

Notice that the schedule arrives at the salvage value of $15,000 exactly.

✓ **Checkpoint**   Help at

Make a sum of the years-digits depreciation schedule for the delivery vans using a useful life of 4 years. Use the same total salvage value of $15,000.

EXAMPLE 6   **Comparing Different Types of Depreciation**

You own a fitness center and purchase all new equipment for $600,000. The useful life of the equipment is 10 years, and the salvage value is $50,000. Graphically compare the three methods of depreciation: (a) straight-line, (b) double declining-balance, and (c) sum of the years-digits.

SOLUTION

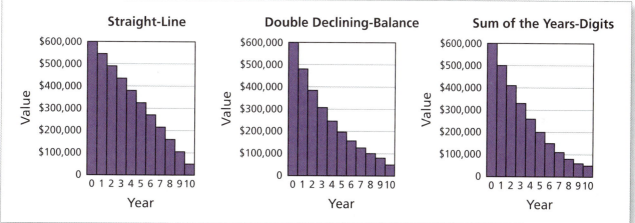

a. Straight-line depreciation is the simplest method. With this method, you expense the *same amount* of depreciation each year.

b. Double declining-balance depreciation is a method of depreciation in which you expense the *same rate* (of the undepreciated value) each year. It is a method of *accelerated depreciation* because the amount of depreciation taken each year is greater during the earlier years of an asset's life. It has the disadvantage that the last year (or last few years) needs to be adjusted to arrive at the salvage value. This is an example of exponential decay.

c. Sum of the years-digits depreciation is similar to double declining-balance depreciation. It is an accelerated depreciation, but it has the advantage that it will always arrive at the salvage value exactly (except possibly for round-off error amounting to a few cents).

✓ **Checkpoint**   Help at *Math*.and*Y⊖U*.com

Suppose you own a business. Which method of depreciation do you prefer?

4.4 Exercises

 Hair Salon The owner of a hair salon buys new equipment. In Exercises 1 and 2, make a straight-line depreciation schedule for the equipment. *(See Examples 1 and 2.)*

1.

Cost: $1100
Salvage value: $400
Useful life: 7 years

2.

Cost: $14,500
Salvage value: $2500
Useful life: 10 years

 Barbershop The owner of a barbershop buys new equipment. A graph of the straight-line depreciation schedule for the equipment is shown. In Exercises 3–6, use the graph. *(See Examples 1 and 2.)*

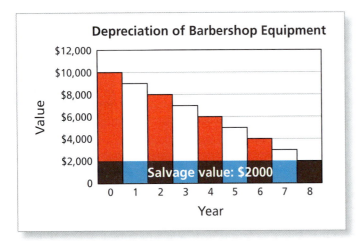

In medieval times, barbers were also surgeons. The red and white stripes on a barber pole represent bloody bandages.

3. What is the value of the equipment after 5 years?

4. What is the value of the equipment after 7 years?

5. When a business sells equipment at a price greater than its value, the U.S. Internal Revenue Service collects taxes on the difference. This is called *depreciation recapture*. Find the taxable amount for each year in which depreciation recapture could occur for a selling price of $6000.

6. When a business sells equipment at a price less than its value, the difference is tax deductible. Find the tax deductible amount for each year in which a selling price of $4000 would cause a loss.

Construction Equipment The owner of a construction company buys new equipment. In Exercises 7 and 8, make a double declining-balance depreciation schedule for the equipment. *(See Examples 3 and 4.)*

7.

Cost: $50,000
Salvage value: $10,000
Useful life: 5 years

8.

Cost: $60,000
Salvage value: $12,000
Useful life: 8 years

Construction Company The owner of a construction company buys new equipment. A graph of the double declining-balance depreciation schedule for the equipment is shown. In Exercises 9–14, use the graph. *(See Examples 3 and 4.)*

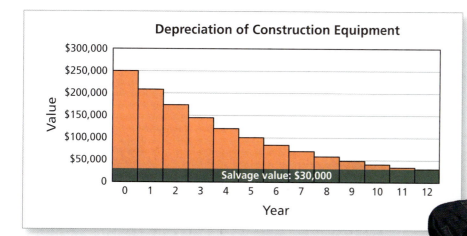

9. What is the value of the equipment after 3 years?

10. What is the value of the equipment after 5 years?

11. How much depreciation did the company expense in year 8?

12. How much depreciation did the company expense in year 12?

13. Using double declining-balance depreciation, how much more of the value did the company expense during the first 6 years when compared to straight-line depreciation?

14. Is there any year in which the value of the equipment using double declining-balance depreciation is greater than the value of the equipment using straight-line depreciation?

Kitchen Equipment The owner of a restaurant buys new kitchen equipment. In Exercises 15 and 16, make a sum of the years-digits depreciation schedule for the equipment. *(See Example 5.)*

15.

Cost: $4500
Salvage value: $0
Useful life: 5 years

16.

Cost: $400
Salvage value: $50
Useful life: 7 years

Restaurant The owner of a restaurant buys new equipment. The salvage value of the equipment is $25,000. The graph shows three methods that the restaurant can use to depreciate the equipment. In Exercises 17–20, use the graph. *(See Example 6.)*

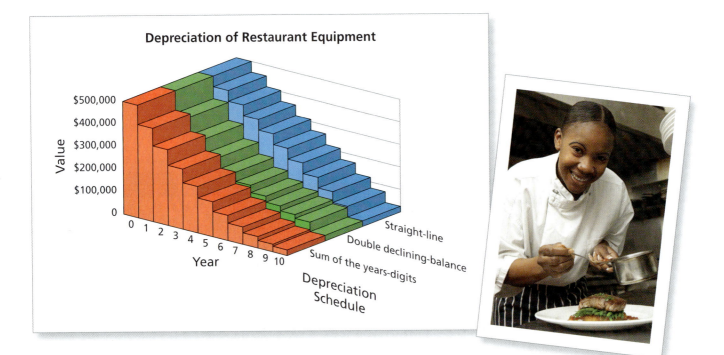

17. Which method depreciates the value the most during the first 3 years?

18. Which method depreciates the value the most during years 4 through 10?

19. Using sum of the years-digits depreciation, how much more of the value did the restaurant expense during the first 4 years when compared to straight-line depreciation?

20. Suppose the restaurant sells the equipment for $250,000 after 4 years. Describe the restaurant's gain or loss using each method.

▶ Extending Concepts

DATA **MACRS** **In Exercises 21–26, use the information below.**

You can also use the modified accelerated cost recovery system (MACRS) to depreciate assets. One method available under MACRS is double declining-balance depreciation with a switch to straight-line depreciation.

Under this method, you compare the annual rates of depreciation for double declining-balance depreciation and straight-line depreciation each year. When the rate for straight-line depreciation is greater than or equal to the rate for double declining-balance depreciation, you switch to straight-line depreciation. The annual rate of depreciation for straight-line depreciation is given by

$$\text{Annual rate of depreciation} = \frac{1}{\text{remaining useful life}}.$$

In the exercises, you should use the half-year convention, which begins depreciation halfway through the year that an asset is put into service. An asset with a useful life of n years will have a depreciation schedule with $n + 1$ years because you start and end the useful life halfway through the year. For straight-line depreciation, the remaining useful life in year 2 is $n - \frac{1}{2}$. After year 2, the remaining useful life decreases by 1 each year. For both methods, the depreciation deductions in the first year and last year are reduced by half. (*Note:* Salvage value is not used under MACRS.)

21. Using the method described above, in which year do you switch to straight-line depreciation for an asset with the given useful life?

    a. 5 years   **b.** 10 years

22. Using the method described above, in which year do you switch to straight-line depreciation for an asset with the given useful life?

    a. 3 years   **b.** 7 years

23. You purchase a $4000 copy machine. The useful life of the machine is 5 years. Make a depreciation schedule using the method described above.

24. You purchase a $10,000 knitting machine. The useful life of the machine is 5 years. Make a depreciation schedule using the method described above.

25. You purchase a $24,000 digital printing press. The useful life of the machine is 7 years. Make a depreciation schedule using the method described above.

26. You purchase a $3000 ticket booth. The useful life of the booth is 7 years. Make a depreciation schedule using the method described above.

4.3–4.4   Quiz

DATA **Wastewater Filtration** In Exercises 1–4, use the information below.

A factory discharges wastewater into a river. The initial concentration of pollutants in the wastewater is 2000 parts per billion (ppb). Before entering the river, the water must pass through multiple filters to reduce the concentration to an acceptable level of 10 ppb.

1. A single filter removes 70% of pollutants in the wastewater.

   a. What is the concentration of pollutants after passing through three filters?

   b. How many filters must the water pass through to meet the acceptable level?

2. A single filter removes half the pollutants in the wastewater.

   a. What is the concentration of pollutants after passing through four filters?

   b. How many filters must the water pass through to meet the acceptable level?

   c. Is it possible to remove all the pollutants? Explain your reasoning.

3. After the water passes through 1 filter, the concentration of pollutants is 900 ppb. Each filter removes the same percent of pollutants. How many filters must the water pass through to meet the acceptable level?

4. After passing through 2 filters, the concentration of pollutants is 320 ppb. Each filter removes the same percent of pollutants. What is the concentration of pollutants after passing through four filters?

DATA 5. **Water Treatment Equipment** A factory owner buys water treatment equipment for $10,000. Make a straight-line depreciation schedule for the equipment using a useful life of 5 years and a salvage value of $1000.

DATA 6. **Water Treatment Equipment** A factory owner buys water treatment equipment for $25,000. The useful life of the equipment is 10 years, and the salvage value is $0. The factory owner uses double declining-balance depreciation. What is the value of the equipment after 5 years?

Depreciation Schedule	
Year	Value
0	$20,000
1	$15,000
2	$11,000
3	$8000
4	$6000
5	$5000

DATA 7. **Financial Records** A factory owner buys $20,000 worth of water treatment equipment. The owner depreciates the equipment over 5 years, and the salvage value is $5000. The depreciation schedule for the equipment is shown. What depreciation method did the owner use? Explain your reasoning.

Chapter 4   Summary

Section Objectives

How does it apply to you?

Section 1

Make a table showing exponential growth.	→	You can analyze a quantity that increases by the same percent each time period. *(See Examples 1 and 2.)*
Draw a graph showing exponential growth.	→	You can visualize exponential growth. *(See Examples 3 and 4.)*
Find an exponential growth rate.	→	You can determine by what percent a quantity is increasing. *(See Examples 5 and 6.)*

Section 2

Use a consumer price index.	→	You can describe the change in prices for consumer goods over time. *(See Examples 1 and 2.)*
Use a graph to interpret a consumer price index.	→	You can visualize the trends in a consumer price index over time. *(See Examples 3 and 4.)*
Compare inflation to the value of the dollar.	→	You can understand the relationship between inflation and the value of the dollar. *(See Examples 5 and 6.)*

Section 3

Make a table and graph showing exponential decay.	→	You can determine how much of a quantity that decays exponentially remains after a period of time. *(See Examples 1 and 2.)*
Calculate and use half-life.	→	You can determine the concentration of a drug in your body and use carbon dating to determine the age of a fossil. *(See Examples 3 and 4.)*
Find an exponential decay rate.	→	You can determine by what percent a quantity is decreasing. *(See Examples 5 and 6.)*

Section 4

Use straight-line depreciation.	→	Straight-line depreciation reduces the value of an item by the same amount each year. *(See Examples 1 and 2.)*
Use double declining-balance depreciation.	→	Double declining-balance depreciation reduces the value of an item by the same percent each year. *(See Examples 3 and 4.)*
Use sum of the years-digits depreciation.	→	Sum of the years-digits depreciation expenses more of the purchase price in the early years. *(See Examples 5 and 6.)*

Chapter 4 Review Exercises

Section 4.1

DATA **Apportionment** **In Exercises 1–6, use the information below.**

Apportionment is the proportional distribution of the 435 members of the U.S. House of Representatives based on the population of each state. In 1960, each member of the House represented about 415,000 people. On average, this number has increased by 11.5% each decade.

1. Write a formula that can be used to model the exponential growth of the population per representative every decade since 1960. (*Note: n* represents decades.)

2. Make a table showing the estimated population per representative every decade from 1960 through 2010.

3. Use the formula from Exercise 1 to project the population per representative every decade from 2020 through 2100.

4. Graph the estimated and projected populations from Exercises 2 and 3.

5. Suppose the actual populations per representative for 2050 and 2060 are 1,100,000 and 1,230,000, respectively. What is the rate of growth per decade? How do these numbers compare to your model?

6. Use the map to estimate the population of California.

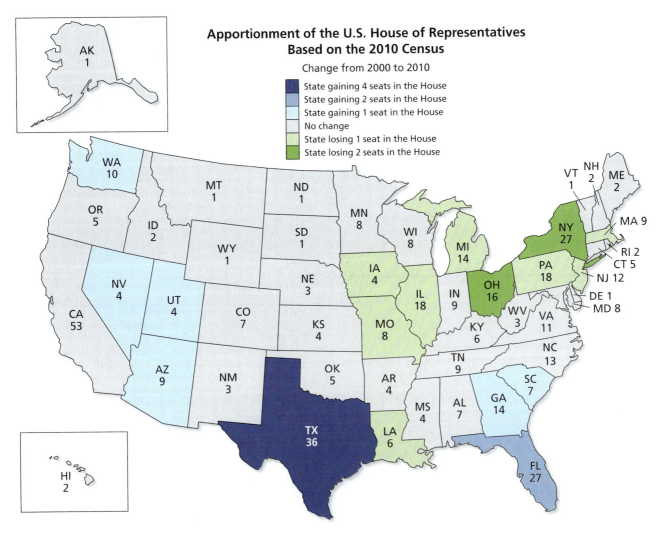

Apportionment of the U.S. House of Representatives Based on the 2010 Census

Change from 2000 to 2010

- State gaining 4 seats in the House
- State gaining 2 seats in the House
- State gaining 1 seat in the House
- No change
- State losing 1 seat in the House
- State losing 2 seats in the House

AK 1

WA 10
MT 1
ND 1
MN 8
VT 1
NH 2
ME 2

OR 5
ID 2
SD 1
WI 8
NY 27
MA 9

WY 1
IA 4
MI 14
RI 2
CT 5

NV 4
UT 4
CO 7
NE 3
IL 18
IN 9
OH 16
PA 18
NJ 12

CA 53
KS 4
MO 8
KY 6
WV 3
VA 11
DE 1
MD 8

AZ 9
NM 3
OK 5
AR 4
TN 9
NC 13

MS 4
AL 7
GA 14
SC 7

TX 36
LA 6

HI 2

FL 27

Section 4.2

Natural Gas The table shows the natural gas price index of natural gas delivered to residential consumers in the United States. Consumers are charged per thousand cubic feet. In Exercises 7–14, use the table. (*Note:* The base year is 1983.)

National Gas Price Index					
Year	CPI	Year	CPI	Year	CPI
1966	—	1981	70.8	1996	104.6
1967	17.2	1982	85.3	1997	114.5
1968	17.2	1983	100.0	1998	112.5
1969	17.3	1984	101.0	1999	110.4
1970	18.0	1985	101.0	2000	128.1
1971	19.0	1986	96.2	2001	158.9
1972	20.0	1987	91.4	2002	130.2
1973	21.3	1988	90.3	2003	158.9
1974	23.6	1989	93.1	2004	177.4
1975	28.2	1990	95.7	2005	209.6
1976	32.7	1991	96.0	2006	226.6
1977	38.8	1992	97.2	2007	215.8
1978	42.2	1993	101.7	2008	229.2
1979	49.2	1994	105.8	2009	200.3
1980	60.7	1995	100.0	2010	—

7. Was inflation of natural gas the highest in the 1970s, 1980s, or 1990s? Explain your reasoning.

8. During which years did the price of natural gas decrease?

9. Graph the natural gas price index from 1967 through 2009. Use the graph to describe the rate of inflation for natural gas.

10. Compare the rate of inflation for natural gas to the rate of inflation represented by the consumer price index.

11. In 2009, the average price of natural gas per thousand cubic feet was $12.14. What was the price in 1970?

12. In 2006, the average price of natural gas per thousand cubic feet was $13.73. What was the price in 1983?

13. Would you rather pay a natural gas bill of $115.50 in 2004 or $92.35 in 1990? Explain your reasoning.

14. Do you think using a price index to compare prices between years is better than using a dollar amount? Explain your reasoning.

Section 4.3

Medicine **In Exercises 15–18, use the information below.**

Subjects in a study receive daily doses of a medicine for 10 weeks. In the first week, they receive 100 milligrams each day. Each week thereafter, the dosage decreases 10% from the previous week.

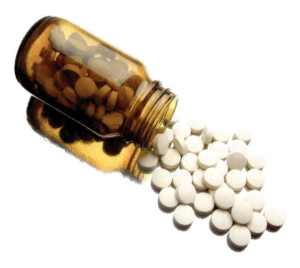

15. Write a formula that represents the daily dosage during each week.

16. Make a table showing the daily dosage during the 10 weeks.

17. Sketch a graph showing the decrease in the daily dosage during the 10 weeks.

18. The half-life of the medicine is 24 hours. How much of the medicine received in a dose during week 10 remains in the patient's bloodstream after 24 hours?

Technetium **In Exercises 19–22, use the information below.**

Most technetium is man-made in nuclear reactors. There are many isotopes of technetium, two of which are technetium-99 and technetium-99m. The half-life of technetium-99 is 210,000 years.

19. How much of a 120-gram sample of technetium-99 will remain after 630,000 years?

20. A vial contains 10 grams of technetium-99m. Use the spreadsheet to determine how long it will take the 10 grams of technetium-99m to decay to about 2.1 grams.

	A	B
	Years	**Grams Remaining**
1		
2	0	10.0
3	1.5	8.4
4	3	7.1
5	4.5	5.9
6	6	5.0
7	7.5	4.2

21. What is the half-life of technetium-99m? Explain your reasoning.

22. A solution containing technetium-99m is often injected into a patient to help diagnose problems in the body. Explain why a hospital or veterinary clinic should not order excessive amounts of technetium-99m with the intent of placing it in storage.

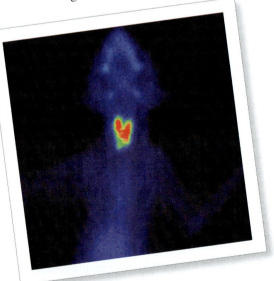

This technetium scan helped veterinarians diagnose this cat with hyperthyroidism.

Section 4.4

 Outdoor Water Park You own an outdoor water park and you purchase new equipment. In Exercises 23 and 24, make a straight-line depreciation schedule for the equipment.

23.

Cost: $2500
Salvage value: $500
Useful life: 5 years

24.

Cost: $4000
Salvage value: $1200
Useful life: 7 years

 Outdoor Water Park You own an outdoor water park and you purchase new equipment. A graph of the double declining-balance depreciation schedule for the equipment is shown. In Exercises 25–28, use the graph.

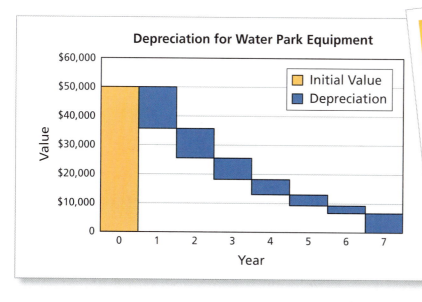

25. How much depreciation did you expense in year 1?

26. How much depreciation did you expense in year 7?

27. Using double declining-balance depreciation, how much more of the value did you expense during the first 4 years when compared to straight-line depreciation?

28. Suppose you had used sum of the years-digits depreciation instead of double declining-balance depreciation. How much more did you expense during the first 4 years when compared to straight-line depreciation?

5 The Mathematics of Taxation

5.1 Flat Tax & Political Philosophy

▶ Calculate a flat income tax.

▶ Identify types of taxes.

▶ Analyze an indirect tax.

5.2 Graduated Income Tax

▶ Calculate a graduated income tax.

▶ Analyze a graduated income tax system.

▶ Compare a graduated income tax with a flat income tax.

5.3 Property Tax

▶ Calculate a property tax.

▶ Analyze assessments and tax credits.

▶ Analyze exemptions for property tax.

5.4 Social Security & Payroll Taxes

▶ Calculate Social Security & Medicare taxes.

▶ Evaluate the benefits of Social Security.

▶ Analyze the viability of Social Security.

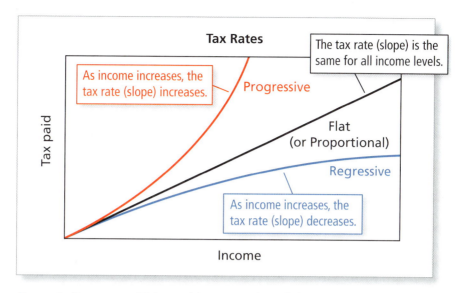

Tax Rates

As income increases, the tax rate (slope) increases.

Progressive

The tax rate (slope) is the same for all income levels.

Flat (or Proportional)

Regressive

As income increases, the tax rate (slope) decreases.

Tax paid

Income

Example 3 on page 204 provides examples of the taxes shown above. Is a sales tax of 5% regressive, flat, or progressive?

5.1 Flat Tax & Political Philosophy

▶ Calculate a flat income tax.
▶ Identify types of taxes.
▶ Analyze an indirect tax.

Calculating a Flat Income Tax

A **flat tax** (short for **flat rate tax**) is a tax system with a constant tax rate. In the United States, the federal income tax is not a flat tax. People with higher incomes not only pay more income tax, they pay a higher tax rate (see Section 5.2). As of the writing of this text, seven states have a flat income tax.

- Colorado 4.63%
- Indiana 3.40%
- Michigan 4.35%
- Utah 5.00%
- Illinois 5.00%
- Massachusetts 5.30%
- Pennsylvania 3.07%

Additionally, there are seven states with no income tax: Alaska, Florida, Nevada, South Dakota, Texas, Washington, and Wyoming.

EXAMPLE 1 Calculating a Flat Income Tax

Two people who live in Utah have the following taxable incomes.

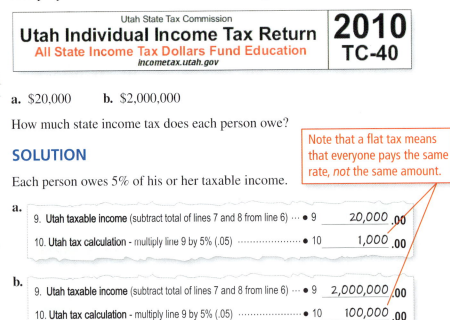

Utah State Tax Commission
Utah Individual Income Tax Return
All State Income Tax Dollars Fund Education
incometax.utah.gov
2010 TC-40

a. $20,000 **b.** $2,000,000

How much state income tax does each person owe?

SOLUTION

Each person owes 5% of his or her taxable income.

Note that a flat tax means that everyone pays the same rate, *not* the same amount.

a.

9. Utah taxable income (subtract total of lines 7 and 8 from line 6) ··· ● 9 20,000 .00
10. Utah tax calculation - multiply line 9 by 5% (.05) ·················· ● 10 1,000 .00

b.

9. Utah taxable income (subtract total of lines 7 and 8 from line 6) ··· ● 9 2,000,000 .00
10. Utah tax calculation - multiply line 9 by 5% (.05) ·················· ● 10 100,000 .00

✓ **Checkpoint** Help at *Math*.and**YOU**.com

Is the following logic valid? Explain your reasoning.

Flat income taxes are not fair because the rich should have to pay more income taxes than the poor.

Taxable Income: Taxable Income:
$20,000 $2,000,000

EXAMPLE 2   **Determining Taxable Income**

The Internal Revenue Service (IRS) estimates that about $290 billion in income goes unreported each year in the United States. Which of the following do you think is part of your taxable income for the IRS?

a. $20 per week you receive for babysitting your niece

b. $500,000 you win in a state lottery

c. $40 per night you receive in tips as a food server

d. $1000 monthly payment by your employer for your medical insurance

e. $15 per month interest you receive from a savings account

f. $50,000 interest you receive from municipal bonds

g. $250,000 profit you receive from selling your home

h. $30,000 inheritance you receive from your spouse

i. $2000 per month you receive for child support from your former spouse

SOLUTION

IRS regulations are complicated and open to interpretation. In addition, there are annual changes to IRS regulations as to what constitutes taxable income. If an IRS audit claims that you owe more tax, you can hire an accountant and an attorney to challenge the audit. With that said, here is one tax accountant's opinion regarding the above list.

a. Babysitting: This is a wage and is taxable.

b. Lottery winnings: Lottery and gambling winnings are taxable.

c. Tips: Tips are taxable.

d. Employer-provided medical insurance: At this time, this is *not* taxable.

e. Interest from savings accounts: This type of interest is taxable.

f. Interest from municipal bonds: This interest is generally *not* taxable.

g. Profit from the sale of a home: This is generally *not* taxable.

h. Inheritance: Inheritance from a spouse is generally *not* taxable.

i. Child support: This is *not* taxable.

✓ **Checkpoint**   Help at *Math*.and**Y☺U**.com

Income taxes were imposed at various times in history, generally because of national emergencies. With the adoption of the 16th Amendment of the U.S. Constitution in 1913, the government established the present form of federal income tax. Some people, such as presidential candidate Ron Paul, have claimed that the amendment is unconstitutional. What do you think?

> "Ron Paul supports the elimination of the income tax and the Internal Revenue Service (IRS). He asserts that Congress had no power to impose a direct income tax and has called for the repeal of the 16th Amendment to the Constitution, which was ratified on February 3, 1913."
>
> "End the Income Tax, Abolish the IRS," *RonPaul.com*, 2009

Ron Paul is a congressman for the 14th Congressional District of Texas. His libertarian positions on political issues have often clashed with both Republican and Democratic party leaders. Paul ran for president of the United States in 1988 and 2008.

Identifying Types of Taxes

A **regressive tax** is a tax that takes a smaller proportion of an income as the income rises. In other words, it is a tax that affects people with low incomes more than people with high incomes. A **progressive tax** (or *graduated tax*) is a tax that takes a larger proportion of an income as the income rises (see Section 5.2). So, based on tax rates and income, taxes can be classified in three ways.

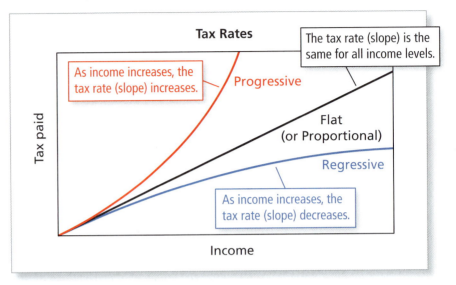

Tax Rates

The tax rate (slope) is the same for all income levels.

As income increases, the tax rate (slope) increases. — Progressive

Flat (or Proportional)

Regressive

As income increases, the tax rate (slope) decreases.

Tax paid

Income

EXAMPLE 3 **Identifying Types of Taxes**

Determine whether each tax is regressive, flat, or progressive.

a. Medicare tax of 1.45% of gross personal income

b. Sales tax of 5%   **c.** Federal income tax

SOLUTION

a. The upper wage limit for the Medicare tax was eliminated in 1994, meaning that the tax is now applied to all wages. The rate is the same for all wage earners, so it is an example of a flat tax.

b. Sales tax is commonly used as an example of a regressive tax. Suppose, for example, that a family earning $50,000 a year buys a new car for $20,000 and pays $1000 in sales tax. This is 2% of the family's income. If a family earning $100,000 a year buys the same car and pays $1000 in sales tax, the family is paying only 1% of its income.

c. Federal income tax is a classic example of a progressive tax. People with low taxable incomes pay 10% federal income tax. People with higher taxable incomes pay up to 35% federal income tax.

✓ **Checkpoint**   Help at *Math*.and**Y☺U**.com

Determine whether each tax is regressive, flat, or progressive.

d. An annual automobile registration fee of $35 per car

e. Federal estate tax

f. Capital gains tax of 15% on the profit from the sale of a stock investment

It would be difficult for a country to openly establish a tax policy that attempts to tax poor people more punitively than rich people. Instead, taxes that are considered regressive, such as sales tax, are often levied as a flat tax. It is only in analyzing the consequences of the tax that the tax is determined to be regressive.

EXAMPLE 4   Identifying Types of Taxes

The graph shows the average state and local taxes paid by different income groups. Use the graph to discuss whether the three types of taxes are regressive, flat, or progressive.

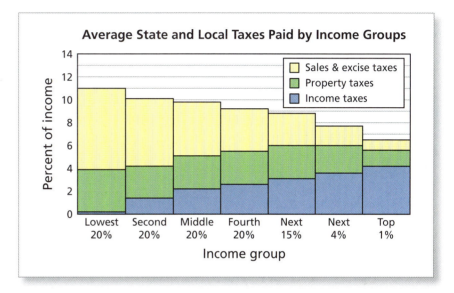

SOLUTION

a. **Sales and excise taxes:** Although these taxes are not explicitly written to be regressive, they end up being classified that way in practice. The graph shows that as your income increases, the percent of your income that goes to sales and excise taxes decreases.

b. **Property taxes:** Although property tax is considered by some people to be regressive, for the vast number of Americans, it is a flat tax. The reason for this is that as a person's income increases and he or she buys a larger home, the assessment of the person's property generally increases.

c. **Income taxes:** Although some states have a flat income tax system, the majority have a progressive income tax system (see Section 5.2). The result is that, on average, people with greater incomes tend to pay a greater percent of their income toward state income tax.

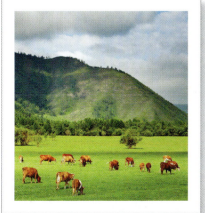

In 2010, there were about 2.2 million farms in the United States. The average farm size was about 418 acres. Farms are often inherited from family members.

✓ **Checkpoint**   Help at

Discuss the following claim. Do you agree or disagree? Explain your reasoning.

> Estate taxes are regressive taxes. They hurt farmers who want to pass their farms on to their children.

Analyzing Indirect Taxes

An **indirect tax** is a tax that increases the price of a product. Consumers actually pay the tax by paying more for the product.

The diagram illustrates an example. The *direct tax* is on an automobile manufacturer. However, the person who ultimately pays the tax is the consumer.

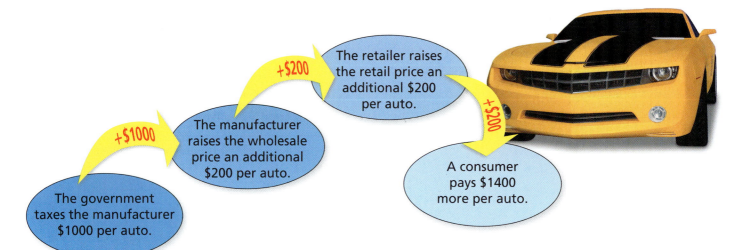

+$1000

The government taxes the manufacturer $1000 per auto.

The manufacturer raises the wholesale price an additional $200 per auto.

+$200

The retailer raises the retail price an additional $200 per auto.

+$200

A consumer pays $1400 more per auto.

The key feature of an indirect tax is that the tax can be shifted to someone else who is farther along in the manufacturing chain.

EXAMPLE 5   Analyzing an Indirect Tax

A store in the United States imports fireworks from China and pays $250, plus an import duty of 2.4%. Is the tax regressive, flat, or progressive? Draw a diagram showing a chain of events that could occur.

SOLUTION

Here is one possible diagram.

+$6

The government taxes the store $6.

The store raises the retail price an additional $4.

+$4

A consumer pays $10 more for the fireworks.

The point is that, regardless of who the tax is levied at *directly*, usually the consumer ends up *indirectly* paying the tax. Excise taxes and sales taxes are generally considered to be regressive.

✓ **Checkpoint**   Help at *Math*.and*YOU*.com

Suppose the federal government levies a $1 per gallon excise tax on diesel fuel. Draw a diagram showing a chain of events that could occur.

Nicholas Gregory Mankiw is an American macroeconomist. From 2003 to 2005, Mankiw was the chairman of President George W. Bush's Council of Economic Advisers. He is a professor of economics at Harvard University.

"Economists have often advocated taxing consumption rather than income, on the grounds that consumption taxes do less to discourage saving, investment and economic growth. . . . The main issue for the soda tax, however, is whether certain forms of consumption should be singled out for particularly high levels of taxation. . . .

Taxes on gasoline can be justified along these lines. Whenever you go out for a drive, you are to some degree committing an antisocial act. You make the roads more congested, increasing the commuting time of your neighbors. You increase the likelihood that other drivers will end up in accidents. And the gasoline you burn adds to pollution. . . .

Taxing soda may encourage better nutrition and benefit our future selves. But so could taxing candy, ice cream and fried foods. Subsidizing broccoli, gym memberships and dental floss comes next. Taxing mindless television shows and subsidizing serious literature cannot be far behind."

"Can a Soda Tax Save Us From Ourselves?," N. Gregory Mankiw

EXAMPLE 6   **Analyzing Possible Effects of Taxes**

It is often assumed that the basic goal behind taxation policies is to raise money that will be spent to help all members of society. Discuss ways in which the following taxes might help the people who are paying the taxes.

a. A property tax that is used to fund local public schools

b. A gasoline excise tax that is used to build and improve roads

c. A tax on fast food that is used to lessen medical insurance premiums

d. A tax on casino gambling that is used to buy prescription medicines for seniors

SOLUTION

a. **Property tax:** Home owners pay property taxes directly, and renters pay property taxes indirectly. The general argument is that everyone benefits from having well-educated members of society, regardless of whether the people who are paying the property tax have children in school.

b. **Gasoline excise tax:** This is an often-cited example of a tax that directly benefits the people who are paying the tax. The argument is that people use gas to travel on roads. They benefit from the tax because they will continue to have new and improved roads to travel on.

c. **Tax on fast food:** There is a growing interest in this type of tax. The general argument is that fast food contributes to obesity, which contributes to increased medical expenses and higher insurance premiums. The idea behind this tax is that people who eat fast food should contribute more toward their own medical expenses.

d. **Tax on casino gambling:** Casino patrons are often 65 or older. So, it could be argued that the tax benefits many of the people who pay the tax.

✓ **Checkpoint**   Help at

Suppose you are a member of a state legislature that is voting on starting a special tax on soda. Would you vote for or against the bill? Explain your reasoning.

5.1 Exercises

State Income Tax In Exercises 1–4, determine how much state income tax the person owes. *(See Example 1.)*

1. A person who lives in Indiana has a taxable income of $49,000.

7. **State Taxable Income** Subtract line 6 from line 5 _____	7 _____ . 00
8. State income tax: multiply line 7 by 3.4% (.034)	8 _____ . 00
(if answer is less than zero, leave blank) _____	

2. A person who lives in Indiana has a taxable income of $2,500,000.

7. **State Taxable Income** Subtract line 6 from line 5 _____	7 _____ . 00
8. State income tax: multiply line 7 by 3.4% (.034)	8 _____ . 00
(if answer is less than zero, leave blank) _____	

3. A person who lives in Michigan has a taxable income of $25,000.

16. **Taxable income.** Subtract line 15 from line 14. If line 15 is greater than line 14, enter "0" .... 16.	00
17. **Tax.** Multiply line 16 by 4.35% (0.0435) .... 17.	00

4. A person who lives in Michigan has a taxable income of $60,000.

16. **Taxable income.** Subtract line 15 from line 14. If line 15 is greater than line 14, enter "0" .... 16.	00
17. **Tax.** Multiply line 16 by 4.35% (0.0435) .... 17.	00

Taxable Income In Exercises 5–8, use the Internet to determine whether the income is taxable by the federal government. *(See Example 2.)*

5. $100 gift card you receive from your parents

6. $2000 per year scholarship you receive for college tuition

7. $250 bonus you receive from your employer

8. $5000 prize you receive from a game show

Taxes **In Exercises 9–12, determine whether the tax is regressive, flat, or progressive. Explain your reasoning.** *(See Example 3.)*

9. Social Security tax of 6.2% on earnings up to $106,800

10. 17% income tax on all income over $12,500

11. 19% income tax on all income

12. 7% sales tax

Connecticut State and Local Taxes **In Exercises 13–16, use the display.** *(See Example 4.)*

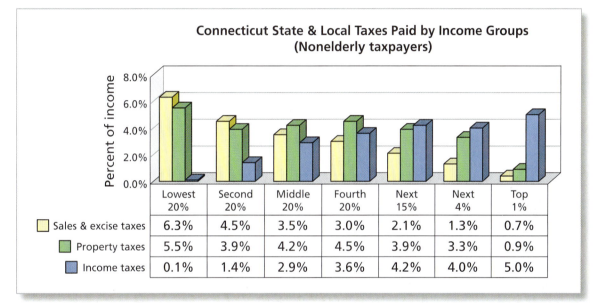

Connecticut State & Local Taxes Paid by Income Groups
(Nonelderly taxpayers)

	Lowest 20%	Second 20%	Middle 20%	Fourth 20%	Next 15%	Next 4%	Top 1%
Sales & excise taxes	6.3%	4.5%	3.5%	3.0%	2.1%	1.3%	0.7%
Property taxes	5.5%	3.9%	4.2%	4.5%	3.9%	3.3%	0.9%
Income taxes	0.1%	1.4%	2.9%	3.6%	4.2%	4.0%	5.0%

13. The income for a family in the middle 20% is $58,000. How much, in dollars, does this family pay for each type of tax?

14. The income for a family in the top 1% is $3,160,000. How much, in dollars, does this family pay for each type of tax?

15. Determine whether the three types of taxes are regressive, flat, or progressive in Connecticut. Explain your reasoning.

16. Are taxes, as a whole, regressive, flat, or progressive in Connecticut? Explain your reasoning.

Connecticut has the second-highest property taxes per person in the United States, behind only New Jersey. It is also the most recent state to adopt an income tax (1991).

Hawaii's General Excise Tax **In Exercises 17–21, use the information below.** *(See Example 5.)*

The state of Hawaii does not have a sales tax. Instead, it has a general excise tax. Unlike a sales tax, the general excise tax applies to businesses, not customers. Hawaii taxes retailers at a rate of 4% on gross income from business transactions.

17. Complete the diagram. Explain your reasoning.

A retailer buys a box of leis for $15 per lei.

The retailer doubles the price of each lei.

The retailer charges an extra 4.166% per lei to cover the tax.

The consumer pays $_____ more per lei because of the tax.

18. Complete the diagram. Explain your reasoning.

A retailer buys a surfboard for $470.

The retailer marks up the price of the surfboard by $30.

The retailer charges an extra 4.166% to cover the tax.

The consumer pays $_____ more because of the tax.

19. A retailer purchases $150 worth of food. Draw a diagram showing a chain of events that could occur. Is the tax regressive, flat, or progressive? Explain.

20. A retailer purchases $70 worth of medicine. Draw a diagram showing a chain of events that could occur. Is the tax regressive, flat, or progressive? Explain.

21. A business passes on the general excise tax to the consumer as a charge that is a percent of the retail price. The charge is also subject to the general excise tax. The business cannot charge more than 4.166% of the retail price because consumer protection laws prohibit businesses from passing on an amount that exceeds the general excise tax on a transaction. Explain why the percent is 4.166%, not 4%.

22. **Oahu Surcharge** The island of Oahu adds an extra 0.5% surcharge to the general excise tax to fund its mass transit system. Discuss ways in which the tax might help the people who are paying it. *(See Example 6.)*

▶ Extending Concepts

Deductions and Credits In Exercises 23–28, use the information below. Assume the country has a flat income tax rate of 14%.

A *tax deduction* is an amount that is subtracted from your gross income before the tax rate is applied to determine your income tax liability. A *tax credit* is an amount that is subtracted from your income tax liability after the income tax rate has been applied.

23. A taxpayer has a gross income of $45,000 and tax deductions of $10,000. How much income tax does the taxpayer owe?

24. A taxpayer has a gross income of $65,600 and tax deductions of $12,000. How much income tax does the taxpayer owe?

25. A taxpayer has a taxable income of $68,500 and tax credits of $3000. How much income tax does the taxpayer owe?

26. A taxpayer has a gross income of $76,400, tax deductions of $11,000, and tax credits of $2000. How much income tax does the taxpayer owe?

27. Which saves a taxpayer more, a tax deduction of $1000 or a tax credit of $1000? Explain your reasoning.

28. Sketch a graph comparing income tax as a percentage of income for a 14% flat tax on all income and a 14% flat tax on all income over $12,000. Explain how tax deductions can make a flat tax more progressive.

Negative Income Tax In Exercises 29 and 30, use the information below. Assume the country has a flat income tax rate of 14%.

Under a negative income tax system, deductions are subtracted from your gross income to determine your taxable income. If the difference is positive, the flat income tax rate is applied to determine how much income tax you owe. If the difference is negative, you receive a check from the government equal to the absolute value of the difference times the flat tax rate. Let the deduction for an adult be $14,000 and the deduction for a dependent be $7000.

29. A family of 2 adults and 2 dependents earns $40,000.

   a. Does the family owe income tax or receive a check from the government? What is the amount?

   b. What is the cutoff amount for the family to qualify for a check from the government?

30. A family of 2 adults and 3 dependents earns $72,000.

   a. Does the family owe income tax or receive a check from the government? What is the amount?

   b. What is the cutoff amount for the family to qualify for a check from the government?

5.2 Graduated Income Tax

▶ Calculate a graduated income tax.

▶ Analyze a graduated income tax system.

▶ Compare a graduated income tax with a flat income tax.

Calculating a Graduated Income Tax

A **graduated tax** (or *progressive tax*) is a tax with a rate that increases as the taxable amount increases. In the United States, the federal income tax is a graduated tax, as are the income taxes in 36 states.

The table shows the graduated income tax for the taxable income (after deductions) of a single person in 2010. The rate for each taxable income bracket is called the **marginal tax rate.** The overall rate of tax that a person pays on his or her entire taxable income is called the **effective tax rate.**

Taxable Income	Marginal Tax Rate
$0–$8375	10%
$8376–$34,000	15%
$34,001–$82,400	25%
$82,401–$171,850	28%
$171,851–$373,650	33%
$373,651+	35%

EXAMPLE 1   **Calculating a Graduated Income Tax**

Use the table above to find the income tax and the effective tax rate for each taxable income.

a. $67,850   **b.** $1,000,000

SOLUTION

a.

	A	B	C
1	Taxable Income	Marginal Tax Rate	Tax
2	$8,375.00	10%	$837.50
3	$25,625.00	15%	$3,843.75
4	$33,850.00	25%	$8,462.50
5	**$67,850.00**		**$13,143.75**
6			
7			
8			

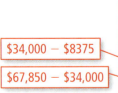
$34,000 − $8375
$67,850 − $34,000

The effective tax rate is $\frac{13,143.75}{67,850.00} \approx 19.4\%$.

b.

	A	B	C
1	Taxable Income	Marginal Tax Rate	Tax
2	$8,375.00	10%	$837.50
3	$25,625.00	15%	$3,843.75
4	$48,400.00	25%	$12,100.00
5	$89,450.00	28%	$25,046.00
6	$201,800.00	33%	$66,594.00
7	$626,350.00	35%	$219,222.50
8	**$1,000,000.00**		**$327,643.75**

The effective tax rate is $\frac{327,643.75}{1,000,000.00} \approx 32.8\%$.

✓ **Checkpoint**    Help at Math.andYOU.com

Find the income tax and the effective tax rate for a taxable income of $92,500.

The three graduated tax tables on these two pages are for single taxpayers. In many cases, married couples pay less income tax than if they stayed single. This occurs primarily in lower taxable income brackets.

EXAMPLE 2	Calculating a Graduated Income Tax

As of the writing of this text, California's income tax rates were the second highest of all state income tax rates. Use the table to find the state income tax for a taxable income of $200,000. What was the effective tax rate?

Taxable Income	Marginal Tax Rate
$0–$7124	1.25%
$7125–$16,890	2.25%
$16,891–$26,657	4.25%
$26,658–$37,005	6.25%
$37,006–$46,766	8.25%
$46,767–$1,000,000	9.55%
$1,000,001+	10.55%

SOLUTION

DATA

	A	B	C
1	Taxable Income	Marginal Tax Rate	Tax
2	$7,124.00	1.25%	$89.05
3	$9,766.00	2.25%	$219.74
4	$9,767.00	4.25%	$415.10
5	$10,348.00	6.25%	$646.75
6	$9,761.00	8.25%	$805.28
7	$153,234.00	9.55%	$14,633.85
8	**$200,000.00**		**$16,809.77**

The effective tax rate was $\dfrac{16,809.77}{200,000.00} \approx 8.4\%$.

✓ **Checkpoint**   Help at *Math*.and**Y☺U**.com

As of the writing of this text, Hawaii's income tax rates were the highest of all state income tax rates. Use the table to find the state income tax for a taxable income of $200,000. What was the effective tax rate?

Taxable Income	Marginal Tax Rate
$0–$2400	1.40%
$2401–$4800	3.20%
$4801–$9600	5.50%
$9601–$14,400	6.40%
$14,401–$19,200	6.80%
$19,201–$24,000	7.20%
$24,001–$36,000	7.60%
$36,001–$48,000	7.90%
$48,001–$150,000	8.25%
$150,001–$175,000	9.00%
$175,001–$200,000	10.00%
$200,001+	11.00%

Analyzing a Graduated Income Tax System

EXAMPLE 3 Analyzing a Graduated Tax System

Use the table and the circle graph to analyze the income tax system in the United States.

Group	Adjusted Gross Income
Top 1%	$380,354+
Next 4%	$159,619−$380,353
Next 5%	$113,799−$159,618
Next 15%	$67,280−$113,798
Next 25%	$33,048−$67,279
Bottom 50%	$0−$33,047

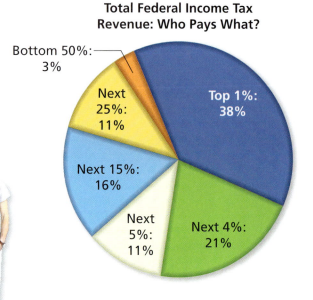

Total Federal Income Tax Revenue: Who Pays What?

Bottom 50%: 3%
Next 25%: 11%
Next 15%: 16%
Next 5%: 11%
Top 1%: 38%
Next 4%: 21%

Of all wage earners in the United States, about half pay no federal income tax. This is not to say that they have no payroll deductions. They do pay federal Social Security tax and federal Medicare tax.

SOLUTION

The income tax system in the United States is a classic progressive system. The top 25% of all wage earners pay 86% of the total income tax revenue. The bottom 75% of all wage earners pay only 14% of the total income tax revenue.

 **Checkpoint**

Help at *Math*.and**Y♥U**.com

The circle graph shows how the adjusted gross income earned in the United States is divided among the different income groups.

Compare each income group's percent of the adjusted gross income earned with the percent of the total income tax revenue it pays. What can you conclude?

Adjusted Gross Income Earned: Who Gets It?

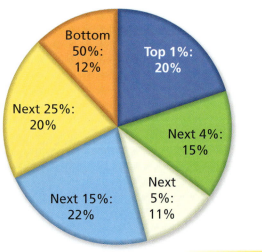

Bottom 50%: 12%
Next 25%: 20%
Next 15%: 22%
Next 5%: 11%
Top 1%: 20%
Next 4%: 15%

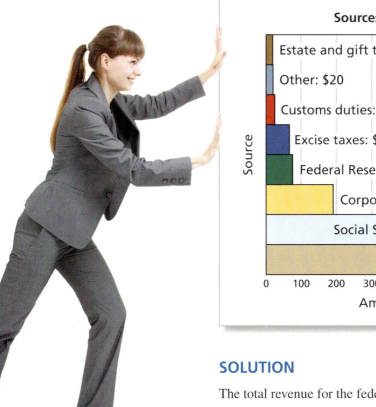

| EXAMPLE 4 | **Analyzing Federal Tax Revenue** |

What percent of the U.S. federal tax revenue comes from individual income taxes?

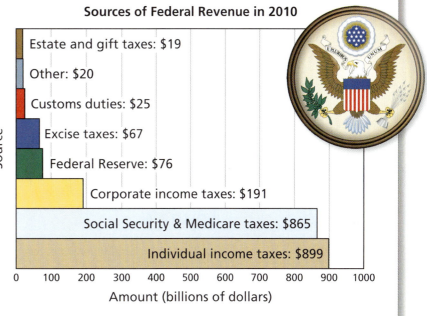

Sources of Federal Revenue in 2010

- Estate and gift taxes: $19
- Other: $20
- Customs duties: $25
- Excise taxes: $67
- Federal Reserve: $76
- Corporate income taxes: $191
- Social Security & Medicare taxes: $865
- Individual income taxes: $899

Source / Amount (billions of dollars)

SOLUTION

The total revenue for the federal government in 2010 was

$$899 + 865 + 191 + 76 + 67 + 25 + 20 + 19 = \$2{,}162 \text{ billion}$$
$$= \$2{,}162{,}000{,}000{,}000.$$

The percent of this that came from individual income taxes was

$$\frac{899{,}000{,}000{,}000}{2{,}162{,}000{,}000{,}000} \approx 41.6\%.$$

 **Checkpoint**

 Help at *Math*.and**YOU**.com

DATA Deciding how to display data is a field in mathematics and statistics called information design. You have already looked at several traditional styles of information design in this text, including bar graphs and circle graphs. In Section 9.1, you will look at several nontraditional styles.

a. Enter the data in Example 4 into a spreadsheet. Make a circle graph and label each section with its title, amount, and percent.

b. Use the spreadsheet to create a third graph of the data.

c. Which type of graph do you prefer? Explain.

Comparing Graduated and Flat Income Tax Systems

EXAMPLE 5 **Comparing Possible Systems**

The total personal income in the United States is estimated to be about $13 trillion. The federal income tax revenue is about $1 trillion, and there are about 150 million taxpayers.

a. What percent of the total personal income is paid to federal income tax?

Discuss the following options for raising this amount of income tax revenue.

b. **Assessment:** Every taxpayer pays the same amount.

c. **Flat tax on total income:** Every taxpayer pays the same rate.

d. **Graduated tax on taxable income:** Continue the current U.S. system.

e. **Graduated tax on total income:** Eliminate deductions and loopholes.

SOLUTION

a. The total federal income tax revenue is

$$\frac{1,000,000,000,000}{13,000,000,000,000} \approx 0.0769$$

$$= 7.69\%$$

of the total personal income in the United States.

b. For each taxpayer to pay the same amount, the amount would have to be

$$\frac{1,000,000,000,000}{150,000,000} \approx \$6667.$$

This is not feasible. People with low incomes cannot pay this amount without extreme hardship.

c. The rate necessary to raise the revenue is the rate found in part (a). If every taxpayer pays 7.69% of his or her *total* income, the federal government will raise $1 trillion in income tax.

d. The current graduated income tax system has been developing since 1913. It contains thousands of pages of tax code. With the many deductions and exemptions, a significant amount of income in the United States goes untaxed.

e. This is not a description of the current income tax code in the United States. Taxpayers can subtract deductions from their total income to reduce their tax liability. There are many deductions, such as some types of medical expenses, certain types of interest payments, contributions to charities, and deductions for dependents. If these deductions are removed, it will no longer be possible for one of the wealthiest Americans, Warren Buffett, to claim that he pays only 17.7% tax on an annual income of approximately $46 million.

Warren Buffett is an American investor and industrialist. He is consistently ranked among the world's wealthiest people. Buffett is an outspoken critic of the American income tax system.

 **Checkpoint**

 Help at *Math*.and**YOU**.com

Write a detailed description of an income tax system that you think would be fair to all citizens in the United States.

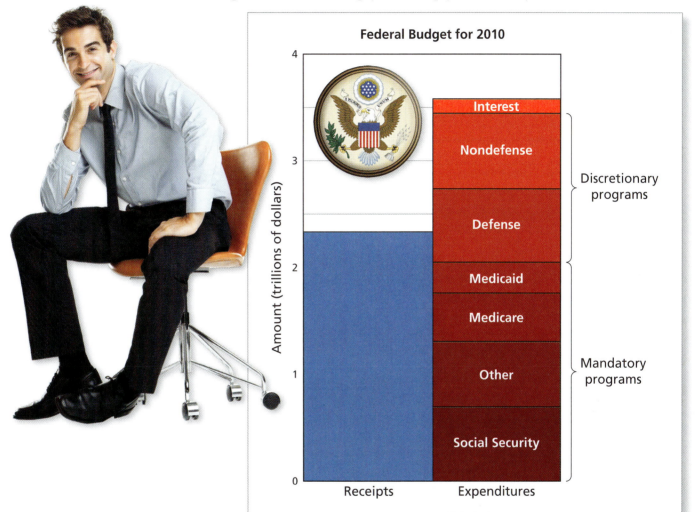

EXAMPLE 6   **Analyzing the Deficit**

Use the information graphic to estimate how much additional income tax revenue the U.S. government would have to raise to eliminate its annual deficit. What percent would each taxpayer have to pay in a flat tax system?

Federal Budget for 2010

Amount (trillions of dollars)

Interest
Nondefense
Defense
— Discretionary programs

Medicaid
Medicare
Other
Social Security
— Mandatory programs

Receipts   Expenditures

SOLUTION

The expenses were projected to be about $3.58 trillion. The income was projected to be about $2.3 trillion. So, the deficit was projected to be about $1.28 trillion. With a total personal income of about $13 trillion, each taxpayer would have to pay an additional

$$\frac{1{,}280{,}000{,}000{,}000}{13{,}000{,}000{,}000{,}000} \approx 9.8\%$$

to cover the deficit. With the rate from Example 5(a), this implies that the federal government could balance its budget by charging each American taxpayer a flat tax rate of about 17.5%.

 **Checkpoint**

 Help at *Math*.and**YOU**.com

Suppose you are given the task of balancing the federal budget without raising taxes. Which expenses would you eliminate to balance the budget? Explain your reasoning.

5.2 Exercises

Joint Return Tax Rates The table shows the marginal tax rates for a married couple filing jointly. In Exercises 1–4, use the table. *(See Examples 1 and 2.)*

Married Filing Jointly in 2010	
Taxable Income	**Marginal Tax Rate**
$0–$16,750	10%
$16,751–$68,000	15%
$68,001–$137,300	25%
$137,301–$209,250	28%
$209,251–$373,650	33%
$373,651+	35%

For tax purposes, your marital status is determined on the last day of the tax year. Even if you get married on December 31, you can file jointly for the whole year.

1. You and your spouse have a taxable income of $58,750. How much do you pay in income tax? What is the effective tax rate?

2. You and your spouse have a taxable income of $75,000. Compare your income tax with that of a single taxpayer who has the same taxable income.

3. You and your spouse have an adjusted gross income of $141,500. You subtract a standard deduction of $11,400 and 2 personal exemptions of $3650 each to determine your taxable income. How much do you pay in income tax? What is the effective tax rate?

4. You have an adjusted gross income of $214,000, and your spouse has no income.

   a. You subtract a standard deduction of $11,400 and 2 personal exemptions of $3650 each to determine your taxable income. How much do you pay in income taxes? What is the effective tax rate?

   b. The standard deduction for a taxpayer filing "single" is $5700, and the personal exemption is $3650. Suppose you are single and subtract the standard deduction and personal exemption to determine your taxable income. How much more do you pay in income tax than in part (a)? How much higher is the effective tax rate?

5. **Married Filing Separately** The table shows the marginal tax rates for a married couple filing separately. You and your spouse have a taxable income of $54,680. Do you pay less tax to file all the income for yourself under "married filing separately" rather than "married filing jointly"? Explain your reasoning. *(See Examples 1 and 2.)*

6. **Effective Tax Rate** Do incomes in the same bracket have the same effective tax rate? Explain your reasoning. *(See Examples 1 and 2.)*

Married Filing Separately in 2010	
Taxable Income	**Marginal Tax Rate**
$0–$8375	10%
$8376–$34,000	15%
$34,001–$68,650	25%
$68,651–$104,625	28%
$104,626–$186,825	33%
$186,826+	35%

South Carolina State Income Tax In Exercises 7 and 8, use the circle graphs. *(See Example 3.)*

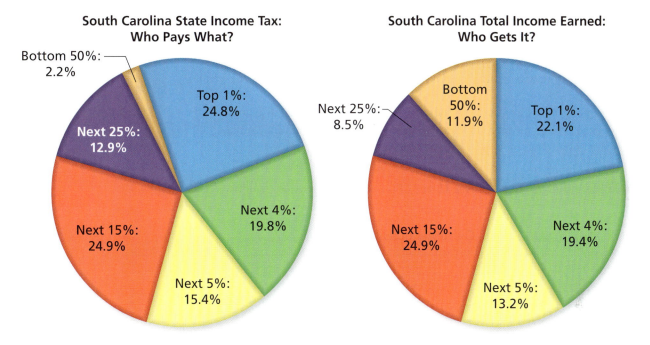

South Carolina State Income Tax:
Who Pays What?

Bottom 50%: 2.2%
Top 1%: 24.8%
Next 25%: 12.9%
Next 4%: 19.8%
Next 15%: 24.9%
Next 5%: 15.4%

South Carolina Total Income Earned:
Who Gets It?

Next 25%: 8.5%
Bottom 50%: 11.9%
Top 1%: 22.1%
Next 15%: 24.9%
Next 4%: 19.4%
Next 5%: 13.2%

7. Use the circle graphs to analyze South Carolina's state income tax.

8. Use the six income groups to compare South Carolina's state income tax to the federal income tax.

South Carolina State and Local Tax Revenue In Exercises 9–12, use the bar graph.
(See Example 4.)

9. What percent of South Carolina's state and local tax revenue comes from individual income taxes?

10. What percent of South Carolina's state and local tax revenue comes from sales taxes?

11. Compare South Carolina's sources of state and local tax revenue with the federal government's sources of tax revenue.

12. Enter the data into a spreadsheet.

    a. Make a circle graph and label each section with its title, amount, and percent.

    b. Use the spreadsheet to create a third graph of the data.

    c. Which type of graph do you prefer? Explain.

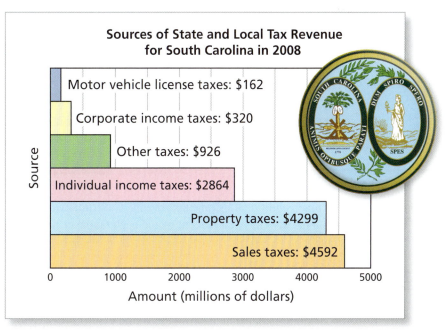

Sources of State and Local Tax Revenue
for South Carolina in 2008

Motor vehicle license taxes: $162
Corporate income taxes: $320
Other taxes: $926
Individual income taxes: $2864
Property taxes: $4299
Sales taxes: $4592

Source

0   1000   2000   3000   4000   5000
Amount (millions of dollars)

California State Income Tax In Exercises 13 and 14, use the information below. *(See Example 5.)*

In 2008, the total personal income in California was about $1.61 trillion. The budgeted income tax revenue was about $58 billion. You are a member of a committee analyzing California's state income tax.

13. Discuss using a flat tax on total income to raise this amount of income tax revenue.

14. Discuss using a graduated tax on total income to raise this amount of income tax revenue. Would the marginal tax rates be higher or lower than California's current graduated tax on taxable income? Explain your reasoning.

California State Budget In Exercises 15 and 16, use the information graphic. *(See Example 6.)*

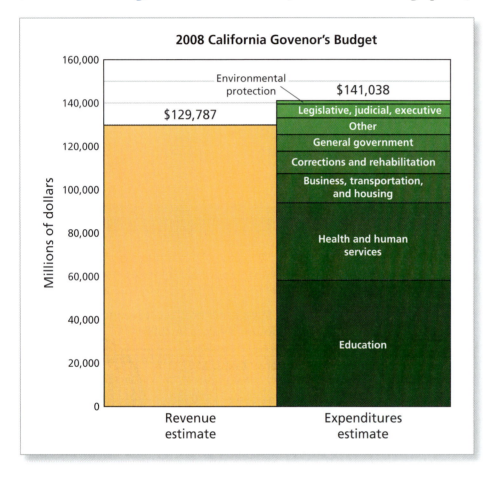

15. Estimate how much additional income tax revenue California would have to raise to eliminate its annual deficit. What percent would each taxpayer have to pay in a flat tax system?

16. In the 2008 California state budget, sales taxes were projected to account for 27% of the state's revenue. Suppose the state eliminated its sales taxes.

   a. How much additional income tax revenue would California have to raise to match its original revenue estimate? What percent would each taxpayer have to pay in a flat tax system?

   b. How much additional income tax revenue would California have to raise to eliminate its annual deficit? What percent would each taxpayer have to pay in a flat tax system?

▶ Extending Concepts

DATA **Long-Term Capital Gains** In Exercises 17–20, use the table and the information below.

Long-term capital gains, such as profit from the sale of stock held for more than one year, are taxed at different rates than ordinary income. For instance, if you have $25,000 of ordinary income and $10,000 of income from long-term capital gains, then your income from $0 to $25,000 is taxed at the marginal rates for ordinary income and your income from $25,001 to $35,000 is taxed at the marginal rates for long-term capital gains. The income brackets are filled by ordinary income first, and then by long-term capital gains beginning on the first dollar after ordinary income.

2010 Marginal Tax Rates for Single Taxpayers		
Taxable Income	Ordinary Income	Long-term Capital Gains
$0–$8375	10%	0%
$8376–$34,000	15%	0%
$34,001–$82,400	25%	15%
$82,401–$171,850	28%	15%
$171,851–$373,650	33%	15%
$373,651+	35%	15%

17. A taxpayer has $30,000 of taxable ordinary income and $5000 of taxable income from long-term capital gains. Find the income tax and the effective tax rate.

18. A taxpayer has $100,000 of taxable ordinary income and $75,000 of taxable income from long-term capital gains. Find the income tax and the effective tax rate.

19. A taxpayer has a taxable income of $1,000,000.

   a. Find the income tax and the effective tax rate when 10% of the taxable income is from long-term capital gains.

   b. Find the income tax and the effective tax rate when 50% of the taxable income is from long-term capital gains.

   c. What happens to the effective tax rate as the percent of taxable income from long-term capital gains increases? Explain your reasoning.

20. Using the graph, explain how the taxes paid as a percent of AGI can be lower for a taxpayer in a higher income bracket than for a taxpayer in a lower income bracket.

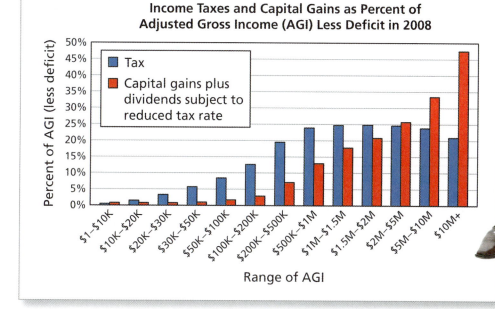

5.1–5.2   Quiz

Alabama State and Local Taxes In Exercises 1 and 2, use the display.

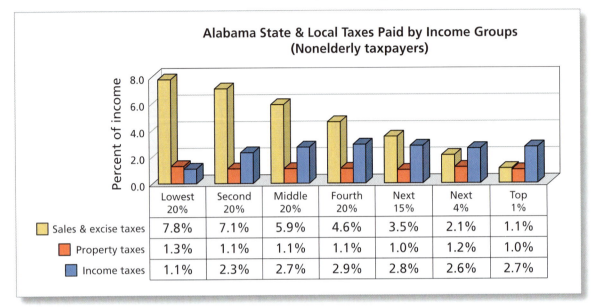

Alabama State & Local Taxes Paid by Income Groups (Nonelderly taxpayers)

	Lowest 20%	Second 20%	Middle 20%	Fourth 20%	Next 15%	Next 4%	Top 1%
Sales & excise taxes	7.8%	7.1%	5.9%	4.6%	3.5%	2.1%	1.1%
Property taxes	1.3%	1.1%	1.1%	1.1%	1.0%	1.2%	1.0%
Income taxes	1.1%	2.3%	2.7%	2.9%	2.8%	2.6%	2.7%

1. The income for a family in the second 20% is $25,000. How much, in dollars, does this family pay for each type of tax?

2. Decide whether each type of tax is regressive, flat, or progressive in Alabama. Explain your reasoning.

   a. Sales and excise taxes   **b.** Property taxes

Alabama State Income Tax In Exercises 3–5, use the information below.

Alabama has a graduated income tax. The table shows the marginal tax rates for the "single" and "married filing jointly" statuses.

Single	Marginal Tax Rate	Married Filing Jointly
$0–$500	2%	$0–$1000
$501–$3000	4%	$1001–$6000
$3001+	5%	$6001+

3. Find the state income tax and the effective tax rate for a single taxpayer with a taxable income of $30,000.

4. Find the state income tax and the effective tax rate for a married couple filing jointly with a taxable income of $100,000.

5. Analyze the following statement.

   "Although Alabama's income tax is essentially flat, the federal income tax is still progressive."

   Who Pays? A Distributional Analysis of the Tax Systems in All 50 States, 3rd Edition, Institute on Taxation and Economic Policy

Math & the American Dream
PROJECT: Finding a Tax Freedom Day

1. You graduate from college and accept a job in San Francisco. You have a taxable income of $105,000. Find the tax you pay in each category. Then find the percent of your total taxable income that you pay in taxes.

   Federal income tax (Use the rates on page 212.) _____

   California state income tax (Use the rates on page 213.) _____

   Sales tax of 9.5% on $35,000 for a new car _____

   Sales tax of 9.5% on $24,000 in other purchases _____

   Social Security tax of 6.2% _____

   Medicare tax of 1.45% _____

   Gasoline excise tax of $0.35 per gallon
   (30 gallons per week total) _____

   Property tax of 1.164% on $950,000 for a mortgaged home _____

   Indirect taxes paid in higher prices of goods purchased $5000

   All other taxes (direct and indirect) $3500

2. *Tax Freedom Day* is the day in the year that the average American has earned enough to pay all the taxes that he or she must pay for the year. Find the Tax Freedom Day using the information in Exercise 1.

3. Use the *My Tax Freedom Day Calculator** at *Math.andYou.com* to calculate your actual Tax Freedom Day.

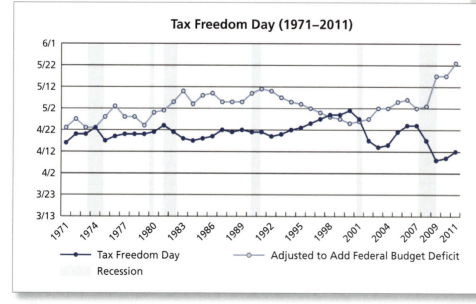

Tax Freedom Day (1971–2011)

— Tax Freedom Day   —○— Adjusted to Add Federal Budget Deficit
Recession

*Provided by *MyTaxFreedomDay.com*

5.3 Property Tax

▶ Calculate a property tax.

▶ Analyze assessments and tax credits.

▶ Analyze exemptions for property tax.

Calculating a Property Tax

In the United States, a **property tax** (or *millage tax*) is a tax levied by city and county governments on real and personal property to generate operating revenues to pay for such public services as schools and emergency services.

Property tax codes vary by state, county, and city. Here are some common characteristics of property taxes.

● Property taxes are generally assessed on land, buildings, and homes, not on other forms of real property, such as jewelry or boats.

● Property taxes are often expressed in **mills** but can also be expressed as percents.

● Market value is an estimate of the price for which a home could sell. **Assessed value** is a percent (called the assessment level) of a property's market value. Property tax is calculated on assessed value.

$$\begin{array}{c}\boxed{\text{Property tax}} = \boxed{\begin{array}{c}\text{Tax rate} \\ \text{(decimal form)}\end{array}} \times \underbrace{\boxed{\begin{array}{c}\text{Assessment level} \\ \text{(decimal form)}\end{array}} \times \boxed{\text{Market value}}}_{\text{Assessed value}}\end{array}$$

Writing a Tax Rate in Decimal Form

A tax rate of 1 mill means $1 of taxes per $1000 of assessed value. To change a tax rate to decimal form, divide by 1000.

Example: 235 mills $= \dfrac{235}{1000} = 0.235$ (decimal form)

EXAMPLE 1   Calculating a Property Tax

There are huge differences in property tax rates throughout the United States. Find the property tax for a $500,000 home in each city.

a. Bridgeport, CT: Tax rate: 38.7 mills, Assessment level: 70%

b. Cheyenne, WY: Tax rate: 71 mills, Assessment level: 9.5%

SOLUTION

a. Property tax $= 0.0387(0.70)(500,000) = \$13,545.00$

b. Property tax $= 0.071(0.095)(500,000) = \$3372.50$

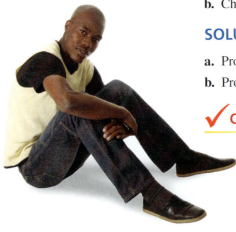

✓ **Checkpoint**

Help at *Math*.andY⊙U.com

Find the property tax for a $400,000 home in each city.

c. Denver, CO: Tax rate: 6.8 mills, Assessment level: 8%

d. Manchester, NH: Tax rate: 17.4 mills, Assessment level: 100%

You can see from Example 1 that the tax rate of a property tax does not by itself tell you whether the tax is high or low. To decide whether a property tax is high or low, use the tax rate and the assessment level to determine the **effective rate** of a property tax. Effective rates are usually expressed as percents.

State and Local Property Taxes Per Capita by Rank

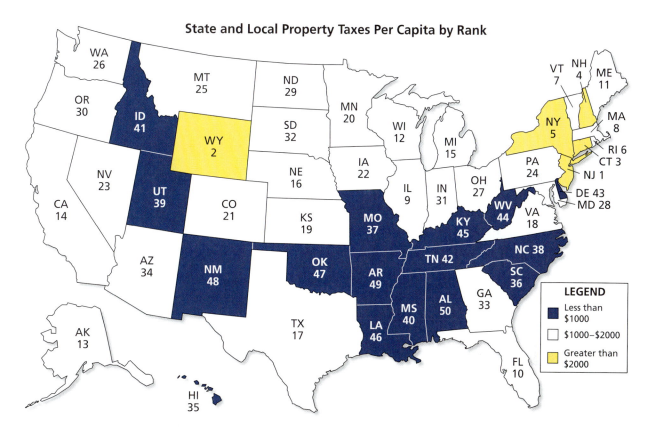

EXAMPLE 2   **Calculating Effective Rates**

Find the effective property tax rate in each city.

a. Omaha, NE: Tax rate: 21.3 mills, Assessment level: 96%

b. Birmingham, AL: Tax rate: 80.2 mills, Assessment level: 10%

SOLUTION

a. Effective rate = 0.0213(0.96) ≈ 0.02 = 2%

b. Effective rate = 0.0802(0.10) ≈ 0.008 = 0.8%

✔ **Checkpoint**   Help at *Math*.and**Y☺U**.com

Find the effective property tax rate in each city.

c. Newark, NJ: Tax rate: 27.4 mills, Assessment level: 59.7%

d. Honolulu, HI: Tax rate: 3.4 mills, Assessment level: 100%

e. New York, NY: Tax rate: 167 mills, Assessment level: 3.7%

Analyzing Assessments and Tax Credits

The term *assessed value* of a home or of land can be confusing to people. Do not confuse the terms *assessed value* and *market value*.

- **Market Value:** When you apply for a home mortgage, an appraiser determines the market value of your home. This appraised *market value* is the amount that the appraiser thinks the home could sell for. Home appraisers are in private business. They are not government employees.

- **Assessed Value:** Your local government conducts assessments of your property. A government assessor determines the market value and then calculates the assessed value. You pay property taxes based on your assessed value. Properties are generally assessed only once every several years, unless you make a considerable home improvement. If you do not agree with the government's assessment of your home, you do have the opportunity to appeal.

EXAMPLE 3   **Appealing an Assessed Value**

You move into a new home and receive the following invoice.

PROPERTY TAX INVOICE

To: John and Jane Doe
1234 Main Street
Anytown, NJ 01234

Assessed Value	Tax Rate
$350,000.00	185 mills

Property Tax Due:	$64,750.00

From talking to your neighbors, you think that the assessed value is too high. Two of your neighbors live in comparable homes and have assessed values of $250,000. Is it worth your time and effort to appeal the assessed value?

SOLUTION

Using your current tax bill, your property tax is

Property tax = 0.185(350,000)

= $64,750.

If you are successful in getting the assessed value lowered to $250,000, the property tax will be

Property tax = 0.185(250,000)

= $46,250.

You could save $18,500 each year. It is probably worth appealing.

✓ **Checkpoint**   Help at

You take out a mortgage for $243,000. Your monthly loan payment is $1073.64. You must also pay for home insurance and property tax. Home insurance costs $650 annually. The tax rate is 358 mills, and the assessed value of your home is $85,050. What is your total monthly payment?

It is relatively common for municipalities to issue **credits** for property tax. Credits are subtracted after taxes have been calculated. One common credit is the homestead tax credit. To qualify as a homestead, a property must be the primary residence of the homeowner.

EXAMPLE 4   **Applying a Tax Credit**

The assessed value of a homeowner's primary residence is $75,000, and the property tax rate is 40 mills. The homeowner receives a homestead tax credit according to the graph. How much does the homeowner pay for property tax?

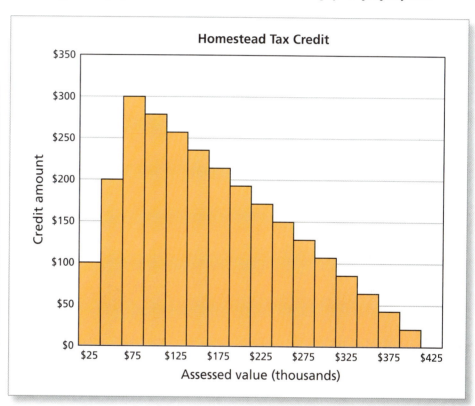

Study Tip

Homestead exemption laws have three main characteristics.

1. They prevent the forced sale of a home to meet creditor demands.
2. They provide a surviving spouse with shelter.
3. They provide an exemption from property taxes.

SOLUTION

The graph shows that the homestead tax credit is $300 for a property with an assessed value of $75,000. So, the property tax is

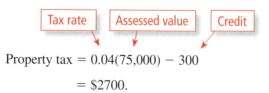

Property tax = 0.04(75,000) − 300

= $2700.

The homeowner pays $2700 for property tax.

✔ **Checkpoint**   Help at *Math*.and**Y☺U**.com

The assessed value of a homeowner's primary residence is $125,000, and the property tax rate is 70 mills. The homeowner receives a homestead tax credit according to the graph. How much does the homeowner pay for property tax?

Analyzing Property Tax Exemptions

It is relatively common for municipalities to have **exemptions** for property tax. Exemptions apply to the assessed value of a property. Here are some typical (full or partial) exemptions.

- Homeowners who are 65 or older
- Homeowners with limited incomes
- Homeowners on full-time disability
- Homeowners who are veterans
- Homeowners who are in the clergy
- Property owned by government agencies
- Property owned by schools
- Property owned by religious organizations
- Property owned by nonprofit organizations

EXAMPLE 5 Estimating Lost Tax Revenue

Suppose that all property in the United States is subject to property tax. Estimate the additional tax that could be collected.

SOLUTION

Here is one way to do this. Use the following information from the U.S. Census Bureau. It shows the total annual tax revenue for state and local governments in the United States.

> **Study Tip**
>
> To find the estimate of $176 billion in Example 5, use the equation
>
> $$70\% \text{ of } \begin{pmatrix} \text{Total} \\ \text{property} \\ \text{tax} \end{pmatrix} = \$410 \text{ billion}$$
>
> to determine that the total property tax is about $586 billion. This implies that the lost revenue is about $176 billion.

Suppose that 30% of the property in the United States is tax exempt. You can conclude that about $176 billion a year is lost to property tax exemptions.

 **Checkpoint**

The United States' policy of "separation of church and state" stems from the First Amendment to the Constitution, which reads, "Congress shall make no law respecting an establishment of religion, or prohibiting the free exercise thereof. . .." Some people argue that allowing religious organizations to be exempt from property tax is a violation of the First Amendment. What do you think?

Study Tip

Property tax codes vary a lot throughout the United States. The time to check your local codes is *before* you move into a municipality. Exemptions also vary, even in neighboring municipalities.

EXAMPLE 6   **Analyzing the Value of an Exemption**

Use the information below to determine how much the veteran pays for property tax and saves from the military exemption.

- A person is 30 years old and has been discharged from 8 years of service in the U.S. military.
- The person buys a home whose assessed value is $75,000.
- The tax rate is 65 mills.
- The municipality has a 15% exemption (up to $5000) for veterans.

SOLUTION

The amount of the exemption is

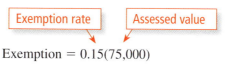

$$\text{Exemption} = 0.15(75,000)$$
$$= \$11,250.$$

This exceeds the maximum, so the exemption is $5000. The property tax is

$$\text{Property tax} = 0.065(75,000 - 5000)$$
$$= \$4550.$$

The tax savings from the $5000 reduction in the assessed value is

$$\text{Tax savings} = 0.065(5000)$$
$$= \$325.$$

The veteran pays $4550 for property tax and saves $325 from the exemption.

 **Checkpoint**

Help at *Math*.and*Y*☺*U*.com

Use the information below to determine how much the veteran pays for property tax and saves with the military exemptions from age 26 to age 40.

- A person is 26 years old and has been discharged from 8 years of service in the U.S. military.
- The person buys a home whose assessed value is $36,800.
- The tax rate is 95 mills.
- The municipality has a 10% exemption (up to $4000) for veterans.

5.3 Exercises

New York Property Taxes In Exercises 1–6, find (a) the property tax for a $200,000 home in the town or city and (b) the effective property tax rate in the town or city. *(See Examples 1 and 2.)*

1. Lewiston
2. Jamestown
3. Fulton
4. Bolton
5. Saratoga Springs
6. Cohoes

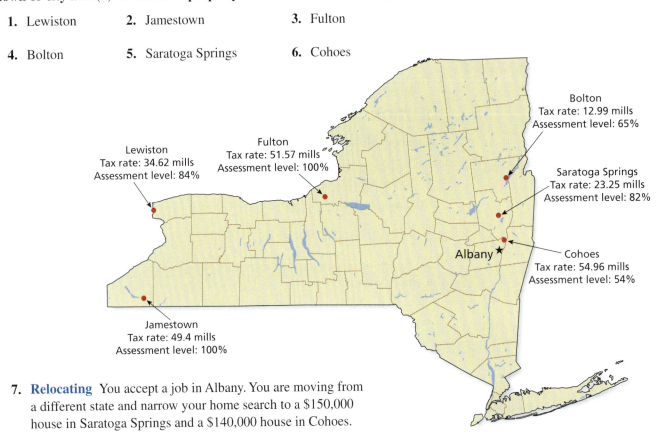

Bolton
Tax rate: 12.99 mills
Assessment level: 65%

Lewiston
Tax rate: 34.62 mills
Assessment level: 84%

Fulton
Tax rate: 51.57 mills
Assessment level: 100%

Saratoga Springs
Tax rate: 23.25 mills
Assessment level: 82%

Albany

Cohoes
Tax rate: 54.96 mills
Assessment level: 54%

Jamestown
Tax rate: 49.4 mills
Assessment level: 100%

7. Relocating You accept a job in Albany. You are moving from a different state and narrow your home search to a $150,000 house in Saratoga Springs and a $140,000 house in Cohoes.

  a. Which house has higher property taxes? How much higher are the taxes?

  b. The drive to work from the home in Saratoga Springs takes twice as much time as the drive to work from the house in Cohoes. Based on driving time, property taxes, and home price, which house do you prefer? Explain.

8. Village of Lakewood A village near Jamestown claims that the owner of a $90,000 house pays a total of $3243.60 in property taxes, of which $642.60 goes to the village.

  a. What is the combined tax rate?

  b. How much does the owner of a $120,000 house in the village pay in property taxes? How much goes to the village?

  c. The circle graph shows how village tax dollars are spent. How much of the village taxes in part (b) go toward snow removal? police?

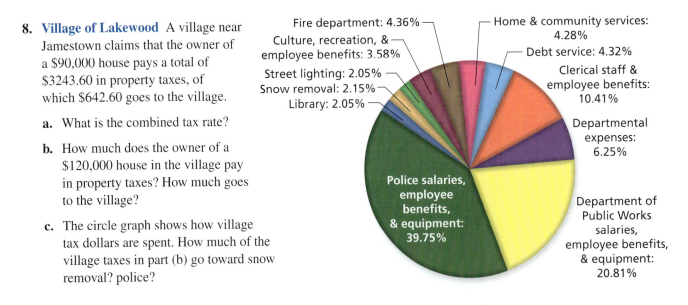

Fire department: 4.36%
Culture, recreation, & employee benefits: 3.58%
Street lighting: 2.05%
Snow removal: 2.15%
Library: 2.05%
Home & community services: 4.28%
Debt service: 4.32%
Clerical staff & employee benefits: 10.41%
Departmental expenses: 6.25%
Police salaries, employee benefits, & equipment: 39.75%
Department of Public Works salaries, employee benefits, & equipment: 20.81%

Property Tax Bills Property taxes in Anytown consist of town, county, and school taxes. The town and county taxes are due in May. The school taxes are due in October. In Exercises 9–13, use the property tax bills shown. Assume that Clarice and Fred are subject to the same tax rates. *(See Examples 3 and 4.)*

ANYTOWN
PROPERTY TAX BILL   May 15, 2012

PROPERTY ADDRESS
Clarice Mason
27411 Route 8
Anytown, MN 01234

PROPERTY VALUATION
The assessor estimates the Assessed Value of this property as
$112,500.
If you feel your property is overvalued, please see the instructions
in the booklet *How to File a Complaint on Your Assessment*.
To obtain a copy of this booklet, contact your assessor's office.

PROPERTY TAXES	TAX RATE	TAX DOLLARS
Town	8.0 mills	$900.00
County	3.6 mills	$405.00
		TAX DUE $1305.00

ANYTOWN
PROPERTY TAX BILL   October 15, 2012

PROPERTY ADDRESS
Fred Alum
3500 Yawkee Hill Rd
Anytown, MN 01234

PROPERTY VALUATION
The assessor estimates the Assessed Value of this property as
$50,000.
If you feel your property is overvalued, please see the instructions
in the booklet *How to File a Complaint on Your Assessment*.
To obtain a copy of this booklet, contact your assessor's office.

PROPERTY TAXES	TAX RATE	TAX DOLLARS
School	18.4 mills	$920.00
		TAX DUE $920.00

9. How much is Clarice's school tax bill? How much more does she pay for town and county taxes than Fred?

10. Clarice plans to sell her home within the next 2 years. Her neighbor's home is similar and has an assessed value of $100,000. Is it worth Clarice's time and effort to appeal the assessed value shown on the property tax bill? Explain.

11. Fred owns a home similar to three other homes on Yawkee Hill Road. The assessed values of the other 3 homes are $47,000, $49,000, and $60,000. Should Fred appeal the assessed value shown on the property tax bill? Explain.

12. After an appeal, the assessed value of Clarice's home decreases to $100,000. She receives a homestead tax credit according to the graph on page 227. How much does she pay in property tax?

13. Fred does not appeal the assessment. He receives a homestead tax credit according to the graph on page 227. How much does he pay in property tax?

14. **Raising Taxes** A municipality can increase taxes by increasing the tax rate. Describe another method municipalities can use to raise property taxes.

Tax Breaks **The circle graph shows the types of property taxes collected in a state in 1 year. In Exercises 15–17, use the circle graph.** *(See Example 5.)*

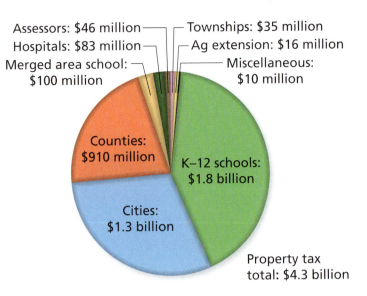

Assessors: $46 million — Townships: $35 million
Hospitals: $83 million — Ag extension: $16 million
Merged area school: $100 million — Miscellaneous: $10 million
Counties: $910 million
K–12 schools: $1.8 billion
Cities: $1.3 billion
Property tax total: $4.3 billion

15. What percent of the property taxes went to K–12 schools?

16. Without the military exemption, the state would have collected about 0.06% more in property taxes. Estimate the total military exemption.

17. Without the homestead tax credit, the state would have collected about 2.3% more in property taxes. Estimate the total property taxes that would have been collected without the homestead tax credit.

Veteran Exemption **In Exercises 18 and 19, use the information below.** *(See Example 6.)*

In New York, veterans of the following conflicts can receive property tax exemptions on their primary residence, as detailed in the table.

EXEMPTION RATE FOR ELIGIBLE VETERANS		
Requirements for Eligible Veterans	Percentage Assessed Value Reduction	Maximum Exemption Rate Reduction
Served during a specified period of war (See list)	15%	Max of $4140 for Class 1; $24,300 for Classes 2 & 4
Served in a combat zone	Plus 10%	Max of $6900 for Class 1; $40,500 for Classes 2 & 4
Disabled (Gold Star parents are not eligible for this portion of the vet's exemption)	Assessed value multiplied by 50% of the vet's disability rating	Max of $13,800 for Class 1; $81,000 for Classes 2 & 4

- Persian Gulf War
- Korean War
- World War I
- Vietnam War
- World War II
- Mexican Border Period

18. A veteran of the Korean War did not serve in a combat zone and is not disabled. He applies for the military exemption. The assessed value of his Class 1 home is $150,000. How much does he save on property taxes when the tax rate is 85 mills?

19. A nondisabled veteran of the Persian Gulf War who served in a combat zone applies for the military exemption. The assessed value of her Class 2 property is $140,000. How much does she save on property taxes when the tax rate is 32 mills?

20. **Solar and Wind Power** Texas offers a property tax exemption on qualifying wind and solar power energy devices, such as the solar hot water system shown. These devices generally increase the assessed value of a home. This increase is the amount of the exemption. *(See Example 6.)*

   a. After installation of the hot water system, the assessed value of a home increases by $15,000 to a total of $175,000. How much is the exemption?

   b. The county tax rate is 9 mills. How much does the exemption save the homeowner in county taxes?

SOLAR HOT WATER SYSTEM

SOLAR HEAT COLLECTOR
HOT WATER FROM COLLECTOR
COLD WATER TO COLLECTOR
HOT WATER TO HOUSE
COLD WATER FROM STREET
SOLAR STORAGE TANK
CIRCULATING PUMP

▶ Extending Concepts

Calculating Property Tax Rates In Exercises 21–23, use the information below.

Property tax rates are calculated based on the amount of money needed from property taxes, called a *levy*, and the taxable assessed value of all property within a municipality.

$$\text{Property tax rate} = \frac{\text{levy}}{\text{taxable assessed value}} \times 1000 \text{ mills}$$

21. A town levy is $600,000, and the taxable assessed value of all property in the town is $50,000,000. Find the property tax rate.

22. A county levy is $50,000,000, and the taxable assessed value of all property in the county is $1,000,000,000. Find the property tax rate.

23. A town levy is $750,000, and the taxable assessed value of all property in the town is $80,000,000. Find the property tax for a home with an assessed value of $50,000.

Equalization In Exercises 24–28, use the information below.

School districts often consist of several municipalities that may have different levels of assessment. Each municipality pays its fair share of the school tax based on its total market value rather than its total assessed value. The total market value of each municipality is the total assessed value divided by the municipality's *equalization rate*.

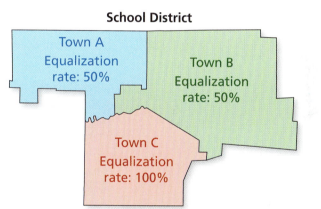

School District

24. The school district shown consists of three towns. The total assessed value of each town is $10 million. Find the total market value of each town.

25. The school district needs to raise $1 million through property tax. Divide the total market value of each town by the total market value of the entire school district to find the percent of the $1 million school tax that each town must pay.

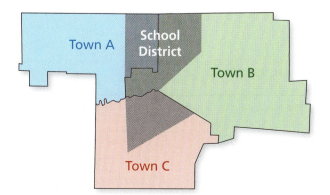

26. Find the school tax levy and the school tax rate for each town. Assume that the taxable assessed value is equal to the total assessed value for each town.

27. Would it be fair or unfair for each town to pay one-third of the school tax levy? Explain.

28. Explain why the school tax rate calculations would be more difficult for a school district organized as shown in the diagram at the left.

5.4 Social Security & Payroll Taxes

▶ Calculate Social Security & Medicare taxes.

▶ Evaluate the benefits of Social Security.

▶ Analyze the viability of Social Security.

Calculating Social Security & Medicare Taxes

Social Security is a social insurance program that is funded through dedicated payroll taxes. The original Social Security Act of 1935 has been broadened to include several programs, including the following.

- Old-Age, Survivors, and Disability Insurance
- Unemployment benefits
- Temporary Assistance for Needy Families
- Health Insurance for the Aged and Disabled (Medicare)
- Grants to States for Medical Assistance Programs (Medicaid)
- State Children's Health Insurance Program (SCHIP)
- Supplemental Security Income (SSI)

Social Security and Medicare Rates

The Social Security and Medicare tax rates for 2010 are as follows.

	Social Security	Medicare
Employees:	6.2% up to $106,800	1.45% (no salary cap)
Employers:	6.2% up to $106,800	1.45% (no salary cap)
Self-Employed:	12.4% up to $106,800	2.90% (no salary cap)

EXAMPLE 1   Calculating Total Payroll Taxes

You live in California and are self-employed. You report a taxable income of $105,000. Estimate your total income tax, including federal, state, Social Security, and Medicare. What percent of your taxable income goes to these four taxes?

SOLUTION

Federal income tax:	$23,109.25	(See page 212.)
California income tax:	$ 7,737.27	(See page 213.)
Social Security tax:	$13,020.00	(12.4% of $105,000)
Medicare tax:	$ 3,045.00	(2.9% of $105,000)
Total:	**$46,911.52**	

You pay 46,911.52/105,000 ≈ 44.7% of your taxable income to these 4 taxes.

 Checkpoint   Help at *Math*.and**Y☺U**.com

Rework Example 1 using taxable incomes of (a) $75,000 and (b) $200,000.

Discuss how your percent toward taxes changes as your income changes.

Study Tip

In reality, it is rare to have the same taxable income for federal, state, Social Security, and Medicare taxes. Taxable income is determined by various deductions, exemptions, and credits dictated by the tax code. To keep calculations simple, the same taxable income is used for all four taxes in Example 1.

In Example 1, the taxpayer is self-employed and must pay both parts of the Social Security and Medicare taxes (for both employee and employer). For people who are not self-employed, the two taxes are shared equally by employee and employer.

"What do I get each month?"   "What do I pay each month?"

EXAMPLE 2   Calculating a Total Compensation Package

You own a business in Massachusetts, which has a flat income tax of 5.3%. You hire a new employee for $60,000 a year. In addition to this salary, you pay a 5% matching contribution to a 401(k) retirement plan, $940 a month for the employee's health insurance, 1.25% for workers' compensation insurance, and $1000 each year for a holiday bonus.

a. What is the *total compensation package* you are paying for this employee?

b. How much does the employee receive each year as "take-home pay"?

SOLUTION

a. Total compensation package:

$60,000.00	Salary	
$ 1,000.00	* Holiday bonus	
$ 3,050.00	* 401(k) matching	5% of $61,000
$11,280.00	* Health insurance	12 months at $940
$ 3,782.00	** Social Security matching	6.2% of $61,000
$ 884.50	** Medicare matching	1.45% of $61,000
$ 762.50	** Workers' compensation	1.25% of $61,000
$80,759.00		

*Optional employee benefits   **Mandated employee benefits*

The total compensation package for this employee is $80,759 per year.

b. Take-home pay: Assume the employee is single, does not have any other taxable income, and claims the standard deduction. The employee must pay income tax on the $1000 holiday bonus but (currently) not on the 401(k) match or the health insurance.

401(k) employee portion:	$ 3,050.00	5% of $61,000
Social Security:	$ 3,782.00	6.2% of $61,000
Medicare:	$ 884.50	1.45% of $61,000
Federal income tax:	$ 8,331.25	Taxable income of $48,600
State income tax:	$ 2,838.15	5.3% of $53,550
Total:	**$18,885.90**	

The employee's take-home pay is 61,000 − 18,885.90 = $42,114.10, which is slightly more than half of the total compensation package.

✔ **Checkpoint**    Help at *Math*.and**Y☺U**.com

Refer to Example 2. Assume the holiday bonus is paid in 12 monthly installments.

c. How much does the employer pay each month?

d. How much does the employee receive each month as take-home pay?

e. How can the difference in these two amounts affect a person's perspective as an employee or as an employer?

Evaluating the Benefits of Social Security

The graph shows the maximum amount of Social Security withholding from 1978 through 2010.

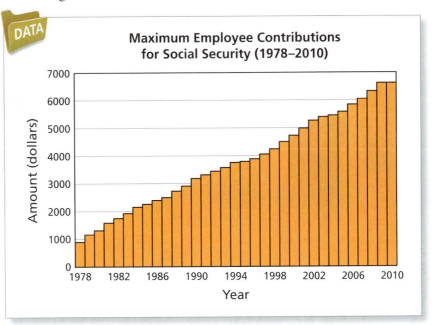

DATA

Maximum Employee Contributions for Social Security (1978–2010)

If you had paid the maximum Social Security tax each year from 1978 through 2010, both you and your employer would have paid about $124,166, for a total of about $248,332 toward your retirement benefits.

EXAMPLE 3 **Evaluating a Social Security Retirement**

Suppose you started your career late in life. You worked for 10 years and paid the maximum employee contribution to Social Security. At age 66 in 2010, you started to receive a retirement benefit of $1220 per month. Compare the amount you put into the system with the amount you will receive.

SOLUTION

Over 10 years, you would have paid about $58,000 into the system. Your employer matched this. If all payments had earned 4% interest each year, the total would have grown to about $140,000. The life expectancy of a 66-year-old is about 20 years. Over the 20 years, you will receive

Total receipts = 20(12)(1220) = $292,800.

Although this comparison is oversimplified, you can still see that the two amounts are very different. In other words, the government will pay you much more than you paid it.

✓ **Checkpoint**

Help at *Math*.and**YOU**.com

People often say, "If I could invest my Social Security tax into a private retirement system, I could end up with a better retirement than the retirement I will receive from Social Security." Analyze this statement. Is it true for all workers? Is it true for any workers?

The age for full retirement benefits used to be 65. That was gradually changed so that now anyone born after 1960 must be 67 years old to receive full retirement benefits. At ages less than 67, the retiree would receive only partial retirement benefits.

EXAMPLE 4   Considering Retirement?

Suppose that you are 62 years old and are considering whether to retire with partial Social Security benefits or wait until you are 67 years old to receive full benefits.

- At age 62, you will receive 75% of $1800 per month.
- At age 67, you will receive 100% of $1900 per month.

How long would you have to live to make waiting for full benefits more economical? Assume a 3% cost-of-living increase each year.

SOLUTION

Use a spreadsheet to analyze this question.

Social Security numbers are used to track workers' earnings over their lifetimes to pay benefits.

1.03*(1350)

1.03*(1900)

	A	B	C	D	E
1	Age	Retire Early	Total Income	Wait Until 67	Total Income
2	62	$1,350.00	$16,200.00	$0.00	$0.00
3	63	$1,390.50	$32,886.00	$0.00	$0.00
4	64	$1,432.22	$50,072.64	$0.00	$0.00
5	65	$1,475.19	$67,774.92	$0.00	$0.00
6	66	$1,519.45	$86,008.32	$0.00	$0.00
7	67	$1,565.03	$104,788.68	$1,900.00	$22,800.00
8	68	$1,611.98	$124,132.44	$1,957.00	$46,284.00
9	69	$1,660.34	$144,056.52	$2,015.71	$70,472.52
10	70	$1,710.15	$164,578.32	$2,076.18	$95,386.68
11	71	$1,761.45	$185,715.72	$2,138.47	$121,048.32
12	72	$1,814.29	$207,487.20	$2,202.62	$147,479.76
13	73	$1,868.72	$229,911.84	$2,268.70	$174,704.16
14	74	$1,924.78	$253,009.20	$2,336.76	$202,745.28
15	75	$1,982.52	$276,799.44	$2,406.86	$231,627.60
16	76	$2,042.00	$301,303.44	$2,479.07	$261,376.44
17	77	$2,103.26	$326,542.56	$2,553.44	$292,017.72
18	78	$2,166.36	$352,538.88	$2,630.04	$323,578.20
19	79	$2,231.35	$379,315.08	$2,708.94	$356,085.48
20	80	$2,298.29	$406,894.56	$2,790.21	$389,568.00
21	81	$2,367.24	$435,301.44	$2,873.92	$424,055.04
22	82	$2.438.26	$464,560.56	$2,960.14	$459,576.72
23	83	$2,511.41	$494,697.48	$3,048.94	$496,164.00

If you live past age 83, you will have received less total retirement income by retiring early. On the other hand, you will have had five additional years of retirement.

✓ Checkpoint

Help at *Math*.andY**O**U.com

Make a double bar graph showing columns C and E. Explain what the graph is showing.

Analyzing the Viability of Social Security

The **economic dependency ratio** is a measure of the portion of a population that is composed of dependents (too young or too old to work).

Economic Dependency Ratio

The economic dependency ratio is the number of people below age 20 or above age 64 divided by the number of people aged 20 to 64.

$$\text{Economic dependency ratio} = \frac{(\text{people } 0\text{–}19) + (\text{people } 65+)}{\text{people } 20\text{–}64}$$

A rising economic dependency ratio is a concern in many countries facing an aging population, because it becomes difficult for pension and social security systems to provide for a significantly older, nonworking population.

EXAMPLE 5 **Analyzing a Graph**

The graph shows the estimated and projected economic dependency ratios in the United States from 1950 through 2080. Discuss the changes in the ratio.

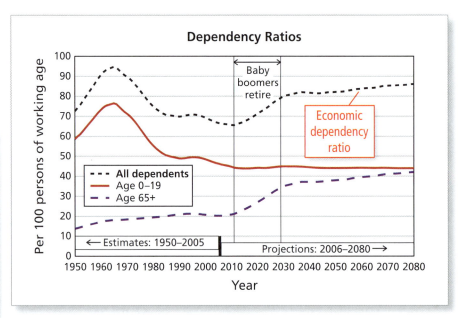

SOLUTION

From 1950 to 1965, the ratio increased. Then, due to a decreasing birth rate, the ratio decreased for many years. Shortly after 2010, the ratio began another period of significant increase as baby boomers (people born in the 20 years following World War II) began to retire. By 2030, the United States should have an economic dependency ratio of 80, a level it has not had since 1980.

 **Checkpoint** Help at *Math*.and*YOU*.com

The graph above generates a basic political philosophy question. That is, in a group, whose responsibility is it to take care of the people who cannot take care of themselves? Explain how different political parties have different answers to this question. What do you think?

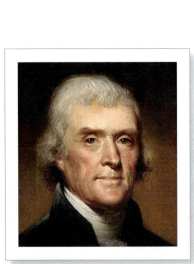

Thomas Jefferson believed that it was the duty of each individual to give to those in need. He thought it was more effective to give entire contributions to local organizations than to divide the contributions among various organizations throughout the country.

The Social Security Board of Trustees has warned that the long-term prospects for the Social Security system are not good. If action is not taken, the Social Security system will only be able to make partial payments in 2037.

EXAMPLE 6   **Looking into the Future**

Explain why the following demographics pose problems for Social Security.

a. Life expectancy: Life expectancy is increasing. In 1940, a 65-year-old man could expect to live another 12 years. In 2010, it was 16 years, and by 2040 it will be 18 years.

b. Birth rate: The birth rate is falling faster than expected, from 23.7 births per 1000 people in 1960 to just 13.5 births per 1000 people in 2009.

c. Elderly population: The elderly portion of the population is expected to rise from 13% in 2010 to 20% by 2050, increasing the number of retirees from 40 million to 88 million.

d. Working-age population: The smaller working-age population and larger elderly population means that while there were 6 workers for each retiree in 1960 and 5 workers for each retiree in 2010, by 2030 there will be just 3 workers to pay the taxes for the benefits of each retiree.

SOLUTION

a. Life expectancy: Longer life expectancy means that people will receive retirement payments from the system for a longer period of time. To help correct this problem, the age for full retirement benefits was raised from 65 to 67.

b. Birth rate: For any country, births are the source of future workers and future taxpayers. When birth rates decline, this source diminishes.

c. Elderly population: The elderly account for the bulk of the Social Security system's expenses. An increase in the percent of elderly people puts a further strain on the system. Over the next 20 years, nearly 80 million baby boomers will become eligible for Social Security.

d. Working-age population: Unlike most private retirement funds, Social Security is a "pay as you go" system. The taxes paid by working people are used to pay current Social Security benefits. As the percent of working people declines, the amount of tax that each worker pays will have to increase, or the amount that is paid out will have to decrease.

When the Social Security Act was signed by President Franklin D. Roosevelt in 1935, the United States was in the midst of the Great Depression. To publicize the new program and its benefits, the Social Security Board began an advertising campaign called "More Security for the American Family."

✓ **Checkpoint**   Help at

Each year, the federal government determines a cost-of-living adjustment (COLA) for recipients of Social Security benefits. Here are a few recent COLA rates.

2005: 4.1%	**2006:** 3.3%	**2007:** 2.3%
2008: 5.8%	**2009:** 0%	**2010:** 0%

About $713 billion in Social Security benefits were distributed in 2010. How much does a COLA rate of 1% cost the federal government in increased Social Security benefits? Discuss the consequences of your answer.

5.4 Exercises

Form W-2 **John Doe works for Company A. In Exercises 1–6, use the incomplete W-2 shown.** *(See Examples 1 and 2.)*

a Employee's social security number 123-00-4567	OMB No. 1545-0008	This information is being furnished to the Internal Revenue Service. If you are required to file a tax return, a negligence penalty or other sanction may be imposed on you if this income is taxable and you fail to report it.	
b Employer identification number (EIN) 12-3456789		**1** Wages, tips, other compensation 36,127.65	**2** Federal income tax withheld 3,597.90
c Employer's name, address, and ZIP code		**3** Social security wages 37,245.00	**4** Social security tax withheld
Company A		**5** Medicare wages and tips 37,245.00	**6** Medicare tax withheld
14 Corporate Drive Anytown, CO 01234		**7** Social security tips	**8** Allocated tips
d Control number 01-234		**9**	**10** Dependent care benefits
e Employee's first name and initial Last name Suff.		**11** Nonqualified plans	**12a** See instructions for box 12 D 1,117.35
John R. Doe 213 Pearl Street Anytown, CO 01234		**13** Statutory employee ☐ Retirement plan ☒ Third-party sick pay ☐	**12b**
		14 Other	**12c**
			12d
f Employee's address and ZIP code			

15 State Employer's state ID number CO 0123-4567	**16** State wages, tips, etc. 36,127.65	**17** State income tax	**18** Local wages, tips, etc. 37,245.00	**19** Local income tax 372.45	**20** Locality name

Form **W-2** Wage and Tax Statement **2010**

Department of the Treasury—Internal Revenue Service

Safe, accurate FAST! Use **e-file**

Copy C—For EMPLOYEE'S RECORDS (See *Notice to Employee* on the back of Copy B.)

1. Box 3 on the W-2 shows the taxable income for Social Security. Box 4 should show the amount of Social Security tax withheld. What amount should be in box 4?

2. Box 5 on the W-2 shows the taxable income for Medicare. Box 6 should show the amount of Medicare tax withheld. What amount should be in box 6?

3. Box 16 on the W-2 shows the taxable income for Colorado. Box 17 should show the amount of state income tax withheld. Colorado has a flat income tax of 4.63%. What amount should be in box 17?

4. John Doe contributes 3% of his salary to a 401(k) retirement plan. This amount is shown in box 12a. How much is John Doe's salary in 2010?

5. What percent of John Doe's salary should be withheld to pay federal, Social Security, Medicare, state, and local taxes?

6. In addition to John Doe's salary, Company A pays a 3% matching contribution to a 401(k) retirement plan and $560 a month for John Doe's health insurance.

    a. What is the total compensation package Company A is paying for John Doe? (*Note:* Worker's Compensation Insurance and Unemployment Insurance are not included for simplicity.)

    b. How much does John Doe receive each year as "take-home pay"?

Social Security Benefit Formula **In Exercises 7–10, use the information below.** *(See Examples 3 and 4.)*

Social Security benefits are based upon lifetime earnings and retirement age. The formula shown is applied to a worker's average indexed monthly earnings (AIME) to arrive at the basic benefit, or primary insurance amount (PIA). This is how much the worker will receive each month after reaching full retirement age. Multiply this amount by 75% to find the estimated monthly retirement benefit at age 62.

2010 Social Security Benefit Formula
1. Multiply the first $761 of the AIME by 90%. $_____
2. Multiply the AIME over $761 and less than or equal to $4586 by 32%. $_____
3. Multiply the AIME over $4586 by 15%. $_____
4. Add 1, 2, and 3. Round down to the next lowest dollar. This is your estimated monthly retirement benefit. $_____

7. A worker's AIME is $3800. Estimate the worker's PIA at full retirement age.

8. A worker's AIME is $6200. Estimate the worker's PIA at full retirement age.

9. A worker's AIME is $4500. Estimate the worker's PIA at age 62.

10. What do you think your AIME will be when you retire? Estimate how much you will receive each month from Social Security at full retirement age. (Neglect inflation.)

Retirement **In Exercises 11 and 12, use the information below taken from a worker's annual Social Security Statement. Assume a 3% cost-of-living increase each year.** *(See Example 4.)*

Your Estimated Benefits

*Retirement You have earned enough credits to qualify for benefits. At your current earnings rate, if you continue working until...

    your full retirement age (66 years), your payment would be about . . . . $ 2,029 a month

    age 70, your payment would be about . . . . $ 2,678 a month

If you stop working and start receiving benefits at...

    age 62, your payment would be about . . . . $ 1,530 a month

*Disability You have earned enough credits to qualify for benefits. If you become disabled right now...

    Your payment would be about . . . . $ 2,067 a month

*Family If you get retirement or disability benefits, your spouse and children also may qualify for benefits.

*Survivors You have earned enough credits for your family to receive survivors benefits. If you die this year, certain members of your family **may** qualify for the following benefits:

    Your child . . . . $ 1,550 a month

    Your spouse who is caring for your child . . . . $ 1,550 a month

    Your spouse who reaches full retirement age . . . . $ 2,067 a month

    Total family benefits cannot be more than . . . . $ 3,617 a month

Your spouse or minor child may be eligible for a special one-time death benefit of $255.

Medicare You have earned enough credits to qualify for Medicare at age 65. Even if you do not retire at age 65, be sure to contact Social Security three months before your 65th birthday to enroll in Medicare.

11. How long would the worker have to live to make waiting for full benefits more economical than retiring at age 62?

12. How long would the worker have to live to make waiting for the benefits at age 70 more economical than retiring at age 62?

Dependency Ratios In Exercises 13–18, use the information and bar graph below
(*See Examples 5 and 6.*)

The economic dependency ratio is also called the total dependency ratio which consists of old-age dependency and youth dependency, as shown in the bar graph. When President Franklin D. Roosevelt signed the Social Security Act in 1935, the United States had a total dependency ratio of 74. Of this, 63 was attributed to youth dependency and 11 to old-age dependency.

13. Why is the total dependency ratio projected to increase from 66 in 2010 to 82 in 2050?

14. Use percent of increase to describe the change in the old-age dependency ratio from 2010 to 2030.

15. What does the bar graph imply about the working-age population from 2010 to 2050?

16. What effects do a rising economic dependency ratio have on a government? Explain your reasoning.

17. The total dependency ratio in 1965 was 95, with a youth dependency of 77 and an old-age dependency of 18. Would the financial burden on the Social Security system be greater in 1965 or 2050? Explain your reasoning.

18. Suppose you are a legislator. How would you propose to fix the imminent financial challenges of the Social Security system?

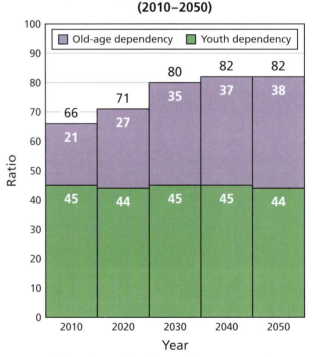

Dependency Ratios for the United States (2010–2050)

Note: Total dependency = ((Population under age 20 + Population aged 65 years and over)/(Population aged 20 to 64 years)) ∗ 100

Old-age dependency = (Population aged 65 years and over/Population aged 20 to 64 years) ∗ 100

Youth dependency = (Population under age 20/Population aged 20 to 64 years) ∗ 100

▶ Extending Concepts

Average Indexed Monthly Earnings In Exercises 19–22, use the information below and the table.

A worker's annual earnings are adjusted, or "indexed," to express the earnings in terms of today's wage levels. The 35 greatest indexed earnings are then averaged to get the average indexed monthly earnings (AIME).

After working for 35 years, a worker retired in 2009 at the age of 62. The table shows the worker's information.

19. Multiply the amounts in column B by the index factors in column C, and enter the results in column D.

20. Find the sum of the amounts in column D. Then divide the sum by 420 (the number of months in 35 years). Round down to the next lowest dollar. This is the worker's AIME.

21. Use the Social Security benefit formula on page 241 to estimate the worker's Social Security monthly retirement benefit at age 62.

22. Explain why the Social Security benefit formula is designed to pay a higher percent for lower AIMEs.

23. **Dependency Ratio** What factors may prevent the economic dependency ratio from being 100% accurate? Explain your reasoning.

24. **Viability of Social Security** Do you think the Social Security system will exist when you retire? Use the concepts in this section to defend your answer.

	A	B	C	D
Year	Maximum Earnings	Actual Earnings (not more than column A)	Index Factor	Indexed Earnings
1975	$14,100	$4,000	4.79	
1976	$15,300	$4,145	4.48	
1977	$16,500	$4,274	4.23	
1978	$17,700	$7,280	3.92	
1979	$22,900	$7,516	3.60	
1980	$25,900	$7,742	3.30	
1981	$29,700	$8,002	3.00	
1982	$32,400	$12,205	2.84	
1983	$35,700	$12,683	2.71	
1984	$37,800	$13,171	2.56	
1985	$39,600	$13,686	2.46	
1986	$42,000	$14,107	2.39	
1987	$43,800	$14,532	2.24	
1988	$45,000	$31,200	2.14	
1989	$48,000	$32,218	2.06	
1990	$51,300	$33,344	1.97	
1991	$53,400	$34,450	1.90	
1992	$55,500	$38,545	1.80	
1993	$57,600	$39,883	1.79	
1994	$60,600	$41,336	1.74	
1995	$61,200	$42,801	1.67	
1996	$62,700	$44,219	1.60	
1997	$65,400	$45,771	1.51	
1998	$68,400	$68,400	1.43	
1999	$72,600	$72,600	1.36	
2000	$76,200	$76,200	1.29	
2001	$80,400	$80,400	1.26	
2002	$84,900	$84,900	1.24	
2003	$87,000	$87,000	1.21	
2004	$87,900	$87,900	1.16	
2005	$90,000	$90,000	1.12	
2006	$94,200	$94,200	1.07	
2007	$97,500	$97,500	1.02	
2008	$102,000	$102,000	1.00	
2009	$106,800	$106,800	1.00	

5.3–5.4   Quiz

Pennsylvania **In Exercises 1–4, use the information and table below.**

You own a $90,000 home in Pennsylvania. The assessment level is 20%. The annual property tax rates on your home are shown.

Taxes	Tax Rate
County	18.1 mills
Township	2.0 mills
School	48.0 mills

1. What is the combined tax rate?

2. What is the property tax for your home?

3. What is the effective property tax rate?

4. County and township taxes are due by June 30. When paid by April 30, there is a 3% discount on county taxes and a 2% discount on township taxes. How much is due when you pay by April 30?

Payroll Taxes **In Exercises 5–7, use the pay stub of a college student residing in Pennsylvania.**

Earnings	Hours	Rate	Current
	84.50	10.00	845.00

Taxes	
Local (flat tax of 1%)	?
Federal	52.00
Social Security	?
Medicare	?
State (flat tax of 3.07%)	?
PA Unemployment	0.67

5. Calculate the local and state taxes.

6. Calculate the Social Security and Medicare taxes.

7. What percent of the earnings go to taxes?

8. **Retirement** A worker's annual Social Security Statement indicates that he would receive $1072 per month by retiring at age 62, or $1938 per month by retiring at age 70. How long would he have to live to make waiting until age 70 more economical? Assume a 3% cost-of-living increase each year.

Chapter 5   Summary

Section Objectives	How does it apply to you?

Section 1

Calculate a flat income tax.	You can determine what income is taxable and calculate income taxes in states with a flat tax. *(See Examples 1 and 2.)*
Identify types of taxes.	You can identify taxes based on how the tax rate changes as income increases. *(See Examples 3 and 4.)*
Analyze an indirect tax.	You can analyze taxes to determine who pays them and who benefits from them. *(See Examples 5 and 6.)*

Section 2

Calculate a graduated income tax.	You can calculate graduated income taxes to compare the taxes paid for different salaries in different states. *(See Examples 1 and 2.)*
Analyze a graduated income tax system.	Analyzing tax systems helps you understand the degree to which income level affects taxation. *(See Example 3.)*
Compare a graduated income tax with a flat income tax.	You can analyze alternative tax systems and determine how they could address the U.S. deficit. *(See Examples 5 and 6.)*

Section 3

Calculate a property tax.	You can compare property taxes in different cities. *(See Example 1.)*
Analyze assessments and tax credits.	You can use the assessed value of a property to appeal assessments and calculate property tax credits. *(See Examples 3 and 4.)*
Analyze exemptions for property tax.	You can calculate property tax savings from exemptions. *(See Example 6.)*

Section 4

Calculate Social Security & Medicare taxes.	You can find how much you and your employer pay for various taxes and benefits. *(See Examples 1 and 2.)*
Evaluate the benefits of Social Security.	You can analyze Social Security benefits to help make retirement decisions. *(See Examples 3 and 4.)*
Analyze the viability of Social Security.	You can analyze Social Security to understand why it may not be available to you when you retire. *(See Examples 5 and 6.)*

Chapter 5   Review Exercises

Section 5.1

Massachusetts State Income Tax **Massachusetts has a flat income tax with a rate of 5.3%. In Exercises 1 and 2, determine how much state income tax the person owes.**

1. A person who lives in Massachusetts has a taxable income of $45,000.

2. A person who lives in Massachusetts has a taxable income of $1,000,000.

Massachusetts State and Local Taxes **In Exercises 3–6, use the display.**

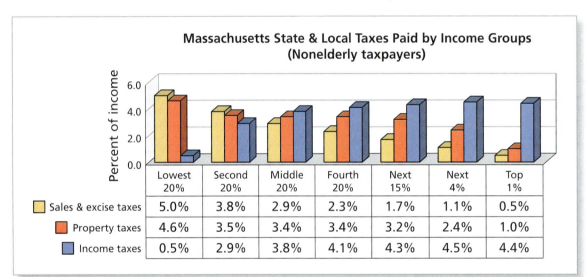

Massachusetts State & Local Taxes Paid by Income Groups
(Nonelderly taxpayers)

	Lowest 20%	Second 20%	Middle 20%	Fourth 20%	Next 15%	Next 4%	Top 1%
Sales & excise taxes	5.0%	3.8%	2.9%	2.3%	1.7%	1.1%	0.5%
Property taxes	4.6%	3.5%	3.4%	3.4%	3.2%	2.4%	1.0%
Income taxes	0.5%	2.9%	3.8%	4.1%	4.3%	4.5%	4.4%

3. The income for a family in the middle 20% is $53,000. How much, in dollars, does this family pay for each type of tax?

4. The income for a family in the top 1% is $2,600,000. How much, in dollars, does this family pay for each type of tax?

5. Determine whether the three types of taxes are regressive, flat, or progressive in Massachusetts.

6. Are taxes, as a whole, regressive, flat, or progressive in Massachusetts? Explain your reasoning.

7. **Alcohol Sales Tax** Massachusetts had a 6.25% sales tax on alcohol from August 2009 to December 2010. Part of the revenue from the tax was used to fund substance abuse programs. This tax was abolished in December 2010. Do you agree with such a tax? Why or why not?

8. **Effects of a Tax** Suppose the revenue from a sales tax on cigarettes is used to offset the costs of a new health insurance law. Discuss ways in which the tax might help the people who are paying the tax.

Section 5.2

 Arizona State Income Tax In Exercises 9–12, use the table.

9. Find the state income tax and the effective tax rate for a taxable income of $18,000.

10. Find the state income tax and the effective tax rate for a taxable income of $45,854.

11. Find the state income tax and the effective tax rate for a taxable income of $75,489.

12. Find the state income tax and the effective tax rate for a taxable income of $214,500.

2010 Arizona State Income Tax for Single Taxpayers	
Taxable Income	**Marginal Tax Rate**
$0–$10,000	2.59%
$10,001–$25,000	2.88%
$25,001–$50,000	3.36%
$50,001–$150,000	4.24%
$150,001+	4.54%

Arizona State and Local Tax Revenue In Exercises 13–16, use the bar graph.

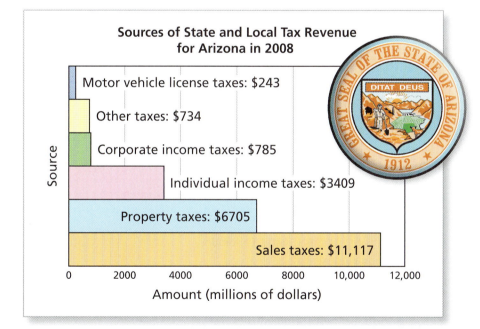

Sources of State and Local Tax Revenue for Arizona in 2008

Motor vehicle license taxes: $243
Other taxes: $734
Corporate income taxes: $785
Individual income taxes: $3409
Property taxes: $6705
Sales taxes: $11,117

Source

Amount (millions of dollars)

13. What percent of Arizona's state and local tax revenue comes from individual income taxes?

14. What percent of Arizona's state and local tax revenue comes from property taxes?

15. Compare the percents of state and local tax revenue that come from sales taxes in Arizona and South Carolina (see Exercise 10 on page 219).

16. In 2008, the total personal income in Arizona was about $224 billion. Discuss using a flat tax to collect the amount of individual income taxes shown in the graph.

Section 5.3

Property Taxes In Exercises 17–20, use the map.

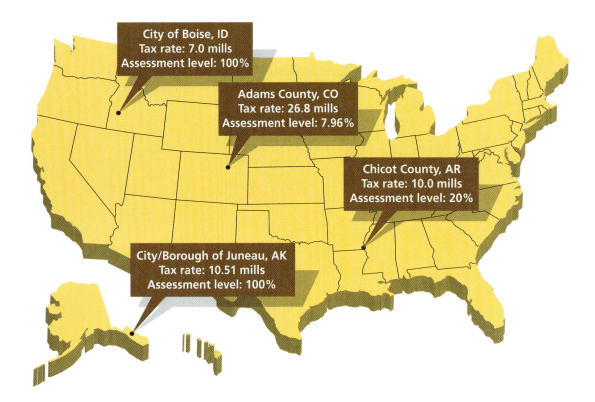

City of Boise, ID
Tax rate: 7.0 mills
Assessment level: 100%

Adams County, CO
Tax rate: 26.8 mills
Assessment level: 7.96%

Chicot County, AR
Tax rate: 10.0 mills
Assessment level: 20%

City/Borough of Juneau, AK
Tax rate: 10.51 mills
Assessment level: 100%

17. Find the county taxes for a $128,000 home in Chicot County, Arkansas.

18. Find the city taxes for a $240,000 home in Juneau, Alaska.

19. Find the effective property tax rate of city taxes in Boise, Idaho.

20. Find the effective property tax rates of county taxes for Adams County, Colorado, and Chicot County, Arkansas. Which county has greater taxes?

21. **Credit** Homeowners in Arkansas can receive a homestead tax credit up to $350. A home with an assessed value of $24,000 is subject to a combined tax rate of 51.8 mills. How much would the homeowner pay in property taxes assuming he or she receives the maximum homestead tax credit?

22. **Exemption** The assessed value of a home in Idaho is $160,000, and the combined tax rate is 10.5 mills. The homeowner receives an exemption of 50% of the assessed value. How much does the homeowner pay in property taxes?

Section 5.4

Social Security **In Exercises 23–30, use the information below.**

Joe and Carolyn have been married for 45 years. In 2010, Joe retired at his full retirement age, 66. During his last year of employment, Joe had an income of $58,000. The Social Security Administration calculates that Joe's average indexed monthly earnings (AIME) is $3226.

Retirement Benefits

A person qualifies for Social Security retirement benefits by earning credits when he or she works and pays Social Security tax. The credits are based on the amount of a worker's earnings. In 2011, a worker received 1 credit for each $1120 of earnings, up to the maximum of 4 credits per year. A worker born in 1929 or later needs 40 credits (10 years of work) to be eligible for retirement benefits.

---

23. During his last year of employment, how much should Joe have paid in Social Security tax?

24. During his last year of employment, how much should Joe have paid in Medicare tax?

25. Use the Social Security benefit formula on page 241 to estimate Joe's primary insurance amount (PIA).

26. Suppose Joe retired at age 62. How long would Joe have to live to make waiting for full benefits more economical? Assume a 3% cost-of-living increase each year.

27. Carolyn has always been a homemaker and has never paid Social Security tax. Will she qualify for Social Security retirement benefits? Explain.

28. Use the Internet to research qualifications for survivors benefits through Social Security. Suppose Joe dies before Carolyn. Will Carolyn qualify for survivors benefits? Explain.

29. How might the economic dependency ratio affect Joe's retirement benefits in the future?

30. How might the economic dependency ratio affect the qualifications for Social Security retirement benefits in the future?

6 The Mathematics of Borrowing & Saving

6.1 Introduction to Lending

▶ Read promissory notes and find due dates.

▶ Find the cost of credit for a loan.

▶ Find the annual percentage rate for a loan.

6.2 Buying Now, Paying Later

▶ Create an amortization table.

▶ Analyze the cost of buying on credit.

▶ Analyze credit in the United States.

6.3 Home Mortgages

▶ Compare rates and terms for a home mortgage.

▶ Analyze the effect of making principal payments.

▶ Compare the costs of buying and renting.

6.4 Savings & Retirement Plans

▶ Find the balance in a savings account.

▶ Find the balance in an increasing annuity.

▶ Analyze a decreasing annuity.

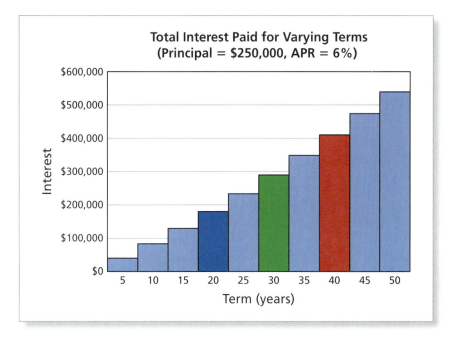

**Total Interest Paid for Varying Terms
(Principal = $250,000, APR = 6%)**

Example 2 on page 275 compares the amount of interest you pay on a home mortgage for terms of 20, 30, and 40 years. Does the interest double when you double the term?

6.1 Introduction to Lending

▶ Read promissory notes and find due dates.

▶ Find the cost of credit for a loan.

▶ Find the annual percentage rate for a loan.

Promissory Notes

At the time you obtain a loan, you are expected to sign a **promissory note** in which you promise to repay the loan. The **term** of a loan is the period of time in which you repay the loan. The amount actually received in a loan is called the **loan proceeds,** and the total amount the borrower must repay the lender is called the total payment, or **total amount due.**

EXAMPLE 1 **Reading a Promissory Note**

Find (a) the term of the loan and (b) the total amount due.

No. __2786__   Boise, Idaho, May 10 , 20 11 $ __1240.00__

_____Ninety days_____ *after date, the undersigned (jointly and severally if more than one) promise(s) to pay to the order of* _____The Lending Bank_____

_____One thousand two hundred forty and no/100_____ *Dollars*

Payable at _____The Lending Bank, Boise, Idaho_____

Each and every party to this instrument, either as maker, endorser, surety, or otherwise, hereby waives demand, notice, protest, and all other demands and notices and assents to any extension of the time of payment or any other indulgence, to any substitution, exchange, or release of collateral and/or to the release of any other party.

Address   __174 Maple Avenue__   *Signed*   *Jane Doe*

__Boise, Idaho__   _____

SOLUTION

a. The term is 90 days.   **b.** The total amount due is $1240.

Total amount due

Term

No. __2786__   Boise, Idaho, May 10 , 20 11 $ __1240.00__

_____Ninety days_____ *after date, the undersigned (jointly and severally if more than one) promise(s) to pay to the order of* _____The Lending Bank_____

_____One thousand two hundred forty and no/100_____ *Dollars*

Payable at   The Lending Bank, Boise, Idaho

✓ **Checkpoint**   Help at *Math*.and*YOU*.com

Promissory notes are negotiable. What does this mean?

Repayment of a loan can occur in many ways. The most common way to repay a loan is with monthly payments. This type of loan is called an **installment loan.** A detailed discussion of installment loans occurs in Section 6.2.

Another way to repay a loan is with a single payment at the end of the term of the loan. Common terms for such loans are 30 days, 60 days, 90 days, 6 months, 1 year, and 18 months.

EXAMPLE 2   Finding the Due Date and Total Amount Due

You are moving from Dallas to Pittsburgh. You plan to sell your home in Dallas and use the money as a down payment on a home in Pittsburgh. Your moving date occurs before your Dallas home sells, so you decide to obtain a 90-day note from a Pittsburgh bank for $10,000.00. The loan takes place on June 5, and the costs include $244.27 for interest and $84.00 in other charges. (a) When is the note due? (b) How much is due at the end of 90 days?

SOLUTION

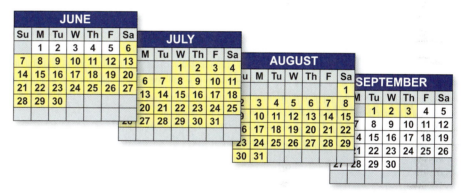

a. The loan occurs on June 5 and is due 90 days from that date.

25 days	Days remaining in June
31 days	Days in July
31 days	Days in August
+ 3 days	Days in September
90 days	

So the note is due on September 3.

b. The total amount due is

10,000.00	Loan proceeds
244.27	Interest
+ 84.00	Other charges
$10,328.27.	Total amount due

✔ **Checkpoint**

Help at *Math*.and**YOU**.com

On March 17, you obtain a 120-day note for $5000. The costs include $250.00 for interest and $77.95 in other charges. When is the note due? How much is due at the end of 120 days?

Cost of Credit

The ability of a person to obtain a loan is called **credit.** When you obtain a loan, you are borrowing someone else's money, and normally you must pay to do so. The **cost of credit** for a loan is the difference between the total amount due and the loan proceeds.

> ### Cost of Credit
>
> Cost of credit = total amount due − loan proceeds

EXAMPLE 3   Finding the Cost of Credit

You purchase a refrigerator for $950.89 plus 6% sales tax. The appliance store offers an installment loan that allows you to pay for the refrigerator by making 12 equal monthly payments of $95.75. What is the cost of credit for this loan?

SOLUTION

Because the loan has 12 equal monthly payments of $95.75, the total amount due is

Total amount due = 12(95.75)   12 monthly payments

= $1149.00.

The loan proceeds are found by adding the cost of the refrigerator to the 6% sales tax.

Loan proceeds = | Cost of refrigerator | + | Sales tax |

= 950.89 + (0.06)(950.89)   6% sales tax

= 950.89 + 57.05

= $1007.94

Finally, the cost of credit for this loan is

Cost of credit = total amount due − loan proceeds

= 1149.00 − 1007.94

= $141.06.

So, by borrowing the money, you pay $141.06 more than if you had paid cash for the refrigerator.

✓ Checkpoint   Help at *Math*.and**Y☺U**.com

What is the cost of credit for each loan?

a. You buy a computer for $1599.99 plus 7% sales tax. The electronics store offers an installment loan that allows you to pay for the computer by making 24 equal monthly payments of $74.20.

b. You borrow $250,000 with a home mortgage. You pay $1342.05 toward the mortgage each month for 30 years.

The federal government requires lending institutions to disclose several pertinent facts about a loan. The disclosure helps the borrower understand the contractual obligations of the loan. These facts must be disclosed *in writing*, *before* the borrower is asked to sign the promissory note. This requirement is specified in the Truth in Lending Act.

The Truth in Lending Act requires lending institutions to classify the cost of credit of every loan into two categories.

1. **Finance charges** such as interest, carrying charges, and service charges

2. Other charges such as insurance premiums, investigation of credit fees, and filing fees

Although the *other charges* may be paid at the time of the loan, they are often incorporated into the loan. Together with the loan proceeds, they make up the **amount financed,** or the **principal.**

EXAMPLE 4   Applying the Truth in Lending Act

You buy a washer and dryer for $1395.00 plus $83.70 in sales tax. You pay $100.70 down and finance the remainder. There is an insurance charge of $60.95 and a finance charge of $122.97. Complete the Truth in Lending disclosure for this loan.

Loan Proceeds	$ _____	Annual Percentage Rate = __8__ %
Other Charges	+ $ _____	
Amount Financed	$ _____	Payable in __24__ payments
Finance Charge	+ $ _____	of $ __65.08__ each.
Total Amount Due	$ _____	

SOLUTION

Insurance   1395.00 + 83.70 − 100.70

Loan Proceeds	$ _1378.00_	Annual Percentage Rate = __8__ %
Other Charges	+ $ _60.95_	
Amount Financed	$ _1438.95_	Payable in __24__ payments
Finance Charge	+ $ _122.97_	of $ _65.08_ each.
Total Amount Due	$ _1561.92_	

✓ Checkpoint

Help at *Math*.and**YOU**.com

You borrow $2000. There is an insurance charge of $89.73. The annual percentage rate is 10%. You make monthly payments of $67.43 for 36 months. Complete a Truth in Lending disclosure for this loan.

Annual Percentage Rate

Interest that is calculated only on the principal is **simple interest.** The rate at which this interest is calculated is the **annual percentage rate** (APR). Interest that is calculated on both the principal *and* accumulated interest is called compound interest, which is discussed in Section 6.2.

> **Study Tip**
>
> When a term is given in days, you can convert it to years by dividing by 365. For example,
>
> $60 \text{ days} = \frac{60}{365} \text{ year.}$

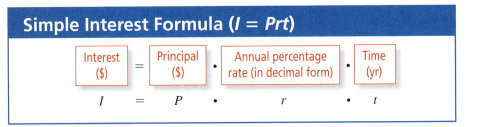

Simple Interest Formula ($I = Prt$)

Interest ($)	=	Principal ($)	·	Annual percentage rate (in decimal form)	·	Time (yr)
I	=	P	·	r	·	t

EXAMPLE 5   Finding Simple Interest

Complete the table showing the interest on $1000 for various terms and rates.

t \ r	4%	8%	12%	16%	20%	24%
60 days						
120 days						
180 days						
240 days						
300 days						
1 year						

SOLUTION

Use a spreadsheet for this type of repetitive calculation.

DATA

	A	B	C	D	E	F	G
1				Rate			
2	Days	4%	8%	12%	16%	20%	24%
3	60	$6.58	$13.15	$19.73	$26.30	$32.88	$39.45
4	120	$13.15	$26.30	$39.45	$52.60	$65.75	$78.90
5	180	$19.73	$39.45	$59.18	$78.90	$98.63	$118.36
6	240	$26.30	$52.60	$78.90	$105.21	$131.51	$157.81
7	300	$32.88	$65.75	$98.63	$131.51	$164.38	$197.26
8	365	$40.00	$80.00	$120.00	$160.00	$200.00	$240.00

$I = Prt$
$= 1000(0.04)\left(\frac{60}{365}\right)$
$= \$6.58$

 **Checkpoint**

Help at *Math*.andY**⊙**U.com

Complete another row in the table for a term of 90 days.

When using the simple interest formula, $I = Prt$, the value of any one of the variables can be unknown. As long as you know the values of three of the variables, you can calculate the value of the fourth variable.

Simple Interest: Related Formulas

Find interest.	Find principal.	Find rate.	Find time.
$I = Prt$	$P = \dfrac{I}{rt}$	$r = \dfrac{I}{Pt}$	$t = \dfrac{I}{Pr}$

Math.and**YOU**.com

You can access a simple interest calculator at *Math.andYou.com*.

EXAMPLE 6   **Finding the Annual Percentage Rate**

Using an ATM, you take a cash advance of $500. The service charge is $3.50. After 45 days, you repay the advance. When you get your credit card statement, you notice that the interest for the "loan" is $17.26. (a) For the interest alone, what is the annual percentage rate? (b) When you add the service charge to the interest, what is the annual percentage rate?

SOLUTION

a. Because you are finding the rate, use the third formula in the table above.

$$r = \frac{I}{Pt}$$

$$= \frac{17.26}{500\left(\frac{45}{365}\right)} \qquad I = \$17.26, \quad P = \$500, \quad t = \frac{45}{365}$$

$$\approx 0.28$$

The annual percentage rate is about 28%.

b. Add the service charge to the stated interest to get $I = \$20.76$.

$$r = \frac{I}{Pt}$$

$$= \frac{20.76}{500\left(\frac{45}{365}\right)} \qquad I = \$20.76, \quad P = \$500, \quad t = \frac{45}{365}$$

$$\approx 0.337$$

The annual percentage rate is about 33.7%.

Study Tip

APR is also called *nominal* APR. The APR found in Example 6(b) is called the *effective* APR, which includes other charges incurred from the loan.

✓ **Checkpoint**   Help at *Math*.and**YOU**.com

You borrow $100 from a friend. You repay the loan in 3 weeks and agree to pay $10 for interest. What is the annual percentage rate for this loan?

6.1 Exercises

Promissory Notes In Exercises 1 and 2, find (a) the term of the loan and (b) the total amount due. *(See Example 1.)*

1.

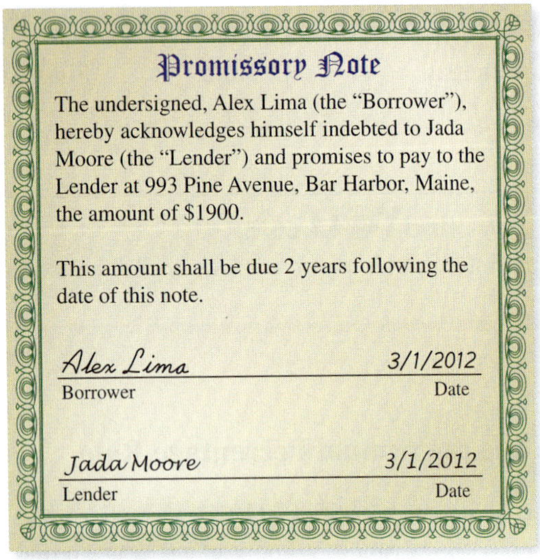

Promissory Note

The undersigned, Alex Lima (the "Borrower"), hereby acknowledges himself indebted to Jada Moore (the "Lender") and promises to pay to the Lender at 993 Pine Avenue, Bar Harbor, Maine, the amount of $1900.

This amount shall be due 2 years following the date of this note.

Alex Lima   3/1/2012
Borrower   Date

Jada Moore   3/1/2012
Lender   Date

2.

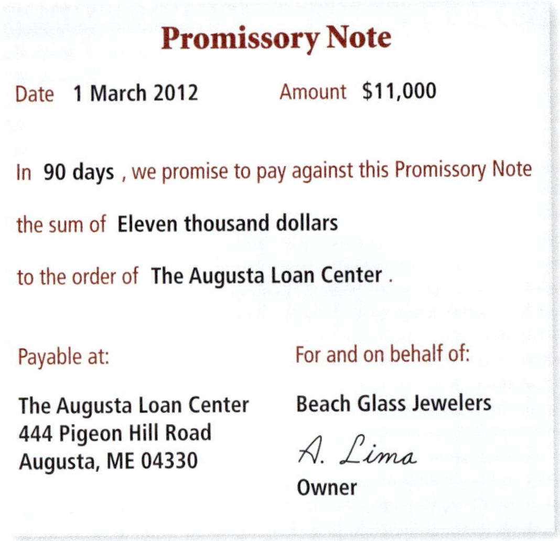

Promissory Note

Date **1 March 2012**   Amount **$11,000**

In **90 days**, we promise to pay against this Promissory Note

the sum of **Eleven thousand dollars**

to the order of **The Augusta Loan Center**.

Payable at: For and on behalf of:

The Augusta Loan Center   **Beach Glass Jewelers**
444 Pigeon Hill Road   *A. Lima*
Augusta, ME 04330   Owner

Due Dates In Exercises 3–5, use the 2012 calendar. *(See Example 2.)*

January

S	M	T	W	T	F	S
1	2	3	4	5	6	7
8	9	10	11	12	13	14
15	16	17	18	19	20	21
22	23	24	25	26	27	28
29	30	31				

February

S	M	T	W	T	F	S
			1	2	3	4
5	6	7	8	9	10	11
12	13	14	15	16	17	18
19	20	21	22	23	24	25
26	27	28	29			

March

S	M	T	W	T	F	S
				1	2	3
4	5	6	7	8	9	10
11	12	13	14	15	16	17
18	19	20	21	22	23	24
25	26	27	28	29	30	31

April

S	M	T	W	T	F	S
1	2	3	4	5	6	7
8	9	10	11	12	13	14
15	16	17	18	19	20	21
22	23	24	25	26	27	28
29	30					

May

S	M	T	W	T	F	S
		1	2	3	4	5
6	7	8	9	10	11	12
13	14	15	16	17	18	19
20	21	22	23	24	25	26
27	28	29	30	31		

June

S	M	T	W	T	F	S
					1	2
3	4	5	6	7	8	9
10	11	12	13	14	15	16
17	18	19	20	21	22	23
24	25	26	27	28	29	30

3. When is the note in Exercise 2 due?

4. On April 19, you obtain a 60-day note for $5000. The costs include $81.37 for interest and a $50.99 service charge. When is the note due? How much is due at the end of 60 days?

5. You obtain a 120-day note for $20,000 to use as a down payment on a home. The loan takes place on January 19. The costs include $789.04 for interest and $134 in other charges. When is the note due? How much is due at the end of 120 days?

6. Leap Year On January 19, 2013, you obtain a 120-day note. Is the due date on the same day of the year as in Exercise 5? Explain. *(See Example 2.)*

Lobstering A fisherman makes purchases for a lobster trapping business. In Exercises 7–9, find the cost of credit for the loan. *(See Example 3.)*

7. The fisherman purchases lobster buoys and paint online for $500. The website offers an installment loan that allows him to pay by making 3 equal monthly payments of $172.25.

8. The fisherman purchases lobster traps online for $1440. The website offers an installment loan that allows him to pay by making 12 equal monthly payments of $127.94.

9. The fisherman purchases a lobster boat for $250,000 plus 5% sales tax. The seller offers an installment loan that allows him to pay by making 60 equal monthly payments of $4988.75.

10. **Cost of Credit** Can you find the cost of credit for the loan in Exercise 1? Exercise 2? Explain. *(See Example 3.)*

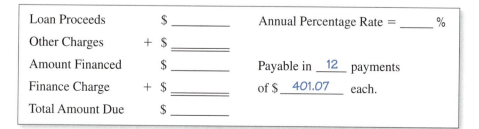

Infrastructure Improvements In Exercises 11 and 12, complete the Truth in Lending disclosure for the loan. *(See Example 4.)*

11. The fisherman pays a contractor to repair his dock. The charges are $1200 for lumber, $60 for sales tax, and $750 for labor. The fisherman pays $260 down and finances the remainder. There is a finance charge of $41.06.

Loan Proceeds	$ _____	Annual Percentage Rate = __8__ %
Other Charges	+ $ _____	
Amount Financed	$ _____	Payable in __6__ payments
Finance Charge	+ $ _____	of $ __298.51__ each.
Total Amount Due	$ _____	

12. The fisherman borrows $4500 to put a new roof on his shop. The other charges are $160. The annual percentage rate is 6%.

Loan Proceeds	$ _____	Annual Percentage Rate = _____ %
Other Charges	+ $ _____	
Amount Financed	$ _____	Payable in __12__ payments
Finance Charge	+ $ _____	of $ __401.07__ each.
Total Amount Due	$ _____	

Pawn Shop In Exercises 13–16, a pawn shop charges simple interest at an annual percentage rate of 40%. *(See Example 5.)*

13. Complete the table showing the interest for various terms and principals.

t \ P	$100	$400	$1000
30 days			
60 days			
180 days			
1 year			

14. You pawn a gold ring and receive a 120-day loan for $500. What is the interest for the loan?

15. You pawn a television and receive a 30-day loan for 40% of the $500 resale value. What is the interest for the loan?

16. You pawn a motorcycle and receive a 180-day loan for 60% of the $2000 resale value. What is the total amount due?

Pawn shops buy and sell merchandise. They also offer loans, using personal belongings as collateral. For example, a pawn shop may offer an individual a loan of $600 in exchange for an antique with a resale value of $1000. This is referred to as *pawning* an item. The individual recovers the antique when the loan is paid in full.

Service Charges A pawn shop includes a service charge according to the table. In Exercises 17–19, find (a) the annual percentage rate for only the interest and (b) the annual percentage rate including the service charge. *(See Example 6.)*

Principal	Under $100	$100.01–$500	Over $500
Service charge	$4	$7	$10

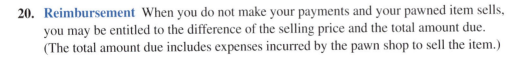

17. You pay $5.59 in interest on a 60-day loan for $85.

18. You pay $336 in interest on a 1-year loan for $700.

19. You pay $10.48 in interest on a 45-day loan for $340.

20. Reimbursement When you do not make your payments and your pawned item sells, you may be entitled to the difference of the selling price and the total amount due. (The total amount due includes expenses incurred by the pawn shop to sell the item.)

a. You do not repay the loan in Exercise 16, and the pawn shop incurs $60 in expenses to sell the motorcycle. Find the sum of the total amount due and the expenses incurred to sell the motorcycle.

b. The motorcycle sells for $1700. How much money should you receive?

▶ Extending Concepts

Ordinary Interest **In Exercises 21–24, use the information below.**

You borrow $4200 from a relative on January 5, 2012, to buy solar panels for your roof and windows with insulated glazing. The annual percentage rate is 8.9%. You agree to repay the loan on May 25, 2012.

21. Simple interest based on a 360-day year in which each month has 30 days is called *ordinary simple interest*. When using ordinary simple interest, you can use the formula below to find the number of days in a term from a month *M*, day *D*, and year *Y* to a later month *m*, day *d*, and year *y*. (*Note:* January = 1, February = 2, etc.)

$$\text{Number of days} = 360(y - Y) + 30(m - M) + (d - D)$$

   a. Use the formula to find the number of days in the term.

   b. Find the total amount due using ordinary simple interest.

22. What is the actual number of days in the term? Find the total amount due using *exact* simple interest, as on page 256.

Low-emissivity (Low-E) glazing on windows helps control heat transfer. These windows may cost 10%–15% more than traditional windows, but can reduce energy loss by 30%–50%.

DATA **23.** Which type of interest costs you more money? Will this always be true? Explain.

24. Do you think it is appropriate to use ordinary simple interest as an approximation of exact simple interest? Explain.

DATA **25.** **Banker's Rule** The *Banker's Rule* is another type of simple interest that is similar to ordinary simple interest. It is based on a 360-day year, but you use the actual number of days in the term when calculating interest. Does this benefit the lender or the borrower? Explain.

26. **Loan Options** You have 3 loan options for borrowing $2500. You will repay the simple interest loan in 180 days.

	Insurance premium	Annual percentage rate	Service charge
Loan A	2% of loan proceeds	25%	$39.50
Loan B	3% of loan proceeds	26%	$0
Loan C	$0	25%	$65.00

   a. Find the interest for each loan.

   b. Find the annual percentage rate of each loan, including the service charge.

   c. Find the total amount due for each loan.

   d. Which loan would you choose? Explain.

6.2 Buying Now, Paying Later

▶ Create an amortization table.

▶ Analyze the cost of buying on credit.

▶ Analyze credit in the United States.

Compound Interest Payments

Interest that is calculated on both the principal *and* accumulated interest is **compound interest.** Compound interest is applied to the vast majority of loans.

Monthly Payment for Installment Loans

The monthly payment M for an installment loan with a principal of P taken out for n months at an annual percentage rate of r (in decimal form) is

$$M = P\left(\frac{r/12}{1 - \left(\frac{1}{1 + (r/12)}\right)^n}\right).$$

Math.and**YOU**.com

You can access a monthly payment calculator at *Math.andYou.com*.

Study Tip

To *amortize* means to decrease an amount gradually or in installments. The schedule of payments for an installment loan is called an *amortization table*.

EXAMPLE 1 **Creating an Amortization Table**

You borrow $1200 for 6 months. The annual percentage rate is 6%.

a. What is the monthly payment?

b. Create an amortization table showing how the balance of the loan decreases.

SOLUTION

0.06/12

a. $M = 1200\left[\dfrac{0.005}{1 - \left(\frac{1}{1.005}\right)^6}\right] = \$203.51$

b.

DATA

	A	B	C	D	E
1	**Payment Number**	**Balance before Payment**	**Monthly Interest**	**Monthly Payment**	**Balance after Payment**
2	1	$1,200.00	$6.00	$203.51	$1,002.49
3	2	$1,002.49	$5.01	$203.51	$803.98
4	3	$803.98	$4.02	$203.51	$604.49
5	4	$604.49	$3.02	$203.51	$404.00
6	5	$404.00	$2.02	$203.51	$202.50
7	6	$202.50	$1.01	$203.51	$0.00

0.005(1200)

✓ **Checkpoint** Help at *Math*.and**YOU**.com

What is the cost of credit for the loan above? Does the cost of credit double when the term doubles? Explain your reasoning.

In the initial repayment of a lengthy installment loan, most of the monthly payment goes toward interest, not principal. This is illustrated in Example 2.

EXAMPLE 2   Using an Amortization Table

You take out a $25,000 loan for a new car. The term is 5 years, and the annual percentage rate is 8%. In 30 months, is the remaining balance one-half of the original loan amount?

SOLUTION

Create an amortization table for the loan. Notice that after 30 payments (out of 60), the remaining balance is *not* one-half of the original loan amount.

	A	B	C	D	E
	Payment	Balance before	Monthly	Monthly	Balance after
1	Number	Payment	Interest	Payment	Payment
2	1	$25,000.00	$166.67	$506.91	$24,659.76
3	2	$24,659.76	$164.40	$506.91	$24,317.25
4	3	$24,317.25	$162.11	$506.91	$23,972.45
30	29	$14,564.11	$97.09	$506.91	$14,154.29
31	30	$14,154.29	$94.36	$506.91	$13,741.74
32	31	$13,741.74	$91.61	$506.91	$13,326.44
59	58	$1,500.68	$10.00	$506.91	$1,003.77
60	59	$1,003.77	$6.69	$506.91	$503.55
61	60	$503.55	$3.36	$506.91	$0.00
62					

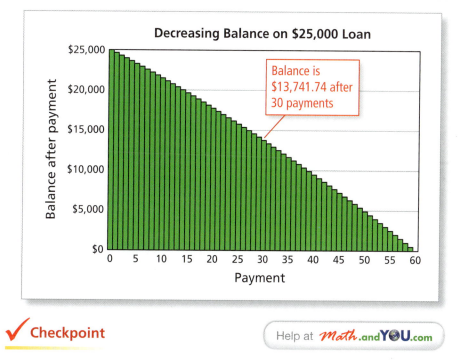

Decreasing Balance on $25,000 Loan

Balance is $13,741.74 after 30 payments

✓ Checkpoint

Help at **Math.andYOU.com**

a. You make all 60 payments on time. How much interest do you pay?

b. How much interest do you pay when the annual percentage rate is 9%? 10%?

The Cost of Buying on Credit

EXAMPLE 3 **Finding the Cost of Buying on Credit**

You use a credit card to purchase a plasma television for $2500. The credit card company sends you a monthly bill, asking for a minimum payment of 3% of the balance or $25, whichever is greater. The annual percentage rate for the unpaid balance is 28%. How long does it take to pay for the television by making only minimum payments? How much interest do you pay?

SOLUTION

The spreadsheet illustrates why you should make more than the minimum payment each month.

	A	B	C	D	E
1	Payment Number	Balance before Payment	Minimum Payment	Interest	Balance after Payment
2	1	$2,500.00	$75.00	$58.33	$2,483.33
3	2	$2,483.33	$74.50	$57.94	$2,466.78
4	3	$2,466.78	$74.00	$57.56	$2,450.33
228	227	$81.93	$25.00	$1.91	$58.84
229	228	$58.84	$25.00	$1.37	$35.21
230	229	$35.21	$25.00	$0.82	$11.03
231	230	$11.03	$11.29	$0.26	$0.00
232	**Total**		**$9,130.26**		

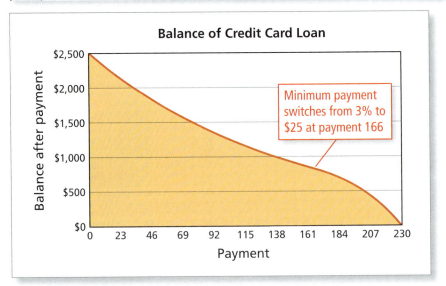

Balance of Credit Card Loan

Minimum payment switches from 3% to $25 at payment 166

This repayment plan takes 230 payments, or 19 years and 2 months. During this time, you pay 9130.26 − 2500.00 = $6630.26 in interest.

✔ **Checkpoint** Help at

Suppose you make a payment of $75 each month.

a. How long does it take to repay the loan?

b. How much interest do you pay?

When you apply for a credit card or any other type of loan, the lending institution will check your credit score. Your credit score is a number from 300 to 850. The higher your score, the better your credit rating.

The two most important things in determining your credit score are (1) your past history of making payments on time and (2) your total indebtedness.

Excellent	760–850
Good	700–759
Fair	620–699
Poor	300–619

Factors Used to Determine Credit Score

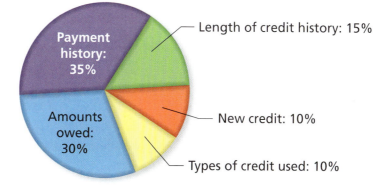

Payment history: 35%
Length of credit history: 15%
Amounts owed: 30%
New credit: 10%
Types of credit used: 10%

EXAMPLE 4   Comparing Rates

You apply for a small business loan for $30,000 for 6 years. If your credit score is 800 or greater, then the annual percentage rate is 6%. If your score is less than 800, then the rate is 8%. How much can you save in interest by having a credit score of 800 or greater?

SOLUTION

6% for 72 months:

$$M = 30{,}000\left[\frac{0.005}{1 - \left(\frac{1}{1.005}\right)^{72}}\right]$$

$$= \$497.19$$

After 72 payments, you will have paid 72(497.19) = $35,797.68, of which $5797.68 is interest.

8% for 72 months:

$$M = 30{,}000\left[\frac{0.00667}{1 - \left(\frac{1}{1.00667}\right)^{72}}\right]$$

$$= \$526$$

After 72 payments, you will have paid 72(526) = $37,872, of which $7872 is interest.

So, if your credit score is 800 or greater, you can save more than $2000 in interest on this loan.

✓ Checkpoint

Help at *Math*.and**YOU**.com

You apply for a debt consolidation loan for $40,000 for 5 years. If your credit score is 760 or greater, then the APR is 5%. If your score is less than 760, then the rate is 8%. How much can you save in interest by having a credit score of 760 or greater?

Analyzing Credit in the United States

The **prime interest rate,** or prime lending rate, is the rate that banks charge their most creditworthy customers.

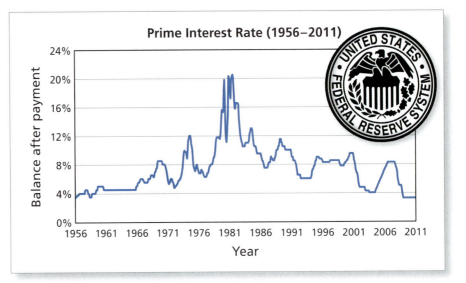

EXAMPLE 5 **Tracking the Prime Interest Rate**

Use the graph above to discuss why interest rates increase and decrease.

SOLUTION

Here are three factors that cause interest rates to fluctuate.

1. **Supply and Demand of Funds:** If the demand for borrowing is higher than the funds that banks have available, then the rates increase. If the demand for borrowing is lower than the available funds, then the rates decrease.

2. **Monetary Policy:** Sometimes the federal government "loosens monetary policy" by printing more money. This causes interest rates to decrease because more money is available to lenders.

3. **Inflation:** Investors want to preserve the buying power of their money. When the inflation rate is high, investors need a higher interest rate to consider lending their money.

✓ **Checkpoint** Help at *Math*.and**Y☺U**.com

Alan Greenspan is an American economist who served as chairman of the Federal Reserve from 1987 to 2006.

Which of the following has had the greatest impact on the prime interest rate?

a. Political party of the president: 1961–1968 (D), 1969–1976 (R), 1977–1980 (D), 1981–1992 (R), 1993–2000 (D), 2001–2008 (R)

b. Country at war: Vietnam War (1965–1973), Persian Gulf War (1990–1991), Iraq War (2003–2010)

c. Inflation rate: See page 162.

Justify your reasoning in words and graphically.

EXAMPLE 6   **Comparing Two Levels of Indebtedness**

The graph shows the total consumer indebtedness of households in the United States from 1952 through 2008. Was the typical household more in debt in 2008 than in 1952? Explain your reasoning.

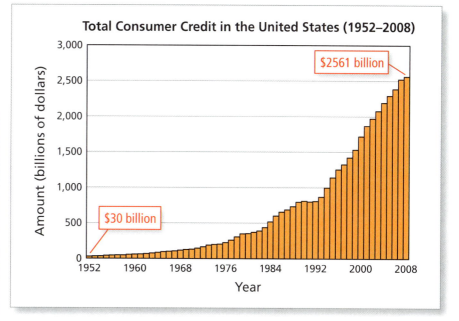

Total Consumer Credit in the United States (1952–2008)

$2561 billion

$30 billion

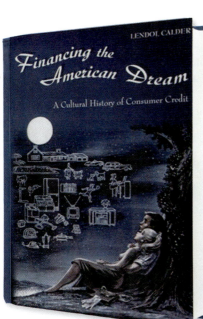

Financing the American Dream: A Cultural History of Consumer Credit by Lendol Calder

"Once there was a golden age of American thrift, when citizens lived sensibly within their means and worked hard to stay out of debt. The growing availability of credit in this century, however, has brought those days to an end—undermining traditional moral virtues such as prudence, diligence, and the delay of gratification while encouraging reckless consumerism. Or so we commonly believe. In this engaging and thought-provoking book, Lendol Calder shows that this conception of the past is in fact a myth."

SOLUTION

At first glance, the answer looks obvious. However, there are two things you need to consider before reaching a conclusion. First, you need to account for the increase in population. Second, you need to account for inflation.

Year	Total Indebtedness	Number of Households	CPI
1952	$30 billion	46 million	26.5
2008	$2561 billion	117 million	215.3

In 1952, the average indebtedness per household was about $700. In 2008, the average indebtedness per household was about $21,900.

So, the average indebtedness per household increased by a factor of about 31. During the same time, the CPI increased by a factor of only about 8.

So, you can conclude that the typical American household was more in debt in 2008 than in 1952.

✓ **Checkpoint**    Help at *Math*.and**YOU**.com

What is your opinion about the increased consumer indebtedness that has occurred in the United States during the past 60 years? Do you think it is a national problem? Or do you agree with historians like Lendol Calder, who believe that the comparison between "now and then" is not as straightforward as it appears at first glance? Explain your reasoning.

6.2 Exercises

DATA **Motorcycles** The tables show loan rates and terms for new and used motorcycle purchases. In Exercises 1–5, use the tables. *(See Examples 1 and 2.)*

NEW MOTORCYCLE	
Loan	**APR**
0 to 24 months	4.24
25 to 36 months	4.49
37 to 48 months	4.74
49 to 60 months	4.99
61 to 72 months	5.49
73 to 84 months	6.74

USED MOTORCYCLE	
Loan	**APR**
0 to 24 months	5.24
25 to 36 months	5.49
37 to 48 months	5.74
49 to 60 months	5.99
61 to 66 months	6.49

1. You buy a brand-new, custom-built chopper for $35,000.

  a. What is the monthly payment on a 60-month loan?

  b. Create an amortization table showing how the balance of the loan decreases.

2. You buy a used motorcycle for $8000.

  a. What is the monthly payment on a 36-month loan?

  b. Create an amortization table showing how the balance of the loan decreases.

3. You buy a used motorcycle for $5000 and choose a 24-month loan. In 12 months, is the remaining balance less than one-half of the original loan amount? If not, when does this occur?

4. You buy a new motorcycle for $18,000 and choose an 84-month loan. What is the cost of credit?

5. You buy a new dirt bike for $13,000.

  a. Find the monthly payments on loans with terms of 2 years, 3 years, 4 years, and 5 years.

  b. Find the cost of credit for each loan. How does the cost of credit change as the term increases?

6. **Double Payments** You are considering a 5-year loan for a motorcycle. Will making double monthly payments decrease the cost of credit by 50%? Given that you can easily make double payments, should you choose a different term? Explain.

DATA **Credit Cards** **You have had a credit card for 2 years. In Exercises 7–11, use the terms and conditions below that apply to your credit card.** *(See Examples 3 and 4.)*

Interest Rates and Interest Charges	
Annual Percentage Rate (APR) for Purchases	**0%** Introductory APR for the first 12 statement Closing Dates following the opening of your account. After that, your Standard APR will be **12.99%** to **20.99%**, based on your creditworthiness when you open your account. This APR will vary with the market based on the Prime Rate.
APR for Balance Transfers	**12.99%** to **20.99%** Standard APR, based on your creditworthiness when you open your account. After that, your APR will vary with the market based on the Prime Rate.
APR for Cash Advances	**24.24%** Standard APR for Direct Deposits, Check Cash Advances, ATM Cash Advances, Bank Cash Advances, Overdraft Protection and Cash Equivilent transactions. This APR will vary with the market based on the Prime Rate.
Penalty APR	None
How to Avoid Paying Interest on Purchases	Your due date is at least 25 days after the close of each billing cycle. We will not charge you any interest on purchases if you pay your entire balance by the due date each month.
For Credit Card Tips from the Federal Reserve Board	To learn more about factors to consider when applying for or using a credit card, visit the website of the Federal Reserve Board at **http://www.federalreserve.gov/creditcard**.

7. You have the lowest possible APR. You use the credit card to purchase 3 nights in a hotel for a total of $350. Your bill also includes $250 from last month. The minimum payment each month is $15. How long does it take to pay the credit card bill by making only the minimum payment each month? How much do you pay in interest?

8. You have the lowest possible APR. You use the credit card to purchase airplane tickets to Australia for $1500. Your bill also includes $591.50 from last month. The minimum payment is either 2% of your statement balance rounded to the nearest whole dollar or $15, whichever is greater. How long does it take to pay the credit card bill by making only the minimum payment each month? How much do you pay in interest?

9. You take a cash advance of $500. How long does it take to pay for the advance by making $50 payments each month? How much do you pay in interest?

10. Your credit card statement is shown. You plan to pay $300 each month. How much more do you pay in interest with the maximum APR than with the minimum APR? (*Note:* There is an interest charge after this month.)

11. You can pay the $1730 credit card balance with a 2-year installment loan that has an APR of 6.5%, or a 1-year installment loan that has an APR of 9%. Which loan has a lesser cost of credit? How much do you save by choosing this loan?

Previous Balance	673.92
Payments	-673.92
Charges	1,730.00
Fees	0.00
Interest Charged	0.00
Credits	0.00
New Balance	**1,730.00**
Minimum Payment Due	**35.00**
	PAY BILL

12. **Paying Debt** Your credit card balance is $500. You can afford to pay $100 each month toward the balance. Should you do this or should you make the minimum payment of $20 each month? Explain your reasoning.

Federal Funds Rate In Exercises 13 and 14, use the graph. *(See Example 5.)*

13. The *federal funds rate* is the interest rate that banks charge each other for overnight loans. Do you think the prime interest rate and the federal funds rate are related? Explain.

14. The APR on a *fixed-rate loan* stays the same during the term of the loan. The APR on an *adjustable-rate loan* generally increases and decreases along with rates set by the Federal Reserve, but it starts out at a lesser rate than a fixed-rate loan. When do you think an adjustable-rate loan would have been cheaper than a fixed-rate loan? Which is more risky? Explain.

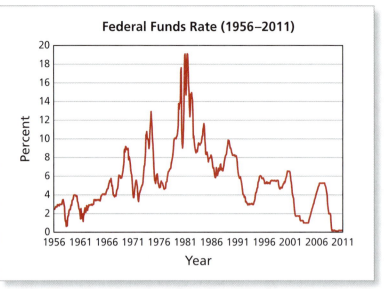

Student Loans In Exercises 15 and 16, use the bar graph. *(See Example 6.)*

15. Taking inflation into account, did the typical graduating senior have more student loan debt in 2004 than in 2000? Explain.

16. Taking inflation into account, did the typical graduating senior have more student loan debt in 2008 than in 1996? Explain.

17. **Credit Card Debt** In 2000, there were 159 million credit card holders with a total of $680 billion in outstanding credit card debt. In 2008, there were 176 million credit card holders with a total of $976 billion in outstanding credit card debt. Did the typical credit card holder have more outstanding credit card debt in 2008 than in 2000? Explain. *(See Example 6.)*

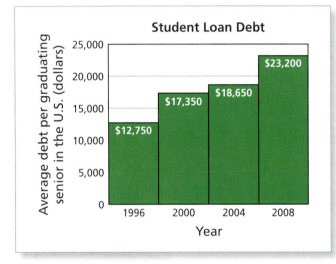

18. **Mortgage Debt** Consumer credit does not include home mortgages. The graph shows consumer credit and home mortgage debt from 1960 through 2010. *(See Example 6.)*

a. There were 53 million households in 1960 and 118 million households in 2010. Did a typical household have more home mortgage debt in 2010 than in 1960?

b. Did home mortgage debt or consumer credit increase at a greater rate since 1980? What may have caused this?

Consumer Credit and Mortgage Debt

$13,830 billion

☐ Home mortgage debt
☐ Consumer credit

$210 billion

▶ Extending Concepts

Credit Cards The tables show all the activity in two credit card accounts for an entire billing period. In Exercises 19–24, use the tables.

Date	Activity	Amount	Balance	
April 1			$300.00	← Starting balance
April 5	Gas	$45.00	$345.00	
April 8	Payment	−$300.00	$45.00	
April 23	Groceries	$70.00	$115.00	
April 30			$115.00	← Ending balance

Date	Activity	Amount	Balance	
July 21			$800.00	← Starting balance
July 29	Car insurance	$110.00	$910.00	
August 3	Gym membership	$40.00	$950.00	
August 3	Payment	−$500.00	$450.00	
August 9	Cell phone	$75.00	$525.00	
August 20			$525.00	← Ending balance

19. When a billing period ends, the ending balance plus any additional charges becomes the starting balance of the next billing period. The account owners have 25 days from the end of the billing period to make a payment. Are the payments shown made on time?

20. Each credit card account has no interest charge as long as the balance of the previous billing period was paid in full. Do the accounts have interest charges for the billing periods shown? Explain.

21. The owner of the Gold Card receives a billing statement for the month of May. The statement includes an interest charge. What can you determine about the amount paid on the April statement?

22. The *daily periodic rate* (DPR) of a credit card is the daily interest rate, which is the APR divided by the number of days in a year. The APR of each credit card is 19.99%. Find the DPR of each credit card.

DATA 23. The *average daily balance* of a credit card account in a billing period is the sum of the balances at the end of each day in the billing period divided by the number of days in the billing period. Find the average daily balance of each account.

24. The interest charge on each credit card account is the product of the average daily balance, the DPR, and the number of days in the billing period. Find the interest charge for the World Card.

6.1–6.2   Quiz

Credit Unions In Exercises 1–6, use the partially completed promissory note for a recreational vehicle (RV) loan.

> ### Promissory Note
>
> I, _Raja Kumar_ , promise to repay My Federal Credit Union the loan amount of _$100,000_ . Repayment is to be made in the form of _60_ equal payments at _6.75%_ interest, or $ _____ payable on the _1st_ of each month, beginning _____ , until the total debt is satisfied.
>
> Signed,
> *Raja Kumar*
> *2/1/2011*

1. Find the term of the loan and the loan proceeds.

2. What is the monthly payment?

3. Find the total amount due and the cost of credit.

4. The first payment is due March 1, 2011. When is the last payment due?

5. The graph shows the decreasing balance of the loan. When is the remaining balance about one-fourth of the original loan amount?

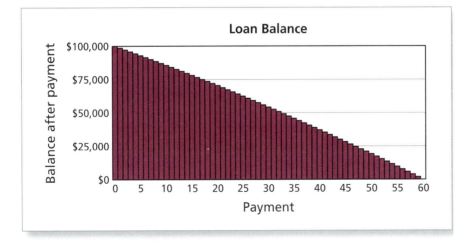

6. The credit union offers Raja a 0.25% reduction in the APR by initiating automatic monthly payments from his checking account. How much does he save over the term of the loan by doing this?

Math & Buying a Car

PROJECT: How Much Car Can I Afford?

1. Use the *How Much Car Can I Afford** calculator at *Math.andYou.com.*

1   Zip Code

Zip Code
(for rates and pricing in your area)

89521   **UPDATE ZIP CODE**

2   Financing

Your Target Monthly Payment ⑦ $ 250

> Enter the maximum monthly payment you can afford.

Loan Term
(months)   60 months

Market Finance Rate ⑦   6.00 %

> Enter the interest rate for the loan.

3   Trade-In and Down Payment

2005 Jeep Grand Cherokee
Limited 4WD 4dr SUV (4.7L 8cyl 5A)
Pricing | Research

SELECT TRADE-IN

> Describe the car you will trade in. The calculator will compute its value.

Value of Your Trade-In ⑦ $ 8796

Amount Owed on Your Trade-In ⑦ $ 0

Cash Down Payment ⑦ $ 0

CALCULATE

Results

Sticker Price Range $18,600 - $21,600

> You can afford a car in this price range.

**Provided by edmunds.com*

2. Enter other zip codes until you find one that gives you different results. Why might other zip codes give you different results?

3. Enter the market finance rates shown. How much is subtracted from the sticker price range each time you increase the rate by 1%?

Market Finance Rate	Sticker Price Range
0.00 %	
1.00 %	
2.00 %	
3.00 %	
4.00 %	
5.00 %	
6.00 %	

6.3 Home Mortgages

▶ Compare rates and terms for a home mortgage.

▶ Analyze the effect of making principal payments.

▶ Compare the costs of buying and renting.

Comparing Rates and Terms for Home Mortgages

A **home mortgage** is an installment loan that is taken out to pay for a home. For most people, it is the largest loan they will ever assume. For this reason, consider the following when purchasing a home.

1. Understand the annual percentage rate for the loan. Is it fixed throughout the term or is it adjustable?

2. Make sure the contract allows you to make extra payments toward the principal.

3. Shop for or negotiate the best possible rate. A difference of even 1% can save you tens of thousands of dollars.

4. Shop for real estate agents. The fee an agent charges you is negotiable.

> **EXAMPLE 1** **Comparing Rates for a Home Mortgage**

You take out a home mortgage for $250,000 for 30 years. Compare the total interest you pay for annual percentage rates of (a) 4% and (b) 6%.

SOLUTION

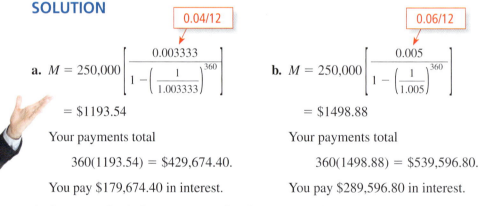

a. $M = 250{,}000 \left[\dfrac{0.003333}{1 - \left(\dfrac{1}{1.003333} \right)^{360}} \right]$ **b.** $M = 250{,}000 \left[\dfrac{0.005}{1 - \left(\dfrac{1}{1.005} \right)^{360}} \right]$

$= \$1193.54$ $= \$1498.88$

Your payments total   Your payments total

    $360(1193.54) = \$429{,}674.40.$ $360(1498.88) = \$539{,}596.80.$

You pay $179,674.40 in interest.   You pay $289,596.80 in interest.

An increase of only 2 percentage points increases the interest that you pay by about $110,000! Can you imagine how much interest people paid in the early 1980s, when home mortgage rates were about 20%?

✓ **Checkpoint**   Help at *Math*.and**YOU**.com

c. In Example 1, do you pay about $110,000 more in interest when the annual percentage rate is 8%? Explain your reasoning.

d. In general, does the amount of interest you pay double when the annual percentage rate doubles? Explain your reasoning.

e. In Example 1, does the amount of interest you pay double when the amount borrowed doubles? Explain your reasoning.

f. In general, does the amount of interest you pay double when the principal doubles? Explain your reasoning.

The median price of a new home in the United States during 2010 was about $221,000.

EXAMPLE 2   **Comparing Terms for a Home Mortgage**

You take out a home mortgage for $250,000 at 6%. Compare the total interest you pay for terms of (a) 20 years, (b) 30 years, and (c) 40 years.

SOLUTION

a. $M = 250{,}000 \left[\dfrac{0.005}{1 - \left(\dfrac{1}{1.005}\right)^{240}} \right] = \$1791.08$ 20 years

Your payments total 240(1791.08) = $429,859.20.
The total interest you pay over 20 years is $179,859.20.

b. $M = 250{,}000 \left[\dfrac{0.005}{1 - \left(\dfrac{1}{1.005}\right)^{360}} \right] = \$1498.88$ 30 years

Your payments total 360(1498.88) = $539,596.80. The total interest you pay over 30 years is $289,596.80.

c. $M = 250{,}000 \left[\dfrac{0.005}{1 - \left(\dfrac{1}{1.005}\right)^{480}} \right] = \$1375.53$ 40 years

Your payments total 480(1375.53) = $660,254.40. The total interest you pay over 40 years is $410,254.40.

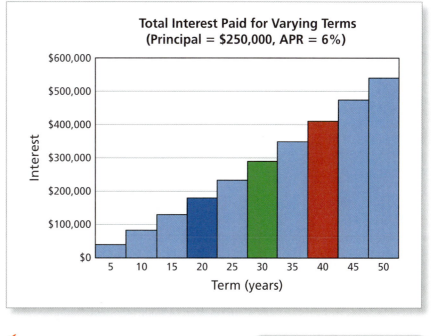

	A	B
	Term	**Monthly**
1	**(years)**	**Payment**
2	5	$4,833.20
3	10	$2,775.51
4	15	$2,109.64
5	20	$1,791.08
6	25	$1,610.75
7	30	$1,498.88
8	35	$1,425.47
9	40	$1,375.53
10	45	$1,340.71
11	50	$1,316.01

This table shows the monthly payment for a mortgage of $250,000 at 6% for varying terms. Notice that increases in the term eventually amount to insignificant reductions in the monthly payment.

✓ **Checkpoint**   Help at *Math*.and**YOU**.com

You take out a home mortgage for $250,000 at 12%. Compare the total interest you pay for the following terms.

d. 20 years   **e.** 30 years   **f.** 40 years

g. Are your answers double those in Example 2? What can you conclude from this?

Analyzing the Effect of Principal Payments

There are three important numbers connected with any home mortgage: market value, principal balance, and equity.

- **Market value** is the amount that the home could sell for.
- **Principal balance** is the amount owed on the mortgage.
- **Equity** is the difference between the market value and the principal balance.

Each month, after paying the interest, each dollar in your payment goes toward decreasing your principal balance and increasing your equity. Even small additional "principal payments" (payments above the normal monthly payment) can have dramatic effects in the overall amount of interest you pay throughout the term of the mortgage.

EXAMPLE 3   Analyzing the Effect of Principal Payments

You take out a home mortgage for $250,000 for 30 years at 6%. Each month, you make the regular payment of $1498.88 plus an additional $50. (a) How much sooner do you pay off the mortgage? (b) How much do you save in interest?

SOLUTION

a. Use a spreadsheet to create an amortization table.

	A	B	C	D	E	F
	Payment Number	**Balance before Payment**	**Interest on Balance**	**Monthly Payment**	**Extra Payment**	**Balance after Payment**
1						
2	1	$250,000.00	$1,250.00	$1,498.88	$50.00	$249,701.12
3	2	$249,701.12	$1,248.51	$1,498.88	$50.00	$249,400.75
4	3	$249,400.75	$1,247.00	$1,498.88	$50.00	$249,098.88
5	4	$249,098.88	$1,245.49	$1,498.88	$50.00	$248,795.50
6	5	$248,795.50	$1,243.92	$1,498.88	$50.00	$248,490.60
328	327	$5,926.39	$29.63	$1,498.88	$50.00	$4,407.15
329	328	$4,407.15	$22.04	$1,498.88	$50.00	$2,880.31
330	329	$2,880.31	$14.40	$1,498.88	$50.00	$1,345.83
331	330	$1,345.83	$6.73	$1,352.56	$0.00	$0.00
332	Total			$494,482.87	$16,450.00	

Instead of taking 360 months, it takes only 330 months, which is 2.5 years sooner.

b. Instead of your payments totaling $539,596.80 [see Example 1(b)], your payments total 494,482.87 + 16,450.00 = $510,932.87, which is a savings of $28,663.93. All of this savings represents interest that you do not have to pay.

✓ Checkpoint

In Example 3, suppose that each month you make the regular payment plus an additional $100. (c) How much sooner do you pay off the mortgage? (d) How much do you save in interest?

Study Tip

Some mortgages do not allow the homeowner to make extra principal payments. Before signing a mortgage contract, make sure the contract allows you to make extra payments whenever you want.

A **balloon mortgage** is a mortgage that is amortized over a period that is longer than the term of the loan, leaving a large final payment due at maturity. The final payment is called a **balloon payment.**

EXAMPLE 4   Analyzing the Cost of a Balloon Mortgage

You take out a 10-year balloon mortgage for $250,000. The monthly payment is equal to that of a 30-year mortgage with an annual percentage rate of 6%. Find the balloon payment and the total interest that you pay.

SOLUTION

From Example 1(b), the monthly payment for a 30-year mortgage for $250,000 with an annual percentage rate of 6% is $1498.88. Use a spreadsheet to amortize the mortgage for 10 years using this monthly payment.

	A Payment Number	B Balance before Payment	C Interest on Balance	D Monthly Payment	E Balance after Payment
1					
2	1	$250,000.00	$1,250.00	$1,498.88	$249,751.12
	2	$249,751.12	$1,248.76	$1,498.88	$249,501.00
	3	$249,501.00	$1,247.51	$1,498.88	$249,249.63
	4	$249,249.63	$1,246.25	$1,498.88	$248,997.00
	5	$248,997.00	$1,244.99	$1,498.88	$248,743.11
	6	$248,743.11	$1,243.72	$1,498.88	$248,487.95
	7	$248,487.95	$1,242.44	$1,498.88	$248,231.51
	8	$248,231.51	$1,241.16	$1,498.88	$247,973.80
	9	$247,973.80	$1,239.87	$1,498.88	$247,714.79
		$247,714.79	$1,238.57	$1,498.88	$247,454.49
		$212,321.49	$1,061.61	$1,498.88	$211,884.22
		$211,884.22	$1,059.42	$1,498.88	$211,444.77
		$211,444.77	$1,057.22	$1,498.88	$211,003.12
	117	$211,003.12	$1,055.02	$1,498.88	$210,559.25
119	118	$210,559.25	$1,052.80	$1,498.88	$210,113.17
120	119	$210,113.17	$1,050.57	$1,498.88	$209,664.86
121	120	$209,664.86	$1,048.32	$1,498.88	$209,214.31
122	**Total**		**$139,079.47**		

Balloon payment

After making 120 payments, you must make a balloon payment of $209,214.31. The total interest that you pay over the 10 years is $139,079.47.

Study Tip

A balloon mortgage lets you make small monthly payments, but at the end of the mortgage you must make a large final payment.

✓ **Checkpoint**

Help at

You take out a 5-year balloon mortgage for $200,000. The monthly payment is equal to that of a 20-year mortgage with an annual percentage rate of 6.5%. Find the balloon payment and the total interest that you pay.

Comparing the Costs of Buying and Renting

Rent   Home Ownership

When you are deciding whether to buy or rent, *and* you have a substantial amount of money in savings, you have two options: (1) use the money as a down payment for a home mortgage or (2) rent a home and use the money as an income investment.

EXAMPLE 5   Comparing Buying and Renting

You take out a home mortgage for $250,000 for 30 years at 6%. After 5 years, you move to a different state and sell the home for $367,850.

Expenses and Savings Related to Buying

Cost of home: $350,000   Realtor's fee: 6% of cost of home
Down payment: $100,000   Home insurance: $1000 per year
Mortgage: $250,000   Property tax: 3% of cost of home per year
Monthly payment: $1498.88   Home repairs: $17,000
Closing costs: 2% of cost of home   Income tax savings (interest): $18,000

Compare the costs of buying the home and renting a comparable home for $2000 per month.

SOLUTION

Here are your expenses and savings with buying.

60 mortgage payments	89,932.80
Closing costs of 2%	7,000.00
Realtor's fee of 6%	21,000.00
Home insurance	5,000.00
Property tax:	52,500.00
Home repairs:	17,000.00
Income tax savings	−18,000.00
Increase in equity	−17,850.00
Total:	**$156,582.80**

If you had rented for 5 years, you would have paid 60(2000) = $120,000 in rent. However, you could also have invested the $100,000 down payment and perhaps earned $16,000 in interest. So, your total cost would have been about $104,000.

So, in this case, the cost of renting is about $53,000 less than the cost of buying.

Home maintenance is often called the "hidden cost of home ownership." You should expect to spend between 1% and 2% of the cost of your home each year on home repairs and maintenance.

✓ Checkpoint

Help at **Math.andYOU.com**

a. Describe other issues that can affect your decision to rent or to buy.

b. Do you agree with the federal income tax policy that allows homeowners a deduction for interest paid on mortgages, but does not allow a comparable deduction for renters? Explain your reasoning.

EXAMPLE 6   **Analyzing a Housing Bubble**

In the graph, the curve shows that around 1998, the prices of homes began to increase abnormally, which is called a *bubble*. Around 2007, the bubble burst. Explain why this caused a record number of foreclosures throughout the United States.

U. S. Home Prices (1970–2010)

Actual

Pre-bubble trend

SOLUTION

At first glance, you would think that a drop in the value of your home would not be much of a concern. However, during the early 2000s, banks were offering *subprime* and *adjustable-rate* mortgages to entice more and more people to buy homes. Here is a hypothetical, but all too common, example.

It is 2005. You have been looking at a home that was sold for $200,000 in 1999. The home is being sold again and is listed for $400,000. That is a compound increase of about 12.2% per year. You feel like you have to take out a mortgage now or you will never be able to afford a home. Your bank offers you a 10% down mortgage with a *subprime* rate of 4%. The rate will be adjusted in 3 years, but at that time, the home should be worth more money and you can refinance, using the equity you have built up.

> Monthly payment: $1718.70   (2005–2008)

It is now 2008. The rate on your mortgage increases to 6.5%, raising your mortgage payment to $2275.44. You still owe about $340,000 on your mortgage but the value of your home has dropped to $300.000. You cannot afford the increased mortgage payment, and you owe more on your home than you can sell it for. What do you do? For many homeowners in 2008, the answer was to let the bank foreclose on the home.

Study Tip

Foreclosure filings include default notices, scheduled auctions, and bank repossessions. In 2010, 2,871,891 properties received a foreclosure filing.

✓ **Checkpoint**   Help at

How are banks affected by foreclosures?

6.3 Exercises

Mortgage Rates You take out a home mortgage. In Exercises 1–4, compare the total interest you pay for the annual percentage rates. *(See Example 1.)*

1. Home mortgage: $140,000 for 30 years

   a. 5%   **b.** 7%

2. Home mortgage: $165,000 for 30 years

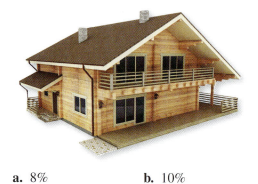

   a. 8%   **b.** 10%

3. Home mortgage: $220,000 for 30 years

   a. 4%   **b.** 6%

4. Home mortgage: $275,000 for 30 years

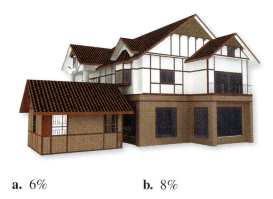

   a. 6%   **b.** 8%

Mortgage Terms You take out a home mortgage. In Exercises 5 and 6, compare the total interest you pay for the terms. *(See Example 2.)*

5. Home mortgage: $200,000 at 5%

   a. 20 years   **b.** 30 years   **c.** 40 years

6. Home mortgage: $180,000 at 6%

   a. 15 years   **b.** 25 years   **c.** 30 years

Amortization Schedules In Exercises 7–10, use a spreadsheet. *(See Example 3.)*

7. You take out a home mortgage for $238,000 for 30 years at 5%. Each month, you make the regular payment of $1277.64 plus an additional $60.

   a. How much sooner do you pay off the mortgage?

   b. How much do you save in interest?

8. You take out a home mortgage for $260,000 for 30 years at 6%. Each month, you make the regular payment of $1558.83 plus an additional $100.

   a. How much sooner do you pay off the mortgage?

   b. How much do you save in interest?

9. You take out a home mortgage for $275,000 for 25 years at 4%.

   a. What is the least amount, to the nearest dollar, that you need to pay in addition to the regular payment to pay off the mortgage 5 years early?

   b. Compare the total interest you pay for the 25-year mortgage to the total interest you pay when you pay off the 25-year mortgage 5 years early.

10. You take out a home mortgage for $190,000 for 20 years at 5%.

   a. What is the least amount, to the nearest dollar, that you need to pay in addition to the regular payment to pay off the mortgage 5 years early?

   b. Compare the total interest you pay for the 20-year mortgage to the total interest you pay when you pay off the 20-year mortgage 5 years early.

   c. Compare the total interest you pay for a 15-year mortgage for the same amount and at the same rate to the total interest you pay when you pay off the 20-year mortgage 5 years early.

Balloon Mortgages In Exercises 11 and 12, use a spreadsheet. *(See Example 4.)*

11. You take out a 5-year balloon mortgage for $150,000. The monthly payment is equal to that of a 15-year mortgage with an annual percentage rate of 5%. Find the balloon payment and the total interest that you pay.

12. You take out a 7-year balloon mortgage for $120,000. The monthly payment is equal to that of a 30-year mortgage with an annual percentage rate of 5.5%. Find the balloon payment and the total interest that you pay.

Buying Versus Renting In Exercises 13 and 14, compare the costs of buying the home and renting a comparable home for $1600 per month. *(See Example 5.)*

13. You take out a home mortgage for $224,000 for 30 years at 5%. After 5 years, you move to a different state and sell the home for $294,280. Assume that if you did not buy the home, you could have invested the down payment and earned $12,000 in interest.

Expenses and Savings Related to Buying

Cost of home: $280,000   Realtor's fee: 6% of cost of home

Down payment: $56,000   Home insurance: $1400 per year

Mortgage: $224,000   Property tax: 2.5% of cost of home per year

Monthly payment: $1202.48   Home repairs: $15,000

Closing costs: 5% of cost of home   Income tax savings (interest): $11,000

14. You take out a home mortgage for $200,000 for 30 years at 4%. After 6 years, you move to a different city and sell the home for $281,540. Assume that if you did not buy the home, you could have invested the down payment and earned $13,000 in interest.

Expenses and Savings Related to Buying

Cost of home: $250,000   Realtor's fee: 5% of cost of home

Down payment: $50,000   Home insurance: $1300 per year

Mortgage: $200,000   Property tax: 1.4% of cost of home per year

Monthly payment: $954.83   Home repairs: $16,000

Closing costs: 3% of cost of home   Income tax savings (interest): $9000

Price-to-Rent Ratio In Exercises 15 and 16, find the price-to-rent ratio.

15. The cost of a home is $162,000. The rent for a comparable home is $800 per month.

16. The cost of a home is $156,000. The rent for a comparable home is $700 per month.

Housing Bubble In Exercises 17–20, use the graph on page 279. *(See Example 6.)*

17. Suppose home prices followed the pre-bubble trend.

    a. What would have been the price of a home in 2006?

    b. How much more was the actual price of a home in 2006?

18. Estimate the percent change in home prices from 2004 to 2010.

19. Estimate the percent decrease in home prices from 2007 to 2009.

20. Compare the percent increase in home prices from 1989 to 1998 to the percent increase in home prices from 1998 to 2007.

▶ Extending Concepts

Discount Points **In Exercises 21 and 22, use the information below.**

Lenders may offer you the option to purchase *discount points* to reduce the interest rate on a loan. One point is equal to one percent of the loan amount.

21. You take out a home mortgage for $150,000 for 30 years.

    a. Compare the monthly payments of mortgage A and mortgage B.

        Mortgage A: 4.5% with 2 points

        Mortgage B: 5% with no points

    b. Suppose you choose mortgage A. How long will it take you to pay off the points with your monthly savings from the lower rate?

22. You take out a home mortgage for $120,000 for 30 years.

    a. Compare the monthly payments of mortgage A and mortgage B.

        Mortgage A: 5.5% with 3 points

        Mortgage B: 6% with 1 point

    b. Suppose you choose mortgage A. How long will it take you to pay off the points with your monthly savings from the lower rate?

23. **Mortgage Reset** You take out a 7-year balloon mortgage for $160,000. The monthly payment is equal to that of a 30-year mortgage with an annual percentage rate of 5.5%. At the end of 7 years, you have the option to reset the mortgage and pay off the remaining balance over the next 23 years with an annual percentage rate of 6.5%.

    a. How much interest do you pay?

    b. How much would you save in interest by taking out a 30-year mortgage with an annual percentage rate of 6%?

24. **Principal and Interest** You take out a home mortgage for $150,000 for 15 years with an annual percentage rate of 6%.

    a. Find the total amount that you pay in interest each year and the total amount that you pay toward the principal each year.

    b. Make a double bar graph that displays the information in part (a). Describe any trends in the graph.

25. **Mortgage Affordability** You can afford to make monthly payments of $600. How large of a home mortgage can you afford at a rate of 5% for a term of 30 years?

26. **Adjustable-Rate Mortgage** You take out a 30-year adjustable-rate mortgage (ARM) for $100,000. The interest rate is 5% for the first 5 years and 8% for the sixth year. What is "ARM reset shock"? How can you avoid it?

6.4 Savings & Retirement Plans

▶ Find the balance in a savings account.

▶ Find the balance in an increasing annuity.

▶ Analyze a decreasing annuity.

Finding the Balance in a Savings Account

When you deposit money into a savings account earning compound interest, the balance in the account grows exponentially. The formula for the balance depends on how often the interest is compounded.

Interest Compounded Monthly

The balance A in a savings account with a principal of P, for n months at an annual percentage rate of r (in decimal form), compounded monthly, is

$$A = P\left(1 + \frac{r}{12}\right)^n.$$

Math.and**YOU**.com

You can access compound interest calculators at *Math.andYou.com*.

EXAMPLE 1   **Comparing Terms for a Savings Plan**

When your daughter is born, your grandparents deposit $5000 into a savings account that earns 4%, compounded monthly.

a. Find the balance in the account when your daughter is 18 years old.

b. Find the balance in the account when your daughter is 26 years old.

SOLUTION

12(18)

a. $A = 5000\left(1 + \dfrac{0.04}{12}\right)^{216} = \$10{,}259.87$ 18th birthday

12(26)

b. $A = 5000\left(1 + \dfrac{0.04}{12}\right)^{312} = \$14{,}121.64$ 26th birthday

> **Study Tip**
>
> The *Rule of 72* is commonly used by investors. It states that the number of years it will take for an investment to double is equal to 72 divided by the interest rate.

✓ **Checkpoint**   Help at *Math*.and**YOU**.com

Suppose your grandparents invest the money into a mutual fund that earns 10%, compounded monthly.

c. Find the balance in the account when your daughter is 18 years old.

d. Find the balance in the account when your daughter is 26 years old.

"...I wish to be useful even after my death, if possible, in forming and advancing other young men, that may be serviceable to their country in both these towns. To this end, I devote two thousand pounds sterling, of which I give one thousand thereof to the inhabitants of the town of Boston, in Massachusetts, and the other thousand to the inhabitants of the city of Philadelphia, in trust, to and for the uses, intents, and purposes herein after mentioned and declared..."

EXAMPLE 2   **Calculating the Consequences of Franklin's Will**

Benjamin Franklin died in 1790. In his will, he left 1000 pounds sterling (about $4444.44) to Boston and to Philadelphia. He expected the money to earn 5% annually for 100 years. Then, about 75% of the money would be distributed for various projects and the remaining 25% would be invested at 5% for another 100 years. The trust funds were set up in 1791.

a. How much did Franklin expect each city to have after 100 years?

b. Despite Franklin's calculations, Boston had only about $100,000 to reinvest after the distributions. This money earned about 4.1% annually for the next 100 years. What was the balance in Boston's fund in 1991?

c. Philadelphia had only about $40,000 to reinvest after the distributions. This money also earned about 4.1% annually for the next 100 years. What was the balance in Philadelphia's fund in 1991?

SOLUTION

a. $A = 4444.44(1 + 0.05)^{100} \approx \$584,449$

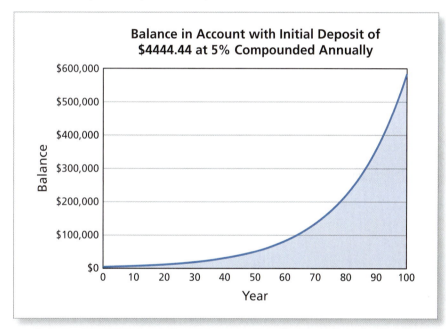

b. In 1991, the balance of Boston's fund was about

$$A = 100,000(1 + 0.041)^{100} \approx \$5,559,976.$$

c. In 1991, the balance of Philadelphia's fund was about

$$A = 40,000(1 + 0.041)^{100} \approx \$2,223,991.$$

So, although Franklin had intended for the two cities to end up with over $38 million, the actual amount they received was about $7.8 million.

✓ **Checkpoint**   Help at *Math*.and**YOU**.com

The CPI from 1774 through 2010 is shown on page 165. Use this index to estimate the value of $5.6 million (in 1991 dollars) at the time of Franklin's death in 1790.

Finding the Balance in an Increasing Annuity

An **increasing annuity** is a savings account in which you make repeated deposits. This type of savings account is often used to create retirement funds.

Balance in an Increasing Annuity

The balance A in an increasing annuity with a monthly deposit of M, for n months at an annual percentage rate of r (in decimal form), compounded monthly is

$$A = M\left[\frac{[1 + (r/12)]^n - 1}{r/12}\right].$$

Math.and**YOU**.com

You can access increasing annuity calculators at *Math.andYou.com*.

EXAMPLE 3 **Creating a Retirement Plan**

You start your working career when you are 22 years old. Each month, you deposit $100 into a pension plan. You continue making deposits into the plan until you are 72 years old. What is the balance when the plan earns

a. 4%, compounded monthly?

b. 6%, compounded monthly?

c. 8%, compounded monthly?

SOLUTION

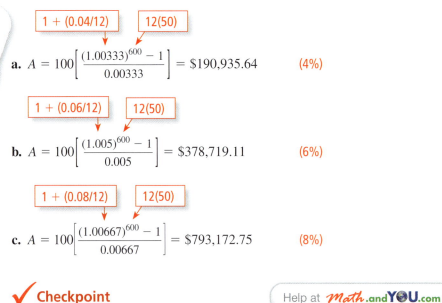

$1 + (0.04/12)$ $12(50)$

a. $A = 100\left[\dfrac{(1.00333)^{600} - 1}{0.00333}\right] = \$190{,}935.64$ (4%)

$1 + (0.06/12)$ $12(50)$

b. $A = 100\left[\dfrac{(1.005)^{600} - 1}{0.005}\right] = \$378{,}719.11$ (6%)

$1 + (0.08/12)$ $12(50)$

c. $A = 100\left[\dfrac{(1.00667)^{600} - 1}{0.00667}\right] = \$793{,}172.75$ (8%)

In 2010, there was an estimated $15 trillion invested in private retirement plans in the United States.

✓ **Checkpoint**   Help at *Math*.and**YOU**.com

Suppose that you continue to make the monthly deposits for only 45 years, instead of 50 years. What is the balance when the plan earns

d. 4%, compounded monthly?   **e.** 6%, compounded monthly?

f. 8%, compounded monthly?

The retirement plan in Example 3 is somewhat unreasonable because it assumes that a person invests the same amount each month for 50 years. In a typical retirement plan, the monthly amount that a person invests increases over time.

EXAMPLE 4   Creating a Retirement Plan

You start your working career when you are 22 years old. Your beginning salary is $40,000 per year. Your employer offers a 401(k) matching retirement plan that amounts to 10% of your salary (5% from you and 5% from your employer). Assume that your salary increases 3% each year and that the 401(k) plan averages 6% annual returns for the life of the plan. What is the balance in your account at age 70?

SOLUTION

The following spreadsheet is oversimplified because it calculates interest annually, instead of monthly. It still gives you the magnitude of the balance after 48 years.

	A	B	C	D
1	**Annual Salary**	**401(k) Contribution**	**Balance in Account**	**Interest Earned**
2	$40,000.00	$4,000.00	$4,000.00	$240.00
3	$41,200.00	$4,120.00	$8,360.00	$501.60
48	$155,801.75	$15,580.17	$1,527,202.89	$91,632.17
49	$160,475.80	$16,047.58	$1,634,882.65	$98,092.96
50	**Total**	**$417,633.58**		**$1,315,342.02**

So, at the end of 48 years, you and your employer will have contributed about $418,000 into the account, and the account will have earned about $1,315,000 in interest, for a total balance of about $1,733,000.

✓ Checkpoint   Help at *Math*.and**YOU**.com

The graph shows the balance in the retirement account in Example 4. Use the graph to estimate the balance in your account at age 59.

Study Tip

The money invested in a 401(k) retirement plan is tax deferred. As such, these plans follow rules and regulations published by the IRS.

Analyzing a Decreasing Annuity

A **decreasing annuity** is an investment that is earning interest, and from which you make regular withdrawals of a fixed amount.

EXAMPLE 5 **Withdrawing from a Retirement Plan**

You retire at age 68. Your 401(k) retirement plan has a balance of $1 million. How much can you withdraw from your account each year?

SOLUTION

One option is to withdraw all of it. Of course, the deposits were tax deferred, so when you withdraw the funds, you will have to pay income tax on the withdrawals. To decide how much you should withdraw for a retirement income, you should consider the following.

• How much interest is the account earning?

• How long do you expect to live?

• What other income sources do you have?

Suppose the account earns 5%, compounded monthly, and you want an income of $60,000 a year. You can use a spreadsheet to determine how many years the account can continue making payments.

	A	B	C	D	E
1	Month Number	Balance before Withdrawal	Monthly Withdrawal	Interest Earned	Balance after Withdrawal
2	1	$1,000,000.00	$5,000.00	$4,145.83	$999,145.83
3	2	$999,145.83	$5,000.00	$4,142.27	$998,288.11
4	3	$998,288.11	$5,000.00	$4,138.70	$997,426.81
5	4	$997,426.81	$5,000.00	$4,135.11	$996,561.92
~	5	$996,561.92	$5,060.00	$4,131.51	$995,363.43
238	237	$658,081.18	$5,000.00	$2,721.17	$655,802.36
239	238	$655,802.36	$5,000.00	$2,711.68	$653,514.03
240	239	$653,514.03	$5,000.00	$2,702.14	$651,216.17
241	240	$651,216.17	$5,000.00	$2,692.57	$648,908.74
242					

After 20 years, you are 88 years old. Your account still has a balance of nearly $650,000, and you have withdrawn a total of $1.2 million from the account.

✓ **Checkpoint**  Help at *Math*.and**Y⊖U**.com

Rework Example 5 using the following conditions.

a. Withdrawals: $70,000 a year   Earned interest: 5%

b. Withdrawals: $60,000 a year   Earned interest: 4%

c. Withdrawals: $100,000 a year   Earned interest: 6%

What would you do? Explain your reasoning.

There are two basic types of retirement plans: (1) defined contribution plans and (2) defined benefit plans. Examples 3, 4, and 5 describe the first type. The following example describes the second type.

EXAMPLE 6   **Analyzing a Defined Benefit Plan**

You are 55 years old and have worked for a government municipality for 30 years. Your defined benefit retirement plan will pay you 2% of your average income for the last 3 years for each year you have worked. Your average annual income during the past 3 years is $85,000. This will increase by 2% each year. Suppose you live to age 85. At what age should you retire to receive the greatest retirement income?

SOLUTION

Use a spreadsheet to analyze the possibilities.

	A	B	C	D	E	F
1	Age at Retirement	Years Worked	Years of Retirement	Average Working Income	Annual Retirement Income	Total Retirement Income
2	55	30	30	$85,000	$51,000	$1,530,000
3	56	31	29	$86,700	$53,754	$1,558,866
4	57	32	28	$88,434	$56,598	$1,584,737
5	58	33	27	$90,203	$59,534	$1,607,412
	59	34	26	$92,007	$62,565	$1,626,679
	60	35	25	$93,847	$65,693	$1,642,320
	61	36	24	$95,724	$68,921	$1,654,107
	62	37	23	$97,638	$72,252	$1,661,804
	63	38	22	$99,591	$75,689	$1,665,162
	64	39	21	$101,583	$79,235	$1,663,927
	65	40	20	$103,615	$82,892	$1,657,832
	66	41	19	$105,687	$86,663	$1,646,601
	67	42	18	$107,801	$90,552	$1,629,944
	68	43	17	$109,957	$94,563	$1,607,565
	69	44	16	$112,156	$98,697	$1,579,152
	70	45	15	$114,399	$102,959	$1,544,384
	71	46	14	$116,687	$107,352	$1,502,926
	72	47	13	$119,021	$111,879	$1,454,431

From the spreadsheet, you can see that for each year you postpone retirement, your annual retirement income increases. However, the *total* amount of retirement income you will receive peaks at a retirement age of 63 years old.

 **Checkpoint**   Help at *Math*.and*YOU*.com

According to the U.S. Bureau of Labor Statistics, only 20% of employees working for private companies have defined benefit retirement plans. In government jobs, however, defined benefit plan coverage is about 4 times greater—about 79%. Why do you think this is true?

In the United States, it is still true that most municipal workers, such as law enforcement employees and firefighters, have defined benefit retirement plans.

6.4 Exercises

Savings Account **In Exercises 1–4, suppose that you deposit $1000 into a savings account.** *(See Example 1.)*

1. The savings account earns 5%, compounded monthly. Find the balance in the account after each time period.

    a. 10 years

    b. 20 years

2. The savings account earns 6%, compounded monthly. Find the balance in the account after each time period.

    a. 10 years

    b. 20 years

3. The savings account earns 5.5%, compounded monthly. Your friend deposits $700 into a savings account that earns 7.5%, compounded monthly. Which account has the greater balance after 15 years?

4. The savings account earns 6.5%, compounded monthly. Your friend deposits $600 into a savings account that earns 8%, compounded monthly. Which account has the greater balance after 40 years?

5. **Purchase of Manhattan** According to legend, in 1626, Peter Minuit purchased Manhattan Island from Native Americans for $24 worth of trade goods. Suppose the $24 had been deposited into a savings account earning 7%, compounded annually. How much would be in the account in 2014? *(See Example 2.)*

6. **Gift for the Future** You deposit $3000 into a savings account that earns 5%, compounded annually, for future generations of your family. How much will be in the account after 200 years? *(See Example 2.)*

7. **Investment by an Ancestor** Suppose that 350 years ago, 1 of your ancestors deposited $1 into a savings account earning 6%, compounded annually. How much would be in the savings account today? *(See Example 2.)*

8. **Compounding a Penny** Suppose that 500 years ago, the equivalent of 1 penny had been deposited into a savings account earning 8%, compounded annually. How much would be in the savings account today? *(See Example 2.)*

Pension Plan **In Exercises 9–12, use the information below.**
(See Example 3.)

You start your working career when you are 22 years old.
Each month, you deposit $50 into a pension plan that
compounds interest monthly. You continue making
deposits into the plan until you are 67 years old.

9. The plan earns 3%. Find the balance in the account.

10. The plan earns 5%. Find the balance in the account.

11. The plan earns 6%.

    a. Find the balance in the account.

    b. Suppose that you deposit $150 each month instead
    of $50. Find the balance in the account.

    c. Compare the account balances in part (a) and
    part (b).

12. The plan earns 7%.

    a. Find the balance in the account.

    b. Suppose that you wait until you are 32 years old to begin making deposits.
    Find the balance in the account.

    c. Compare the account balances in part (a) and part (b).

13. **401(k) Plan** You start your working career when you are 22 years old. Your
beginning salary is $50,000 per year. Your employer offers a 401(k) matching
retirement plan that amounts to 10% of your salary (5% from you and 5% from
your employer). Assume that your salary increases 2% each year and that the
401(k) plan averages 6% annual returns for the life of the plan. *(See Example 4.)*

    a. How much have you and your employer contributed to your 401(k) plan
    when you are 70 years old?

    b. How much interest has your 401(k) plan earned when you are 70 years old?

    c. What is the total balance in your 401(k) plan when you are 70 years old?

14. **401(k) Plan** You start your working career when you are
22 years old. Your beginning salary is $45,000 per year.
Your employer offers a 401(k) matching retirement plan that
amounts to 10% of your salary (5% from you and 5% from your
employer). Assume that your salary increases 3% each year and
that the 401(k) plan averages 8% annual returns for the life of
the plan. *(See Example 4.)*

    a. How much have you and your employer contributed to
    your 401(k) plan when you are 67 years old?

    b. How much interest has your 401(k) plan earned when you
    are 67 years old?

    c. What is the total balance in your 401(k) plan when you are
    67 years old?

DATA   **Retirement Plan** **In Exercises 15–18, use the information below.** *(See Example 5.)*

You retire at age 67. Your 401(k) retirement plan has a balance of $1 million and compounds interest monthly.

15. The account earns 6%, and you want an income of $75,000 a year.

    a. How much have you withdrawn in total from your account after 10 years?

    b. How much interest has the account earned after 10 years?

    c. After 10 years, what is the balance in your account?

16. The account earns 8%, and you want an income of $90,000 a year.

    a. How much have you withdrawn in total from your account after 20 years?

    b. How much interest has the account earned after 20 years?

    c. After 20 years, what is the balance in your account?

17. The account earns 5%. How many years can the account support withdrawals of $60,000 a year?

18. The account earns 5%. How many years can the account support withdrawals of $85,000 a year?

DATA   **Defined Benefit Plan** **In Exercises 19 and 20, use the information below.** *(See Example 6.)*

You are 55 years old and you have worked for a government municipality for 30 years. Your defined benefit retirement plan will pay you 2% of your average income for the last 3 years for each year you have worked. Your average annual income during the past 3 years is $72,000. Suppose you live to age 85.

19. Your salary will increase by 1% each year. At what age should you retire to receive the greatest retirement income?

20. Your salary will increase by 3% each year. At what age should you retire to receive the greatest retirement income?

▶ Extending Concepts

Savings Goals **In Exercises 21 and 22, use the information below.**

For an increasing annuity, the monthly deposit M that you must make for n months, at an annual percentage rate of r (in decimal form), to achieve a balance of A is

$$M = \frac{A\left(\frac{r}{12}\right)}{\left(1 + \frac{r}{12}\right)^n - 1}.$$

21. You start saving for retirement at age 25. You want to have $1 million when you retire in 42 years. You invest in a savings plan that earns 6%, compounded monthly.

   a. How much should you deposit each month?

   b. Suppose you wait until you are 30 to start saving. How much more do you have to deposit each month compared to the amount in part (a)?

22. You want to have $20,000 to help pay for your child's college education in 18 years. You invest in a savings plan that earns 4.8%, compounded monthly.

   a. How much should you deposit each month?

   b. Suppose you want to have the money in 10 years. How much more do you have to deposit each month compared to the amount in part (a)?

Annual Percentage Yield **In Exercises 23–26, use the information below.**

The *annual percentage yield* (APY) is the rate at which an investment increases each year. The formula for the APY of an investment with an annual percentage rate of r that is compounded n times a year is

$$APY = \left(1 + \frac{r}{n}\right)^n - 1.$$

23. Find the APY for an investment that earns 6% for each compounding period.

   a. Daily   **b.** Monthly   **c.** Quarterly

   d. Semiannually   **e.** Annually

24. Find the APY for an investment that earns 7% for each compounding period.

   a. Daily   **b.** Monthly   **c.** Quarterly

   d. Semiannually   **e.** Annually

25. For what compounding period is the APY the same as the APR? Explain your reasoning.

26. Which of the following earns more interest annually?

   a. An investment with an APY of 6%

   b. An investment with an APR of 5.9%, compounded monthly

6.3–6.4   Quiz

Buying a House **In Exercises 1–4, use the information below.**

You want to buy a $100,000 house. You plan to make a
$20,000 down payment.

1. Each month for 5 years, you deposit $300 into a savings account that
   earns 5%, compounded monthly. After 5 years, you use the money in
   the account to make the down payment.

   a. How much is left in the account?

   b. You leave the remaining amount from part (a) in the account.
   Assuming you do not make any more deposits, how much is in
   the account after 14 years?

2. You take out a home mortgage for $80,000 for 30 years. Compare
   the total interest you pay for the annual percentage rates.

   a. 5%   **b.** 7%

3. You take out a home mortgage for $80,000 for 30 years at 5.5%. Each
   month, you make the regular payment of $454.23 plus an additional $50.

   a. How much sooner do you pay off the mortgage?

   b. How much do you save in interest?

4. You take out a home mortgage for $80,000 for 30 years at 6%. After 5 years,
   you move to a different state and sell the home for $140,250.

Expenses and Savings Related to Buying	
Cost of home: $100,000	Realtor's fee: 5% of cost of home
Down payment: $20,000	Home insurance: $600 per year
Mortgage: $80,000	Property tax: 1% of cost of home per year
Monthly payment: $479.64	Home repairs: $6000
Closing costs: 5% of cost of home	Income tax savings (interest): $5000

   Compare the costs of buying the home and renting a
   comparable home for $550 per month. Assume that if
   you did not buy the home, you could have invested the
   down payment and earned $4000 in interest.

Retirement Plan **In Exercises 5 and 6, use the
information below.**

You start your working career when you are 22 years old.
Each month, you deposit $150 into a retirement plan that
earns 6%, compounded monthly. You continue making
deposits into the plan until you are 67 years old.

5. Find the balance in the account.

6. How many years can the account support withdrawals of $70,000 a year?

Chapter 6   Summary

Section Objectives	How does it apply to you?

Section 1

Read promissory notes and find due dates.	➡	It is important to know all the terms and conditions of a loan so that you repay the correct amount on time. *(See Examples 1 and 2.)*
Find the cost of credit for a loan.	➡	When you obtain a loan, you normally must pay a cost of credit in addition to the loan proceeds. *(See Example 3.)*
Find the annual percentage rate for a loan.	➡	The annual percentage rate (APR) is the rate at which interest is calculated. *(See Example 6.)*

Section 2

Create an amortization table.	➡	An amortization table shows how the balance of a loan decreases. *(See Examples 1 and 2.)*
Analyze the cost of buying on credit.	➡	To reduce the amount of interest you pay, you should pay more than the minimum payment on your credit card bill. *(See Example 3.)*
Analyze credit in the United States.	➡	You can compare the indebtedness of households in the United States in different years by using population size and the CPI. *(See Example 6.)*

Section 3

Compare rates and terms for a home mortgage.	➡	The rate and term of a home mortgage affect the size of your monthly payment and the total amount of interest that you pay. *(See Examples 1 and 2.)*
Analyze the effect of making principal payments.	➡	You can pay off your mortgage early and save in interest when you make additional payments. *(See Example 3.)*
Compare the costs of buying and renting.	➡	You should consider all the costs and savings involved in buying and renting to determine what is best for you. *(See Example 5.)*

Section 4

Find the balance in a savings account.	➡	You can find how much is in a savings account after a period of time. *(See Examples 1 and 2.)*
Find the balance in an increasing annuity.	➡	You can find the balance in an account in which you make repeated deposits and determine how much you will have in the account when you retire. *(See Examples 3 and 4.)*
Analyze a decreasing annuity.	➡	You can determine how much you can withdraw annually from an interest-earning account when you retire. You can also determine the age at which you should retire to maximize your retirement income. *(See Examples 5 and 6.)*

Chapter 6 Review Exercises

Section 6.1

Payday Loans Payday loans are short-term loans that are typically due the next time you receive a paycheck. The tables show rates at two different payday loan companies. In Exercises 1–6, use the tables.

You Receive	Fee	Write the Check For
Amount Financed	14-day Finance Charge	Total of Payments
$100.00	$17.65	$117.65
$150.00	$26.48	$176.48
$200.00	$35.30	$235.30
$255.00	$45.00	$300.00

Term: 30 days		
Amount	Fees	Total
$100	$18.62	$118.62
$200	$37.24	$237.24
$300	$55.86	$355.86
$400	$74.48	$474.48
$500	$93.10	$593.10
$600	$111.72	$711.72
$700	$130.34	$830.34
$800	$148.96	$948.96
$900	$167.58	$1067.58
$1000	$186.20	$1186.20

1. You obtain a 30-day payday loan on September 14. What is the due date?

2. You obtain a $100 payday loan with a 14-day term. How much is due at the end of 14 days?

3. Complete the Truth in Lending disclosure for a $255 payday loan with a 14-day term.

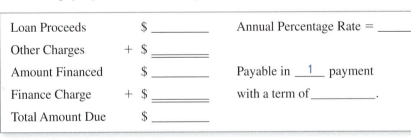

Loan Proceeds	$ _____	Annual Percentage Rate = _____ %
Other Charges	+ $ _____	
Amount Financed	$ _____	Payable in __1__ payment
Finance Charge	+ $ _____	with a term of _____.
Total Amount Due	$ _____	

4. Complete the Truth in Lending disclosure for a $500 payday loan with a 30-day term.

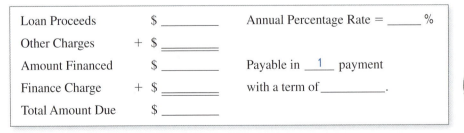

Loan Proceeds	$ _____	Annual Percentage Rate = _____ %
Other Charges	+ $ _____	
Amount Financed	$ _____	Payable in __1__ payment
Finance Charge	+ $ _____	with a term of _____.
Total Amount Due	$ _____	

5. Find the APR for a $200 payday loan with a term of (a) 14 days and (b) 30 days.

6. Does the APR for a 30-day loan change as the amount increases? Explain.

Section 6.2

Home Equity Loans Home equity loans are loans in which a homeowner borrows money using the equity of the home as collateral. The table shows home equity loan rates and terms. In Exercises 7–11, use the table.

Home Equity Loans

Loan Type	APR	Term
Home equity-subordinate mortgage (0–60 months)	5.24	60
Home equity-subordinate mortgage (61–84 months)	5.49	84
Home equity-subordinate mortgage (85–144 months)	5.99	144
Home equity-subordinate mortgage (145–180 months)	6.49	180

7. You choose a term of 144 months for a $50,000 home equity loan. What is the monthly payment?

8. Create an amortization table showing how the balance of the loan decreases.

9. What is the cost of credit for the loan?

10. When is the remaining balance less than one-half of the original loan amount?

11. You receive a 0.25% reduction in APR by initiating automatic monthly payments from your checking account. How much do you save over the term of the loan by doing this?

12. **Comparing Loans** Your credit score of 660 is too low to receive the loan above. You use the Internet to find the home equity loan options shown. You receive the APR discount, and the term is 10 years. How much more do you pay in interest with Radiance Bank than with Sunny Bank?

Results (4)   Calculator

Print

	Loan Type	Lender	Credit Score	APR	Fees and Conditions
☐	$50K home equity loan	**Any Bank** Posted: 03/02/12 Contact	600-850	10.04	0.25% APR disc w/auto debit; $500 fee; $10,000 min
☐	$50K home equity loan	**My Bank** Posted: 03/02/12 Contact	700-850	5.49	
☐	$50K home equity loan	**Sunny Bank** Posted: 03/02/12 Contact	660-850	7.14	0.25% APR disc w/auto debit; $5,000 min
☐	$50K home equity loan	**Radiance Bank** Posted: 03/02/12 Contact	660-850	7.34	0.25% APR disc w/auto debit; $5,000 min

Section 6.3

Home Mortgage In Exercises 13–17, use the information below.

You take out a home mortgage for $170,000 for 30 years at 5%. The regular monthly payment is $912.60.

13. Compare the total interest you pay to the total interest of a home mortgage for $170,000 for 30 years at 7%.

14. Compare the total interest you pay to the total interest of a home mortgage for $170,000 for 15 years at 5%.

DATA 15.   Each month, you make the regular payment of $912.60 plus an additional $50.

   a.   How much sooner do you pay off the mortgage?

   b.   How much do you save in interest?

16. The monthly payment for a 5-year balloon mortgage for $170,000 is equal to that of a 30-year mortgage with an annual percentage rate of 5%. Find the balloon payment and the total interest that you pay.

17. After 5 years, you move to a different state and sell the house for $253,350.

Expenses and Savings Related to Buying	
Cost of home: $205,000	Realtor's fee: 5% of cost of home
Down payment: $35,000	Home insurance: $1000 per year
Mortgage: $170,000	Property tax: 1.5% of cost of home per year
Monthly payment: $912.60	Home repairs: $10,000
Closing costs: 5% of cost of home	Income tax savings (interest): $8000

Compare the costs of buying the home and renting a comparable home for $1200 per month. Assume that if you did not buy the home, you could have invested the down payment and earned $8000 in interest.

18. **Housing Bubble** Describe what happens to the prices of homes during a bubble. What happens when the bubble bursts?

Section 6.4

Savings Account In Exercises 19 and 20, suppose that you deposit $3000 into a savings account that earns 4.5%, compounded monthly.

19. Find the balance in the account after each time period.

    a. 16 years   **b.** 32 years

20. Your friend deposits $2500 into a savings account that earns 6.5%, compounded monthly. Which account has the greater balance after 10 years?

Retirement Plan In Exercises 21 and 22, use the information below.

You start your working career when you are 22 years old. Each month, you deposit $200 into a retirement plan that earns 8%, compounded monthly. You continue making deposits into the plan until you are 67 years old.

21. Find the balance in the account.

22. You want an income of $100,000 a year.

    a. How much have you withdrawn in total from your account after 10 years?

    b. How much interest has the account earned after 10 years?

    c. After 10 years, what is the balance in your account?

23. **Gift for the Future** You deposit $5 into a savings account that earns 6%, compounded annually. You stipulate that the balance will be divided evenly among your living heirs in 500 years. Find the balance in the account after each time period.

    a. 50 years   **b.** 100 years

    c. 150 years   **d.** 200 years

    e. 250 years   **f.** 500 years

24. **Defined Benefit Plan** You are 55 years old and you have worked for a government municipality for 30 years. Your defined benefit retirement plan will pay you 2% of the average income for the last 3 years for each year you have worked. Your average annual income during the past 3 years is $60,000. This will increase by 3.5% each year. Suppose you live to age 85. At what age should you retire to receive the greatest retirement income?

7 The Mathematics of Patterns & Nature

7.1 Linear Patterns

▶ Recognize and describe a linear pattern.

▶ Use a linear pattern to predict a future event.

▶ Recognize a proportional pattern.

7.2 Exponential Patterns

▶ Recognize and describe an exponential pattern.

▶ Use an exponential pattern to predict a future event.

▶ Compare exponential and logistic growth.

7.3 Quadratic Patterns

▶ Recognize and describe a quadratic pattern.

▶ Use a quadratic pattern to predict a future event.

▶ Compare linear, quadratic, and exponential growth.

7.4 Fibonacci & Other Patterns

▶ Recognize and describe the Fibonacci pattern.

▶ Analyze geometric Fibonacci patterns.

▶ Recognize and describe other patterns in mathematics.

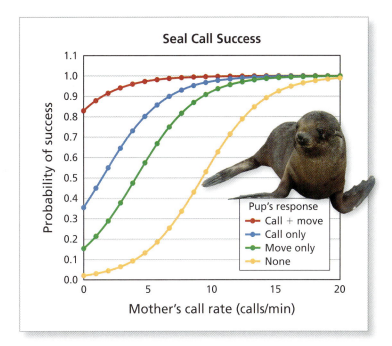

Seal Call Success

Example 6 on page 317 compares a mother seal's call rate with the probability that the mother will relocate her baby after going to sea. What is the mother's and baby's best strategy?

7.1 Linear Patterns

▶ Recognize and describe a linear pattern.

▶ Use a linear pattern to predict a future event.

▶ Recognize a proportional pattern.

Recognizing a Linear Pattern

A sequence of numbers has a **linear pattern** when each successive number increases (or decreases) by the same amount.

EXAMPLE 1 Recognizing a Linear Pattern

Anthropologists use tables like those at the left to estimate the height of a person based on part of the person's skeleton.

a. Does the table relating the length of a man's femur (upper leg bone) to the man's height represent a linear pattern?

b. The femur length of a Roman soldier is 18 inches. What was the height of the Roman soldier?

SOLUTION

a. To determine whether the table represents a linear pattern, find the *differences* between consecutive terms.

	A Femur Length (in.)	B Height (in.)
1	Femur Length (in.)	Height (in.)
2	14	58.32
3	15	60.20
4	16	62.08
5	17	63.96
6	18	65.84
7	19	67.72
8	20	69.60
9	21	71.48
10	22	73.36
11	23	75.24
12	24	77.12

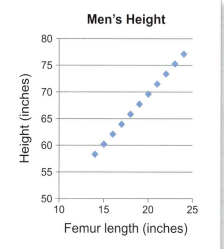

1.88
1.88
1.88
1.88

Men's Height

Each time the femur length increases by 1 inch, the height of the man increases by 1.88 inches. So, the pattern is linear.

b. From the table, an 18-inch femur corresponds to a height of about 66 inches. In other words, the Roman soldier was about 5' 6".

✓ **Checkpoint** Help at *Math*.and**YOU**.com

For women, femur length and height are related as follows.

Height in inches = 1.95(femur length) + 28.7

c. Use a spreadsheet to make a table for this formula.

d. Use the spreadsheet to graph the data in the table and verify that the points on the graph lie on a line.

	A	B
1	Foot Length (inches)	Shoe Size
2	9.30	6.0
3	9.47	6.5
4	9.64	7.0
5	9.81	7.5
6	9.98	8.0
7	10.15	8.5
8	10.32	9.0
9	10.49	9.5
10	10.66	10.0
11	10.83	10.5
12	11.00	11.0
13	11.17	11.5
14	11.34	12.0
15	11.51	12.5
16	11.68	13.0
17	11.85	13.5
18	12.02	14.0
19	12.19	14.5
20	12.36	15.0

EXAMPLE 2   **Recognizing a Linear Pattern**

The table relates a man's shoe size to the length of his foot.

To measure your foot, trace it on a piece of paper. Mark the front and back of your foot. Then measure the length.

a. Does the table represent a linear pattern? Explain.

b. Use a spreadsheet to graph the data. Is the graph linear?

SOLUTION

a. To determine whether the table represents a linear pattern, find the differences between consecutive terms.

Notice that each time the foot length increases by 0.17 (about 1/6) inch, the shoe size increases by a half size. So, the pattern is linear.

b. The points on the graph lie on a line. So, the graph is linear.

 **Checkpoint**   Help at *Math*.and**YOU**.com

Use the table at *Math.andYou.com* that relates a woman's shoe size to the length of her foot.

c. Does the table represent a linear pattern? Explain.

d. Use a spreadsheet to graph the data. Is the graph linear?

Using a Linear Pattern to Predict a Future Event

One common use of linear patterns is predicting future events.

EXAMPLE 3 **Predicting a Future Event**

The graph shows the ages of American women at the time of their first marriage from 1960 through 2010. Use the graph to predict the age in 2020.

SOLUTION

The pattern looks roughly linear. One way to estimate the age in 2020 is to draw a "best-fitting line" to approximate the data. Then use the line to estimate the age in 2020.

From the graph, it appears that the age of women at their first marriage in 2020 will be about 27.5.

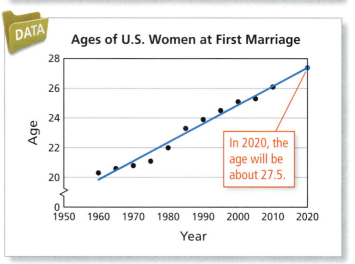

DATA

In 2020, the age will be about 27.5.

Study Tip

The procedure described in Example 3 is called *linear regression*. It is a commonly used procedure in statistics.

✓ **Checkpoint**

Help at *Math*.and**YOU**.com

Use the graph below to predict the marriage rate for women in the United States in 2020. How do these data relate to the data in Example 3?

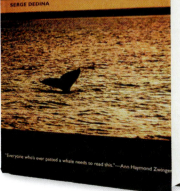

| EXAMPLE 4 | **Describing a Linear Pattern in Nature** |

The figure shows the annual northward migration of Canada geese. Describe any linear patterns that you see in the figure.

SOLUTION

Through the central United States, the migration appears to be moving north at a rate of about 150 miles every 10 days.

Feb 10	Feb 20	Mar 1	Mar 10	Mar 20	Mar 30	Apr 10	Apr 20	Apr 30
0 mi	130 mi	270 mi	420 mi	630 mi	760 mi	900 mi	1040 mi	1200 mi

 **Checkpoint**

Help at Math.andYOU.com

The longest known migration of a mammal is that of the gray whale. It travels the 6000 miles between Baja California, Mexico, and the Bering Sea each spring and fall. Traveling at a rate of 4 miles per hour, how long does it take a gray whale to migrate each spring?

In this book, Serge Dedina discusses the conservation of the gray whale in Baja California, Mexico.

Recognizing a Proportional Pattern

A pattern with two variables is **proportional** when one of the variables is a constant multiple of the other variable. Proportional patterns are also linear.

> **EXAMPLE 5** **Recognizing a Proportional Pattern**

You hang different weights from a spring. You then measure the distance the spring stretches.

a. Describe the pattern. Is the distance the spring stretches proportional to the weight?

b. How much does the spring stretch when you hang 6 pounds from it?

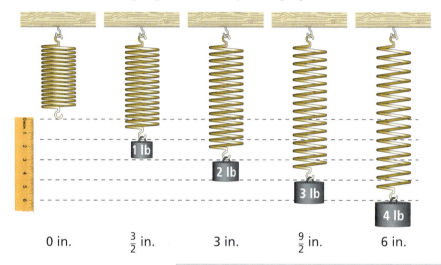

| 0 in. | $\frac{3}{2}$ in. | 3 in. | $\frac{9}{2}$ in. | 6 in. |

Throughout human history, most discoveries have occurred because humans observed patterns. From the patterns, they then formulated laws of nature. The law illustrated in Example 5 is called *Hooke's Law*, after the English scientist Robert Hooke. The law states that the distance a spring stretches is proportional to the weight hanging on the spring.

SOLUTION

a. You can see that the distance the spring stretches is 3/2 times the weight in pounds.

$$\frac{3}{2} \times 0 = 0 \qquad \frac{3}{2} \times 1 = \frac{3}{2}$$

$$\frac{3}{2} \times 2 = 3 \qquad \frac{3}{2} \times 3 = \frac{9}{2}$$

$$\frac{3}{2} \times 4 = 6$$

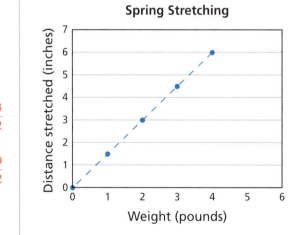

So, the distance the spring stretches *is* proportional to the weight.

b. When you hang 6 pounds from the spring, it will stretch (3/2 × 6), or 9 inches.

✓ Checkpoint

Help at *Math*.andY♥U.com

The distance that a spring stretches depends on its elasticity. Data for a different spring are shown in the table.

c. Is this spring more or less elastic than the spring in Example 5? Explain.

d. How much will this spring stretch when you hang 7 pounds from it?

Weight (pounds)	Distance stretched (inches)
0	0
1	$\frac{3}{4}$
2	$\frac{3}{2}$
3	$\frac{9}{4}$
4	3

EXAMPLE 6   Recognizing a Proportional Pattern

Is the following statement true? Explain your reasoning.

As a human grows, its skull height is proportional to its total height.

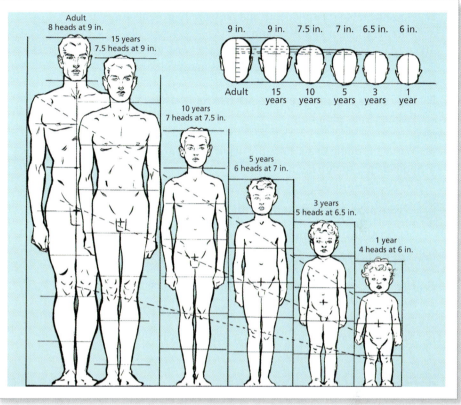

Adult
8 heads at 9 in.

15 years
7.5 heads at 9 in.

9 in. 9 in. 7.5 in. 7 in. 6.5 in. 6 in.

Adult 15 years 10 years 5 years 3 years 1 year

10 years
7 heads at 7.5 in.

5 years
6 heads at 7 in.

3 years
5 heads at 6.5 in.

1 year
4 heads at 6 in.

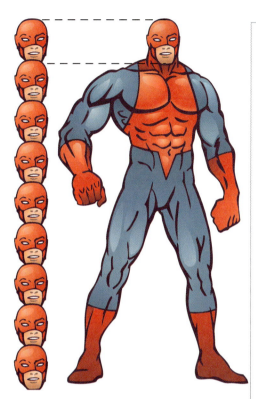

Most adult humans are between 6.5 and 8 heads tall. Comic action heros, however, are often 9 heads tall.

SOLUTION

This is *not* true of humans. A baby's skull height is about one-fourth of its total height. An adult's skull height is only about one-eighth of its total height.

In the graph, notice that the total height is not a constant multiple of the skull height.

Human Body Proportions

Proportional line of 1-to-8 ratio

(graph: x-axis "Skull height (inches)" from 0 to 10; y-axis "Total height (inches)" from 0 to 80)

✓ **Checkpoint**

Help at *Math.andYOU.com*

Baby reptiles are miniature versions of the adults. So, for reptiles, it is true that "as a reptile grows, its skull length is proportional to its total length." Some horned lizards can grow up to a length of 8 inches. Use the photo to estimate the ratio of the lizard's skull length to its total length.

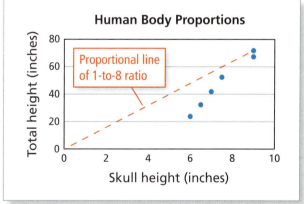

7.1 Exercises

 Freshwater **The table shows the pressures at various depths of freshwater. In Exercises 1–4, use the table.** *(See Examples 1 and 2.)*

Depth (feet)	Pressure (pounds per square inch)
0	14.70
10	19.03
20	23.36
30	27.69
40	32.02
50	36.35
60	40.68
70	45.01
80	49.34
90	53.67
100	58.00

1. Does the table relating depth and pressure represent a linear pattern? Explain your reasoning.

2. Use a spreadsheet to graph the data. Is the graph linear?

3. How much does the pressure increase for every foot of depth? Explain your reasoning.

4. Write a formula that relates the depth in feet to the pressure in pounds per square inch.

 Seawater **For seawater, depth and pressure are related as follows.**

Pressure in pounds per square inch = 0.445(depth in feet) + 14.7

In Exercises 5–8, use this formula. *(See Examples 1 and 2.)*

5. Use a spreadsheet to make a table for the formula. Then graph the data and verify that the points on the graph lie on a line.

6. The recreational diving limit for a scuba diver is 130 feet. Find the pressure at this depth.

7. The wreck of the *Lusitania* lies about 300 feet beneath the Celtic Sea. Find the pressure at this depth.

8. The wreck of the *Titanic* lies about 12,500 feet beneath the Atlantic Ocean. Find the pressure at this depth.

The stern of the *Titanic*, pictured above, was crushed by water pressure as it sank to the bottom of the ocean.

Tree Growth The figure shows the circumference of a tree over a 4-year period. In Exercises 9 and 10, use the figure. *(See Example 3.)*

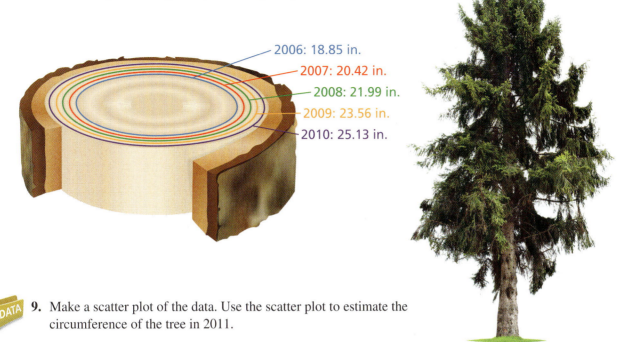

2006: 18.85 in.
2007: 20.42 in.
2008: 21.99 in.
2009: 23.56 in.
2010: 25.13 in.

DATA **9.** Make a scatter plot of the data. Use the scatter plot to estimate the circumference of the tree in 2011.

10. Find the diameter of the tree for each year. Make a scatter plot showing the diameter for each year. Use the scatter plot to predict the diameter of the tree in 2011.
$\left(\text{The formula for the diameter of a circle is } d = \dfrac{C}{\pi}. \right)$

11. Black-and-White Warbler The figure shows the migration of the black-and-white warbler. Describe any linear patterns that you see in the figure. *(See Example 4.)*

12. Distance Traveled A black-and-white warbler flies about 20 miles per day across the United States during migration. Complete the table to estimate the distance that the black-and-white warbler travels between March 30 and each date. *(See Example 4.)*

Day	Distance
March 30	0
March 31	
April 2	
April 6	
April 13	
April 21	
May 1	

May 30
May 20
May 10
May 1
April 20
April 10
March 30
March 20
March 20
March 10
March 10
— Migration line

Absorbance The figure shows light passing through a glass container that contains a substance. Absorbance is a unitless measure of the amount of light that a substance absorbs as light passes through it. In Exercises 13–16, (a) describe the pattern of the absorbance values, (b) make a scatter plot of the data, and (c) predict the next absorbance value in the pattern. *(See Examples 5 and 6.)*

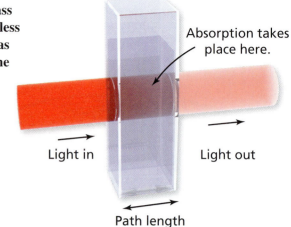

Absorption takes place here.

Light in

Light out

Path length

13. The table shows the absorbance of light with a wavelength of 630 nanometers for a solution of the food dye Blue No. 1 in a 1-centimeter glass container at various concentrations.

Concentration (mg/L)	Absorbance
0	0
1	0.164
2	0.328
3	0.492
4	0.656

14. The table shows the absorbance of light with a wavelength of 625 nanometers for a solution of the food dye Green No. 3 in a 1-centimeter glass container at various concentrations.

Concentration (mg/L)	Absorbance
0	0
2	0.312
4	0.624
6	0.936
8	1.248

15. The table shows the absorbance of light with a wavelength of 527 nanometers for a solution of the food dye Red No. 3 with a concentration of 4 milligrams per liter in glass containers of various path lengths.

Path length (cm)	Absorbance
0	0
0.2	0.088
0.4	0.176
0.6	0.264
0.8	0.352

16. The table shows the absorbance of light with a wavelength of 500 nanometers for a solution of the food dye Red No. 40 with a concentration of 5 milligrams per liter in glass containers of various path lengths.

Path length (cm)	Absorbance
0	0
0.3	0.078
0.6	0.156
0.9	0.234
1.2	0.312

17. Yellow No. 5 The absorbance of light with a wavelength of 428 nanometers for a solution of the food dye Yellow No. 5 in a 1-centimeter glass container is proportional to the concentration of Yellow No. 5. The absorbance of a solution with a concentration of 2 milligrams per liter is 0.106. What is the absorbance of a solution with a concentration of 5 milligrams per liter? *(See Examples 5 and 6.)*

18. Yellow No. 6 The absorbance of light with a wavelength of 484 nanometers for a solution of the food dye Yellow No. 6 with a concentration of 6 milligrams per liter is proportional to the path length. The absorbance of the solution in a 0.5-centimeter glass container is 0.162. What is the absorbance of the solution in a 1.1-centimeter glass container? *(See Examples 5 and 6.)*

▶ Extending Concepts

Linear Regression in Excel In Exercises 19 and 20, use the information below.

You can use Excel to find the best-fitting line for a data set. Enter the data into a spreadsheet. Make a scatter plot of the data. Click on the scatter plot. From the chart menu, choose "Add Trendline." Click on the "Options" tab. Check the box labeled "Display equation on chart." Click "OK." This will add the best-fitting line and its equation to your scatter plot.

19. The data set relates the number of chirps per second for striped ground crickets and the temperature in degrees Fahrenheit.

Chirps per second	Temperature (°F)
20.0	88.6
16.0	71.6
19.8	93.3
18.4	84.3
17.1	80.6
15.5	75.2
14.7	69.7
17.1	82.0

Chirps per second	Temperature (°F)
15.4	69.4
16.2	83.3
15.0	79.6
17.2	82.6
16.0	80.6
17.0	83.5
14.4	76.3

(*Source:* George W. Pierce, *The Song of Insects,* Harvard University Press, 1948)

a. Enter the data into a spreadsheet and make a scatter plot of the data.

b. Graph the best-fitting line on your scatter plot and find its equation.

c. Estimate the temperature when there are 19 chirps per second.

d. Estimate the temperature when there are 22 chirps per second.

20. Data were collected from a sample of 414 infants, grouped by month of birth. The data set relates the average monthly temperature (in degrees Fahrenheit) 6 months after the infants were born and the average age (in weeks) at which the infants learned to crawl.

Average temperature (°F)	Average crawling age (in weeks)
66	29.84
73	30.52
72	29.70
63	31.84
52	28.58
39	31.44

Average temperature (°F)	Average crawling age (in weeks)
33	33.64
30	32.82
33	33.83
37	33.35
48	33.38
57	32.32

(*Source:* Janette Benson, *Infant Behavior and Development,* 1993)

a. Enter the data into a spreadsheet and make a scatter plot of the data.

b. Graph the best-fitting line on your scatter plot and find its equation.

c. Estimate the average crawling age for infants when the average temperature 6 months after they are born is 55°F.

d. Estimate the average crawling age for infants when the average temperature 6 months after they are born is 475°F. Is this temperature reasonable? Is your estimate reasonable? Explain your reasoning.

7.2 Exponential Patterns

▶ Recognize and describe an exponential pattern.

▶ Use an exponential pattern to predict a future event.

▶ Compare exponential and logistic growth.

Recognizing an Exponential Pattern

A sequence of numbers has an **exponential pattern** when each successive number increases (or decreases) by the same percent. Here are some examples of exponential patterns you have already studied in this text.

- Growth of a bacteria culture (Example 1, page 152)
- Growth of a mouse population during a mouse plague (Example 3, page 154)
- Decrease in the atmospheric pressure with increasing height (Example 2, page 175)
- Decrease in the amount of a drug in your bloodstream (Example 3, page 176)

EXAMPLE 1   Recognizing an Exponential Pattern

Describe the pattern for the volumes of consecutive chambers in the shell of a chambered nautilus.

Chamber 7: 1.207 cm³
Chamber 6: 1.135 cm³
Chamber 5: 1.068 cm³
Chamber 4: 1.005 cm³
Chamber 3: 0.945 cm³
Chamber 2: 0.889 cm³
Chamber 1: 0.836 cm³

SOLUTION

It helps to organize the data in a table.

DATA

Chamber	1	2	3	4	5	6	7
Volume (cm³)	0.836	0.889	0.945	1.005	1.068	1.135	1.207

Begin by checking the differences of consecutive volumes to conclude that the pattern is *not linear*. Then find the *ratios* of consecutive volumes.

$$\frac{0.889}{0.836} \approx 1.063 \qquad \frac{0.945}{0.889} \approx 1.063 \qquad \frac{1.005}{0.945} \approx 1.063$$

$$\frac{1.068}{1.005} \approx 1.063 \qquad \frac{1.135}{1.068} \approx 1.063 \qquad \frac{1.207}{1.135} \approx 1.063$$

The volume of each chamber is about 6.3% greater than the volume of the previous chamber. So, the pattern is exponential.

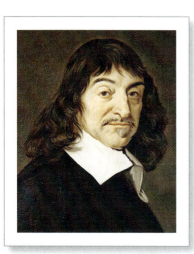

The exponential growth pattern of the chambers in a chambered nautilus was first recorded by the French philosopher René Descartes in 1638.

✓ **Checkpoint**   Help at

Use a spreadsheet to extend the pattern in Example 1 to 24 chambers. Then make a scatter plot of the data and describe the graph.

1000 B.C. Approximate beginning of the Iron Age

2000 B.C. Beginning of the Middle Kingdom in Egypt

3000 B.C. Stonehenge is built in England.

4000 B.C. Civilization begins to develop in Mesopotamia.

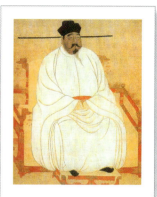

EXAMPLE 2   **Describing an Exponential Pattern**

From 5000 B.C. through 1500 A.D., the population of Earth followed a growth pattern that was roughly exponential. Describe the growth pattern in words.

1 B.C. Augustus Caesar controlled most of the Mediterranean world.

1000 A.D. The Song Dynasty in China had about one-fifth of the world's population.

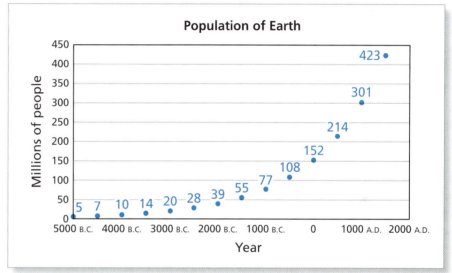

Population of Earth

SOLUTION

Begin by finding the ratios of consecutive populations.

$$\frac{7}{5} = 1.40 \qquad \frac{10}{7} \approx 1.43 \qquad \frac{14}{10} = 1.40 \qquad \frac{20}{14} \approx 1.43 \qquad \frac{28}{20} = 1.40$$

$$\frac{39}{28} \approx 1.39 \qquad \frac{55}{39} \approx 1.41 \qquad \frac{77}{55} = 1.40 \qquad \frac{108}{77} \approx 1.40 \qquad \frac{152}{108} \approx 1.41$$

$$\frac{214}{152} \approx 1.41 \qquad \frac{301}{214} \approx 1.41 \qquad \frac{423}{301} \approx 1.41$$

From these Earth population estimates, you can say that Earth's population was increasing by about 40% every 500 years.

✓ **Checkpoint**

Help at **Math.andYOU.com**

Did the growth pattern described in Example 2 continue through the next 500 years, up through the year 2000? If not, why didn't the pattern continue?

The mission of the U.S. Fish and Wildlife Service is "to work with others to conserve, protect, and enhance fish, wildlife, and plants and their habitats for the continuing benefit of the American people."

Using an Exponential Pattern to Predict a Future Event

EXAMPLE 3   **Predicting a Future Event**

It is estimated that in 1782 there were about 100,000 nesting bald eagles in the United States. By the 1960s, this number had dropped to about 500 nesting pairs. This decline was attributed to loss of habitat, loss of prey, hunting, and the use of the pesticide DDT.

The 1940 Bald Eagle Protection Act prohibited the trapping and killing of the birds. In 1967, the bald eagle was declared an endangered species in the United States. With protection, the nesting pair population began to increase, as shown in the graph. Finally, in 2007, the bald eagle was removed from the list of endangered and threatened species.

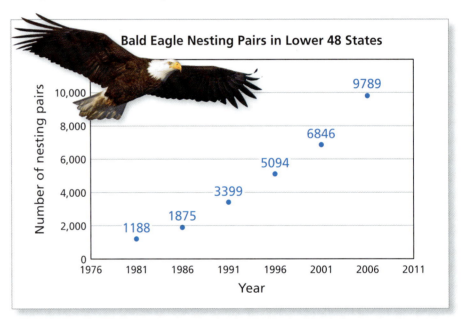

Assume that this recovery pattern continued. Estimate the number of nesting pairs of bald eagles in the lower 48 states in 2011.

SOLUTION

Begin by finding the ratios of consecutive populations.

$$\frac{1875}{1188} \approx 1.58 \qquad \frac{3399}{1875} \approx 1.81 \qquad \frac{5094}{3399} \approx 1.50$$

$$\frac{6846}{5094} \approx 1.34 \qquad \frac{9789}{6846} \approx 1.43$$

From the data, it appears that the population increased by about 50% every 5 years. So, from 2006 to 2011, you can estimate that the population increased to 1.5(9789), or about 14,700 nesting pairs.

 **Checkpoint**

Help at *Math*.and**Y☺U**.com

Suppose the recovery pattern continued for another 5 years. Predict the number of nesting pairs in 2016.

Study Tip

Using a computer and an exponential regression program, you can find that the best estimate for the increase (every 5 years) for the data in Example 3 is 52.8%.

EXAMPLE 4   **Predicting a Future Event**

Discuss the following graph prepared by the World Wildlife Fund. What exponential pattern can you see in the graph?

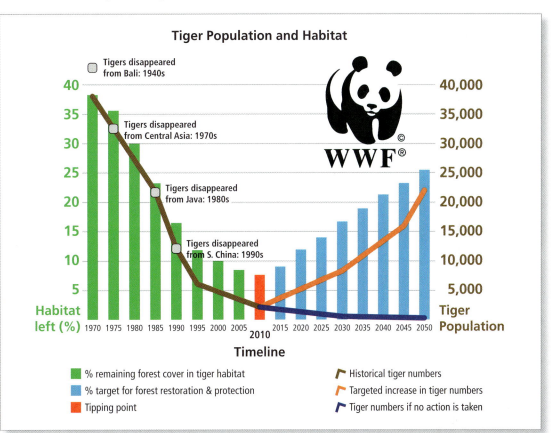

Tiger Population and Habitat

Tigers disappeared from Bali: 1940s

Tigers disappeared from Central Asia: 1970s

Tigers disappeared from Java: 1980s

Tigers disappeared from S. China: 1990s

Habitat left (%)   1970 1975 1980 1985 1990 1995 2000 2005 **2010** 2015 2020 2025 2030 2035 2040 2045 2050   **Tiger Population**

Timeline

■ % remaining forest cover in tiger habitat   ┌ Historical tiger numbers
■ % target for forest restoration & protection   ┌ Targeted increase in tiger numbers
■ Tipping point   ┌ Tiger numbers if no action is taken

SOLUTION

From the graph, the estimated tiger population appears to be decreasing with an exponential pattern, as follows.

1985	1990	1995	2000	2005
22,000	12,500	6000	5000	3800

$$\frac{12,500}{22,000} \approx 0.568 \qquad \frac{6000}{12,500} = 0.480 \qquad \frac{5000}{6000} \approx 0.833 \qquad \frac{3800}{5000} = 0.760$$

Although the rate of decrease in each 5-year period varies, you need to remember that these data are difficult to collect and consequently are only an approximation. Even so, it appears that the tiger population is decreasing by almost 70% every 5 years.

✓ **Checkpoint**   Help at

Estimate the percent of remaining tiger habitat from 1985 through 2010. Describe the pattern.

Comparing Exponential and Logistic Growth

Exponential growth can only occur for a limited time in nature. Eventually, the quantity that is growing reaches physical boundaries. The resulting growth is called **logistic growth.**

Yeasts are single-celled organisms. Most reproduce by asexual budding (splitting to form two new yeast cells). When yeast cells lack oxygen, they die and produce alcohol. This process is called fermentation.

EXAMPLE 5 Comparing Exponential and Logistic Growth

The graph shows the growth of a culture of yeast cells that is introduced into a container of grape juice. Describe the growth.

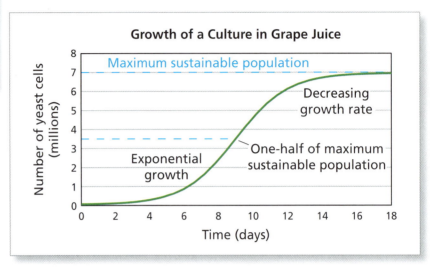

Growth of a Culture in Grape Juice

Maximum sustainable population

Decreasing growth rate

Exponential growth

One-half of maximum sustainable population

Number of yeast cells (millions) vs. Time (days)

SOLUTION

During the exponential growth stage, most of the energy of the yeast culture is devoted to reproducing itself. To do this, it uses the natural sugar that is in the grape juice. Wine fermentation has two stages called aerobic (with oxygen) and anaerobic (without oxygen) fermentations. After a few days in the first stage, most of the sugar and other nutrients in the grape juice are depleted. At this point, the oxygen source is removed and the growth rate of the yeast starts to decrease. Eventually, the yeast cells die (this is not shown in the graph). So, the population is limited by the food and oxygen available.

✓ **Checkpoint**   Help at *Math*.and*YOU*.com

What is your opinion about the sustainable population level of humans on Earth? Do you agree with Thomas Malthus, who predicted that the human population will grow exponentially, creating a permanent class of poor? Explain your reasoning.

Thomas Malthus is known for his theories on population growth. He claimed that populations are checked by famine, disease, and widespread mortality.

"**1.** Population is necessarily limited by the means of subsistence.

2. Population invariably increases where the means of subsistence increase, unless prevented by some very powerful and obvious checks.

3. These checks, and the checks which repress the superior, power of population, and keep its effects on a level with the means of subsistence, are all resolvable into moral restraint, vice and misery." *Thomas Malthus*

EXAMPLE 6   **Comparing Logistic Patterns**

The graph shows four different strategies used by mother and baby fur seals to locate each other after the mother returns from hunting. Discuss the strategies. Which is more effective?

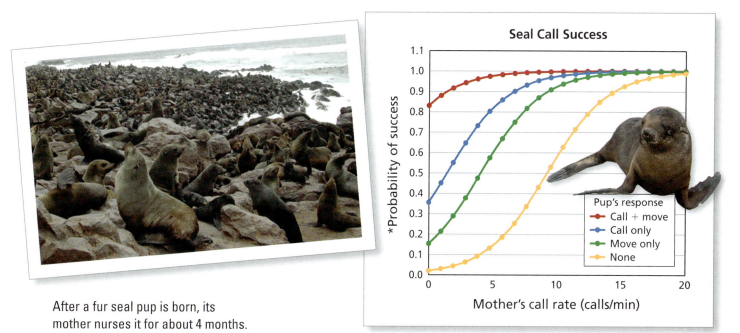

After a fur seal pup is born, its mother nurses it for about 4 months. During this time, the mother makes frequent trips to sea to forage for food. Each time a mother returns from hunting, she has to locate her pup in the colony, which can have thousands of baby and adult seals.

Probability is discussed in Chapter 8.

SOLUTION

Here are some general observations about the graph.

1. In each strategy, the mother has more success locating her pup when she increases her bark rate per minute.

2. As the bark rate increases, the probability of success increases logistically.

3. If the mother calls at a rate of 20 barks per minute, she is almost certain to locate her pup, regardless of the pup's response.

Here are some observations relative to the pup's response.

- **The pup calls and moves.** This is the best strategy for the pup. By calling and moving, there is a good chance that its mother will find it.

- **The pup only calls.** This is the second-best strategy for the pup. By calling, there is still a good chance that its mother can hear it through the noise of the colony.

- **The pup only moves.** This is not a good strategy for the pup. If its mother is calling at the rate of only 5 barks per minute, there is only a 60% chance that its mother will find it.

- **The pup does nothing.** This is the worst strategy for the pup. A pup who is too weak to call or move does not have a good chance of being found.

 Checkpoint

The above graph applies each time the mother goes to sea for food. Explain how the pup's chance for survival changes with multiple trips by the mother.

7.2 Exercises

Water Hyacinth An invasive species of water hyacinth is spreading over the surface of a lake. The figure shows the surface area covered by the water hyacinth over a 3-week period. In Exercises 1–4, use the figure. *(See Examples 1 and 2.)*

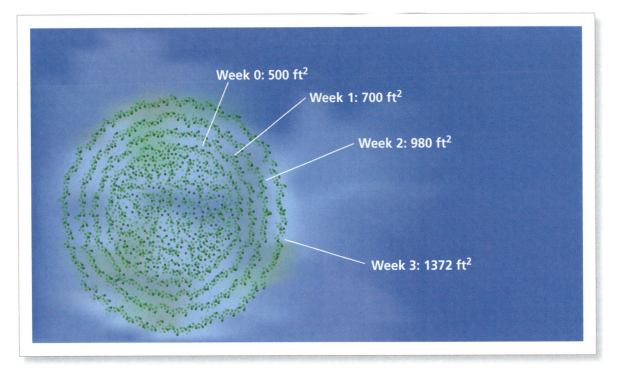

Week 0: 500 ft²

Week 1: 700 ft²

Week 2: 980 ft²

Week 3: 1372 ft²

1. Is the pattern linear? Explain your reasoning.

2. At what rate is the surface area covered by the water hyacinth increasing?

3. Use a spreadsheet to extend the pattern to 20 weeks. Then make a scatter plot of the data and describe the graph.

4. The surface area of the lake is about 800,000 square feet. How many weeks does it take the water hyacinth to cover the entire lake?

5. **Invasive Species** An invasive species of water plant covers 1500 square feet of the surface of a lake. The lake has a surface area of about 2,500,000 square feet. The surface area covered by the plant increases by 60% each week. Make a table and a scatter plot showing the surface area covered by the plant until the plant covers the entire lake. *(See Examples 1 and 2.)*

6. **Invasive Species** Suppose in Exercise 5 that the surface area covered by the plant increases by only 20% each week. How much longer does it take the plant to cover the entire lake? *(See Examples 1 and 2.)*

Rabbits A rabbit population is introduced to a new area. The graph shows the growth of the rabbit population. In Exercises 7–12, use the graph. *(See Example 3.)*

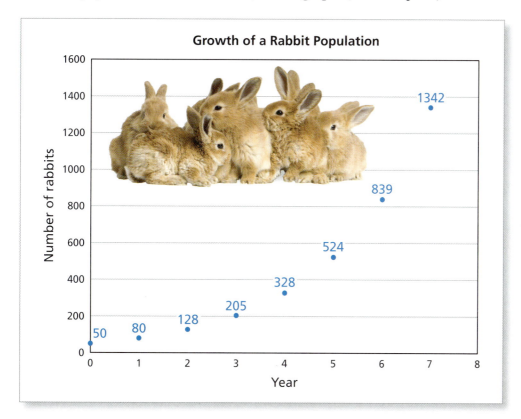

Growth of a Rabbit Population

Number of rabbits vs. Year

- 50
- 80
- 128
- 205
- 328
- 524
- 839
- 1342

7. What does the population for year 0 represent?

8. At what rate is the rabbit population increasing?

9. Suppose the population growth continued for another year. Predict the number of rabbits in year 8.

10. Suppose the population growth continued for another 3 years. Predict the number of rabbits in year 10.

11. When does the rabbit population exceed 3000?

12. When does the rabbit population exceed 6000?

13. **Population Growth** A rabbit population grows exponentially over a 10-year period. The population in year 3 is 150. The population in year 4 is 204. Predict the number of rabbits in year 10. *(See Example 3.)*

14. **Disease Outbreak** The outbreak of a disease causes a rabbit population to decrease exponentially over a 6-year period. The population in year 2 is 1200. The population in year 3 is 960. Predict the number of rabbits in year 6. *(See Example 4.)*

Trout A lake is stocked with 200 trout. The graph shows the growth of the trout population. In Exercises 15–18, use the graph. *(See Example 5.)*

15. What is the maximum sustainable population? Explain your reasoning.

16. Make a table that shows the change in the number of trout for each year. Discuss any trends.

17. Make a table that shows the percent change in the number of trout for each year. Discuss any trends.

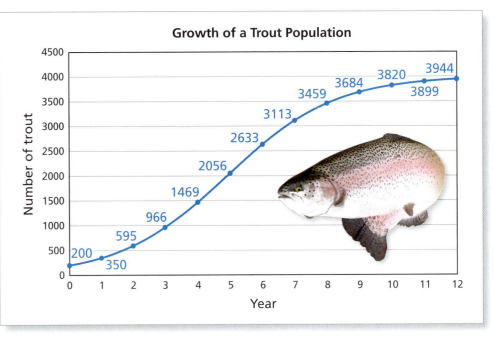

18. Make a table comparing the number of trout for each year in the graph to the number of trout each year if the trout population grew exponentially by 60% each year. Why is exponential growth unrealistic in this situation?

Competing Species The graphs show the growth of the populations of two competing species of fish when they are released into separate ponds and when they are released into the same pond. Assume all the ponds are the same size and have the same resources. In Exercises 19 and 20, use the graphs. *(See Example 6.)*

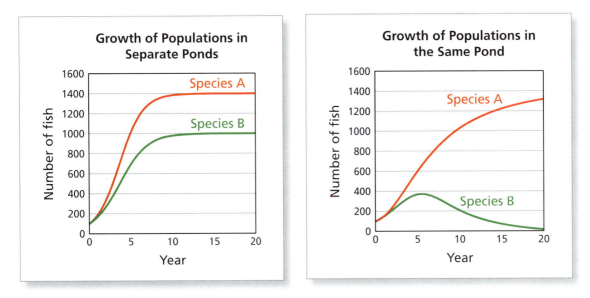

19. Compare the growth of the populations of the two species when they are released into separate ponds.

20. Compare the growth of the populations of the two species when they are released into the same pond.

▶ Extending Concepts

Logistic Growth Rate **In Exercises 21–24, use the information below.**

The formula for the logistic growth rate is

$$r = r_0 \times 1 - \left(\frac{\text{population size}}{\text{maximum sustainable population}} \right)$$

where r_0 is the intrinsic growth rate in decimal form.

21. The population of squirrels in a forest is growing logistically. The intrinsic growth rate is 40% per year, and the maximum sustainable population is 5000. Find the rate at which the population is growing when the population reaches (a) 100, (b) 2000, and (c) 4500.

22. The population of raccoons in a forest is growing logistically. The intrinsic growth rate is 50% per year, and the maximum sustainable population is 2000. Find the rate at which the population is growing when the population reaches (a) 200, (b) 1000, and (c) 1800.

23. What is the growth rate when a population is at the maximum sustainable population? Explain.

24. Describe the growth rate when a population exceeds the maximum sustainable population. Explain what this represents in nature.

DATA **Superexponential Growth** **In Exercises 25 and 26, use the information below.**

A population undergoes *superexponential growth* when the growth rate increases exponentially over time.

25. A population of 100 locusts is introduced to a new area. The initial growth rate is 50% per year. Make a table comparing the population of the locusts over a 10-year period when the growth rate remains constant (exponential) and when the growth rate increases by 20% each year (superexponential). Then make a scatter plot comparing the two data sets.

26. A population of 50 frogs is introduced to a new area. The initial growth rate is 60% per year. Make a table comparing the population of the frogs over a 10-year period when the growth rate remains constant (exponential) and when the growth rate increases by 10% each year (superexponential). Then make a scatter plot comparing the two data sets.

DATA **27. Riddle of the Lily Pad** A single lily pad lies on the surface of a pond. Each day the number of lily pads doubles until the entire pond is covered on day 30. On what day is the pond half-covered?

28. Mutant Plant A mutant strain of water plant covers 100 square feet of the surface of a lake. The lake has a surface area of about 1,000,000 square feet. At the end of each day, the surface area covered by the plant is double what it was at the beginning of the day minus the amount of plant cover that you clear. You can clear 5000 square feet of plant cover in 1 day. On what day do you have to begin clearing the plant cover to stop the plant from spreading across the entire lake?

7.1–7.2   Quiz

Deer The table shows two data sets for the projected growth of a deer population in a forest. In Exercises 1–8, use the table.

Year	Set A	Set B
2012	200	200
2013	224	224
2014	248	251
2015	272	281
2016	296	315
2017	320	352
2018	344	395

The deer population in the United States is estimated at over 20 million. In most states, the population is managed by the state's Department of Fish and Wildlife.

1. Describe the pattern in set A. Then make a scatter plot of the data.

2. Use set A to predict the deer population in 2022.

3. Describe the pattern in set B. Then make a scatter plot of the data.

4. Use set B to predict the deer population in 2022.

5. Using set A, when does the deer population exceed 500?

6. Using set B, when does the deer population exceed 500?

7. Suppose information from a previous study reveals that the deer population was 113 in 2007. Which model fits these data better? Explain your reasoning.

8. Suppose the deer population is growing logistically and the maximum sustainable population is 500 deer. Sketch a graph that illustrates this type of growth for the deer population.

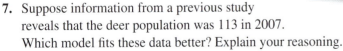

Math & Antibiotics

PROJECT: Are We Running Out of Antibiotics?

"In the future, historians of science may debate whether victory over bacteria was ever within our grasp. But it seems almost certain that the 60 or so years after penicillin came to market will eventually be viewed as just an interlude in the eternal war between us and them. We are multicelled animals of astonishing complexity and delicacy, moving through a world in which they vastly outnumber us. They are single-celled organisms so primitive they lack even a nucleus, marvelously adapted to multiply inside us—under the right circumstances, to consume our flesh and poison us with their waste. For a few decades we gained the upper hand through the use of antibiotics, natural substances that are as toxic to germs as germs are to us. But our ingenuity is in a desperate race against their ability to reproduce. More and more strains of bacteria are developing biological countermeasures to antibiotics—cell membranes that won't let them in, tiny pumps that push them back out, biochemical tweaks that make them harmless. Evolution is a process that has been at work on earth for hundreds of millions of years; modern biological science has been around for less than a century and a half. Which would you bet on?"

Newsweek, Jeneen Interlandi

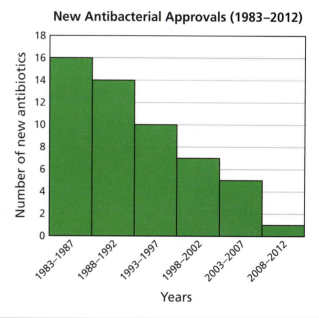

1. The number of new antibiotics that have been approved by the U.S. Food and Drug Administration since 1983 are shown in the graph. What does this pattern show?

2. Use the Internet to find information about the number of "superbugs" that are becoming resistant to all our known forms of antibiotics. Present your findings graphically.

3. What can we do to help slow the number of new strains of bacteria that are resistant to all our known forms of antibiotics?

4. Do you agree with the *Newsweek* article that from the discovery of penicillin through the present is simply a "blip" in time during which we gained the upper hand against bacteria? Explain your reasoning.

7.3 Quadratic Patterns

▶ Recognize and describe a quadratic pattern.

▶ Use a quadratic pattern to predict a future event.

▶ Compare linear, quadratic, and exponential growth.

Recognizing a Quadratic Pattern

A sequence of numbers has a **quadratic pattern** when its sequence of second differences is constant. Here is an example.

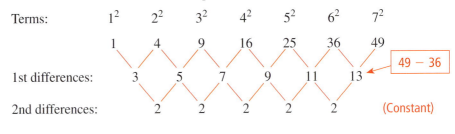

EXAMPLE 1 **Recognizing a Quadratic Pattern**

The distance a hit baseball travels depends on the angle at which it is hit and on the speed of the baseball. The table shows the distances a baseball hit at an angle of 40° travels at various speeds. Describe the pattern of the distances.

Speed (mph)	80	85	90	95	100	105	110	115
Distance (ft)	194	220	247	275	304	334	365	397

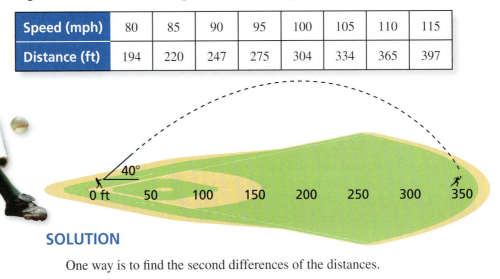

The distance a batter needs to hit a baseball to get a home run depends on the stadium. In many stadiums, the ball needs to travel 350 or more feet to be a home run.

SOLUTION

One way is to find the second differences of the distances.

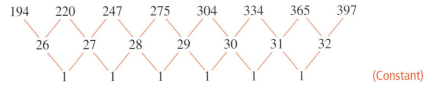

Because the second differences are constant, the pattern is quadratic.

 **Checkpoint**

Help at *Math*.and**Y😀U**.com

In Example 1, extend the pattern to find the distance the baseball travels when hit at an angle of 40° and a speed of 125 miles per hour.

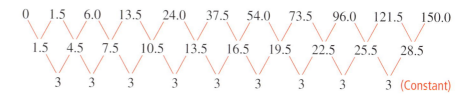

EXAMPLE 2   **Recognizing a Quadratic Pattern**

The table shows the numbers of days an offshore oil well has been leaking and the diameters (in miles) of the oil spill. (a) Describe the pattern of the numbers of days. (b) Use a spreadsheet to graph the data and describe the graph.

Diameter (mi)	0	0.5	1.0	1.5	2.0	2.5	3.0	3.5	4.0	4.5	5.0
Days	0	1.5	6.0	13.5	24.0	37.5	54.0	73.5	96.0	121.5	150.0

The Institute for Marine Mammal Studies in Gulfport, Mississippi, reported that a large number of sea turtles were found dead along the Mississippi coast following the Deepwater Horizon oil spill of 2010.

SOLUTION

a. One way is to find the second differences of the numbers of days.

0   1.5   6.0   13.5   24.0   37.5   54.0   73.5   96.0   121.5   150.0

1.5   4.5   7.5   10.5   13.5   16.5   19.5   22.5   25.5   28.5

3   3   3   3   3   3   3   3   3   (Constant)

Because the second differences are constant, the pattern is quadratic.

b. The graph is a curve that looks something like exponential growth. However, it is not an exponential curve. In mathematics, this curve is called *parabolic*.

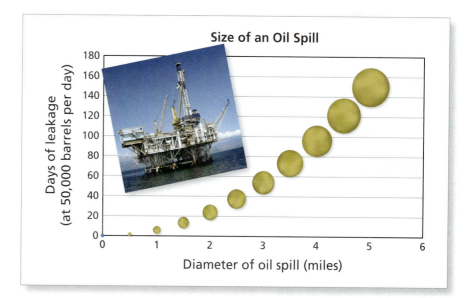

✓ **Checkpoint**   Help at *Math*.and**YOU**.com

Use a spreadsheet to make various graphs, including a scatter plot and a column graph, of the data in Example 1. Which type of graph do you think best shows the data? Explain your reasoning.

Using a Quadratic Pattern to Predict a Future Event

EXAMPLE 3   Predicting a Future Event

The Mauna Loa Observatory is an atmospheric research facility that has been collecting data related to atmospheric change since the 1950s. The observatory is part of the National Oceanic and Atmospheric Administration (NOAA).

The graph shows the increasing levels of carbon dioxide in Earth's atmosphere. Use the graph to predict the level of carbon dioxide in 2050.

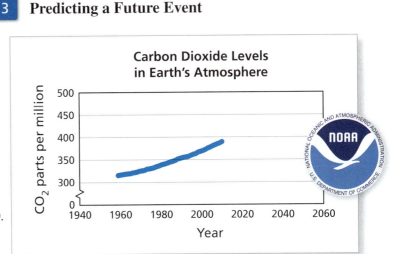

SOLUTION

The graph looks like it has a slight curve upward, which means that the rate of increase is increasing.

Using a linear regression program, the prediction for 2050 is 443 parts per million.

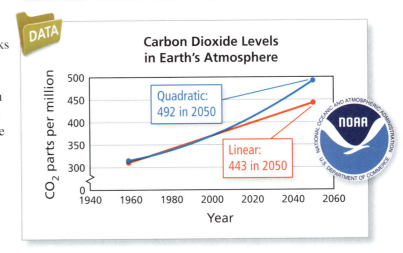

Using a quadratic regression program, the prediction for 2050 is 492 parts per million.

✓ **Checkpoint**   Help at *Math*.and**Y☺U**.com

The graph shows the results of a plant experiment with different levels of nitrogen in various pots of soil. The vertical axis measures the number of blades of grass that grew in each pot of soil. Describe the pattern and explain its meaning.

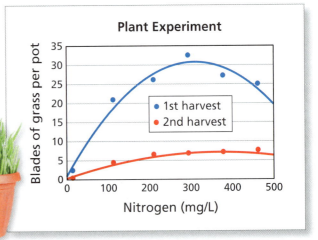

EXAMPLE 4   **Describing Lift for Airplanes**

For a given wing area, the lift of an airplane (or a bird) is proportional to the square of its speed. The table shows the lifts for a Boeing 737 airplane at various speeds.

Speed (mph)	0	75	150	225	300	375	450	525	600
Lift (1000s of lb)	0	25	100	225	400	625	900	1225	1600

a. Is the pattern of the lifts quadratic? Why?

b. Sketch a graph to show how the lift increases as the speed increases.

The Boeing 737 is the most widely used commercial jet in the world. It represents more than 25% of the world's fleet of large commercial jet aircraft.

SOLUTION

a. Begin by finding the second differences of the lifts.

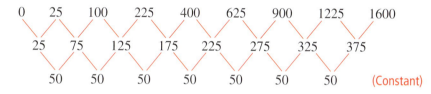

Because the second differences are constant, the pattern is quadratic.

b. Notice that as the speed increases, the lift increases quadratically.

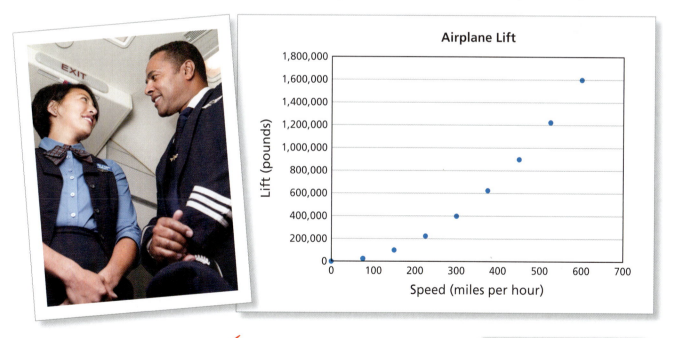

Airplane Lift

✓ **Checkpoint**   Help at *Math*.and**Y♥U**.com

A Boeing 737 weighs about 100,000 pounds at takeoff.

c. Estimate how fast the plane must travel to get enough lift to take flight.

d. Explain why bigger planes need longer runways.

Comparing Linear, Exponential, and Quadratic Models

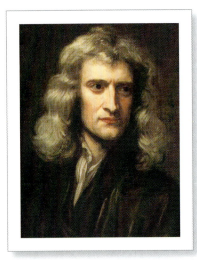

Earth's gravitational attraction was explained by Sir Isaac Newton's Law of Universal Gravitation. The law was published in Newton's *Principia* in 1687. It states that the force of attraction between two particles is directly proportional to the product of the masses of the two particles, and inversely proportional to the square of the distance between them.

EXAMPLE 5 **Conducting an Experiment with Gravity**

You conduct an experiment to determine the motion of a free-falling object. You drop a shot put ball from a height of 256 feet and measure the distance it has fallen at various times.

Time (sec)	0	0.5	1.0	1.5	2.0	2.5	3.0	3.5	4.0
Distance (ft)	0	4	16	36	64	100	144	196	256

Is the pattern of the distances linear, exponential, quadratic, or none of these? Explain your reasoning.

SOLUTION

Begin by sketching a graph of the data.

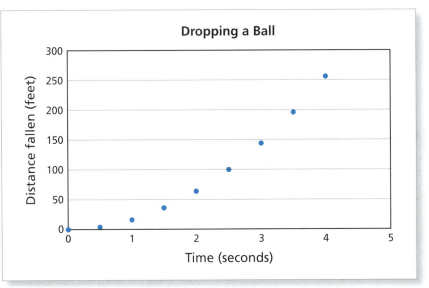

- The pattern is *not* linear because the graph is not a line.
- The pattern is *not* exponential because the ratios of consecutive terms are not equal.
- The pattern *is* quadratic because the second differences are equal.

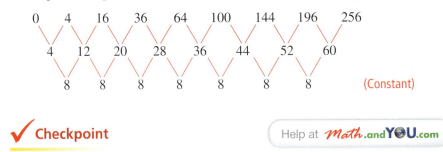

✓ **Checkpoint**

Help at *Math*.and**Y🌐U**.com

A classic problem in physics is determining the speed of an accelerating object. Estimate the speed of the falling shot put ball at the following times. Explain your reasoning.

a. 0 sec   **b.** 1 sec   **c.** 2 sec   **d.** 3 sec   **e.** 4 sec

EXAMPLE 6   **Describing Muscle Strength**

The muscle strength of a person's upper arm is related to its circumference. The greater the circumference, the greater the muscle strength, as indicated in the table.

Circumference (in.)	0	3	6	9	12	15	18	21
Muscle strength (lb)	0	2.16	8.61	19.35	34.38	53.70	77.31	105.21

Is the pattern of the muscle strengths linear, exponential, quadratic, or none of these? Explain your reasoning.

SOLUTION

Begin by sketching a graph of the data.

12 in.

18 in.

A typical upper arm circumference is about 12 inches for women and 13 inches for men.

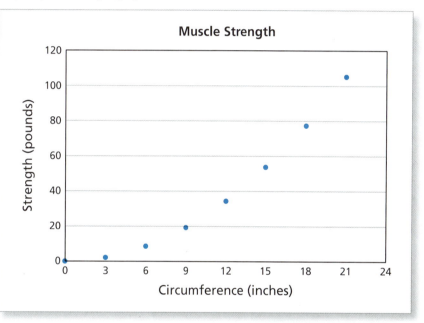

As in Example 5, the pattern is not linear or exponential. By calculating the second differences, you can see that the pattern is quadratic.

$$0 \quad 2.16 \quad 8.61 \quad 19.35 \quad 34.38 \quad 53.70 \quad 77.31 \quad 105.21$$

$$2.16 \quad 6.45 \quad 10.74 \quad 15.03 \quad 19.32 \quad 23.61 \quad 27.90$$

$$4.29 \quad 4.29 \quad 4.29 \quad 4.29 \quad 4.29 \quad 4.29 \qquad \text{(Constant)}$$

✓ **Checkpoint**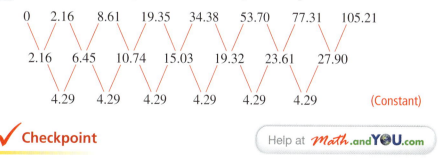

Example 6 shows that the muscle strength of a person's upper arm is proportional to the square of its circumference. Which of the following are also true? Explain your reasoning.

a. Muscle strength is proportional to the diameter of the muscle.

b. Muscle strength is proportional to the square of the diameter of the muscle.

c. Muscle strength is proportional to the cross-sectional area of the muscle.

7.3 Exercises

Football In Exercises 1–3, describe the pattern in the table. *(See Examples 1 and 2.)*

1. The table shows the heights of a football at various times after a punt.

Time (sec)	0	0.5	1	1.5	2	2.5	3
Height (ft)	3	34	57	72	79	78	69

2. The table shows the distances gained by a running back after various numbers of rushing attempts.

Rushing attempts	0	3	6	9	12	15	18
Distance (yd)	0	12.6	25.2	37.8	50.4	63	75.6

3. The table shows the heights of a football at various times after a field goal attempt.

Time (sec)	0	0.5	1	1.5	2	2.5	3
Height (ft)	0	21	34	39	36	25	6

4. **Punt** In Exercise 1, extend the pattern to find the height of the football after 4 seconds. *(See Example 1.)*

5. **Passing a Football** The table shows the heights of a football at various times after a quarterback passes it to a receiver. Use a spreadsheet to graph the data. Describe the graph. *(See Example 2.)*

Time (sec)	0	0.25	0.5	0.75	1	1.25	1.5	1.75	2	2.25	2.5
Height (ft)	6	15	22	27	30	31	30	27	22	15	6

6. **Graph** Use the graph in Exercise 5 to determine how long the height of the football increases.

Stopping a Car In Exercises 7–10, use the graph and the information below. *(See Example 3.)*

Assuming proper operation of the brakes on a vehicle, the minimum stopping distance is the sum of the reaction distance and the braking distance. The reaction distance is the distance the car travels *before* the brakes are applied. The braking distance is the distance a car travels *after* the brakes are applied but *before* the car stops. A reaction time of 1.5 seconds is used in the graph.

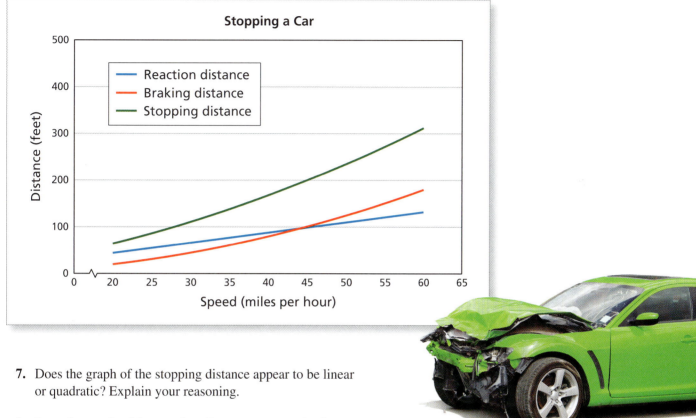

7. Does the graph of the stopping distance appear to be linear or quadratic? Explain your reasoning.

8. Does the graph of the reaction distance appear to be linear or quadratic? Explain your reasoning.

9. Use the graph to predict the stopping distance at 90 miles per hour.

10. The braking distance at 35 miles per hour is about 60 feet. Does this mean that the braking distance at 70 miles per hour is about 120 feet? Explain.

DATA **Slippery Road** The braking distance of a car depends on the friction between the tires and the road. The table shows the braking distance for a car on a slippery road at various speeds. In Exercises 11 and 12, use the table. *(See Example 4.)*

Speed (mph)	20	30	40	50	60	70	80
Distance (ft)	40	90	160	250	360	490	640

11. Is the pattern quadratic? Explain.

12. Graph the data in the table. Compare this graph to the graph above.

Gravity In Exercises 13–16, determine whether the pattern in the table is linear, exponential, quadratic, or none of these. Explain your reasoning. *(See Examples 5 and 6.)*

13. An object is dropped from a height of 50 feet on the moon. The table shows the distances it has fallen at various times.

Time (sec)	0	0.5	1	1.5	2	2.5	3
Distance (ft)	0	$\frac{2}{3}$	$2\frac{2}{3}$	6	$10\frac{2}{3}$	$16\frac{2}{3}$	24

14. An object is dropped from a height of 150 feet on Venus. The table shows the distances it has fallen at various times.

Time (sec)	0	0.5	1	1.5	2	2.5	3
Distance (ft)	0	3.7	14.8	33.3	59.2	92.5	133.2

15. An object is dropped from a height of 300 feet on Mars. The table shows the heights of the object at various times.

Time (sec)	0	1	2	3	4	5	6
Height (ft)	300	293.8	275.2	244.2	200.8	145	76.8

16. An object is dropped from a height of 1600 feet on Jupiter. The table shows the heights of the object at various times.

Time (sec)	0	1	2	3	4	5	6
Height (ft)	1600	1556.8	1427.2	1211.2	908.8	520	44.8

17. Sign of Second Differences Graph the data in Exercises 14 and 15 on the same coordinate plane. Compare the graphs. What appears to be the relationship between the sign of the second differences and the corresponding graph?

18. Moon The moon's gravitational force is much less than that of Earth. Use the table in Exercise 13 and the table in Example 5 on page 328 to estimate how many times stronger Earth's gravitational force is than the moon's gravitational force. Explain your reasoning.

▶ Extending Concepts

Business Data from real-world applications rarely match a linear, exponential, or quadratic model perfectly. In Exercises 19–22, the table shows data from a business application. Determine whether a linear, exponential, or quadratic model *best* represents the data in the table. Explain your reasoning.

19. The table shows the revenue for selling various units.

Units sold	0	40	80	120	160	200
Revenue	$0	$186.30	$372.45	$558.38	$744.24	$930.15

20. The table shows the total cost for producing various units.

Units produced	0	40	80	120	160	200
Total cost	$500.00	$572.05	$627.98	$668.03	$692.10	$700.12

21. The table shows the profit from selling various units.

Units sold	0	40	80	120	160	200
Profit	−$500.00	−$385.75	−$255.53	−$109.65	$52.14	$230.03

22. The table shows the stock price of a company for various years.

Year	2007	2008	2009	2010	2011	2012
Stock price	$21.56	$23.68	$26.08	$28.62	$31.62	$34.79

Activity Fold a rectangular piece of paper in half. Open the paper and record the number of folds and the number of sections created. Repeat this process four times and increase the number of folds by one each time. In Exercises 23–26, use your results.

23. Complete the table.

Folds	1	2	3	4	5
Sections					

24. Graph the data in Exercise 23. Determine whether the pattern is linear, exponential, or quadratic.

25. Write a formula for the model that represents the data.

26. How many sections are created after eight folds?

2 folds
4 sections

7.4 Fibonacci & Other Patterns

▶ Recognize and describe the Fibonacci pattern.
▶ Analyze geometric Fibonacci patterns.
▶ Recognize and describe other patterns in mathematics.

Study Tip

The Fibonacci sequence starts with the numbers 0 and 1. A *general* Fibonacci sequence can start with other numbers, such as 1 and 3.

Characteristics of Fibonacci Patterns

In the **Fibonacci sequence** of numbers, each number is the sum of the 2 previous numbers, starting with 0 and 1.

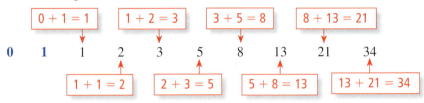

$$0 + 1 = 1 \qquad 1 + 2 = 3 \qquad 3 + 5 = 8 \qquad 8 + 13 = 21$$

0   1   1   2   3   5   8   13   21   34

$$1 + 1 = 2 \qquad 2 + 3 = 5 \qquad 5 + 8 = 13 \qquad 13 + 21 = 34$$

EXAMPLE 1 **Recognizing a Fibonacci Pattern**

Consider a hypothetical population of rabbits. Start with one breeding pair. After each month, each breeding pair produces another breeding pair. The total number of rabbits each month follows the exponential pattern 2, 4, 8, 16, 32, ... Now suppose that in the first month after each pair is born, the pair is too young to reproduce. Each pair produces another pair after it is 2 months old. Describe this pattern.

SOLUTION

Leonardo of Pisa was also known as Leonardo Fibonacci. He was an Italian mathematician who is credited with spreading the Hindu-Arabic numeral system in Europe. He did this through his book *Liber Abaci*, in which he used the Fibonacci sequence as an example.

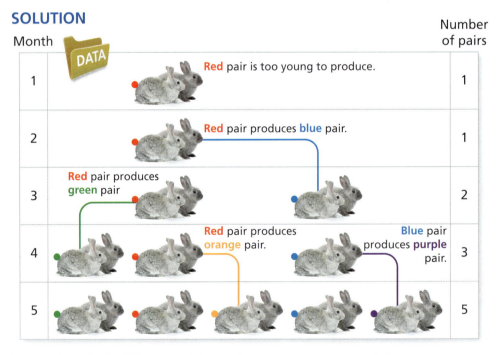

The number of pairs follows the Fibonacci pattern, not an exponential pattern.

✓ **Checkpoint**   Help at *Math*.and**Y❂U**.com

Enter the total number of rabbits for each month in Example 1 (2, 2, 4, 6, 10, . . .) into a spreadsheet. Make a scatter plot of the data. Then compare the scatter plot with the exponential pattern 2, 4, 8, 16, 32, ...

The Fibonacci sequence has captivated people's imaginations for centuries. The Fibonacci Association meets regularly to share ideas about the Fibonacci sequence. One way to describe the sequence is, "It looks at itself, it looks at its most recent past, puts them together, and evolves to the next number."

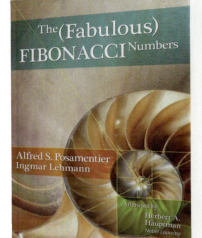

Many people have written about the occurrence of Fibonacci numbers in nature. In *The (Fabulous) Fibonacci Numbers*, math educators Alfred Posamentier and Ingmar Lehmann describe how the Fibonacci numbers occur in dozens of different patterns in the natural world.

EXAMPLE 2   Recognizing a Fibonacci Pattern

Look at the X-ray of the human hand. Describe how the Fibonacci sequence is related to the X-ray.

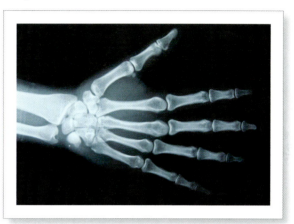

SOLUTION

By looking at the lengths of the bones in the X-ray, you can observe part of the Fibonacci sequence.

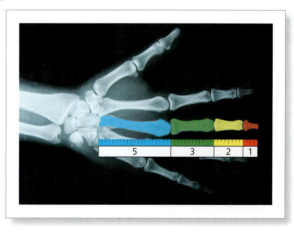

✓ **Checkpoint**

Help at *Math*.and**YOU**.com

The triangle of numbers shown is called Pascal's Triangle, after the French mathematician Blaise Pascal.

a. Describe the pattern in Pascal's Triangle.

b. Describe how the Fibonacci sequence is related to Pascal's triangle.

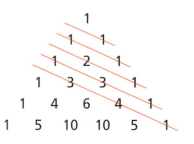

```
          1
        1   1
      1   2   1
    1   3   3   1
  1   4   6   4   1
1   5  10  10   5   1
```

Geometric Fibonacci Patterns

> **EXAMPLE 3** **Drawing the Fibonacci Spiral**

Using only a compass and a ruler, draw the Fibonacci spiral.

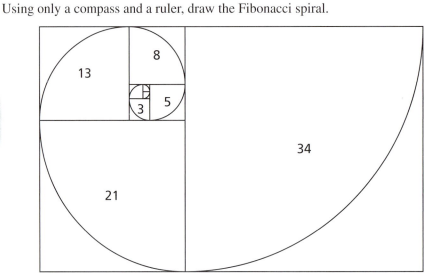

The shell of a chambered nautilus is a spiral. The spiral can be described mathematically as a Fibonacci spiral.

SOLUTION

- Begin by drawing two 1-unit squares.

- Adjoin a 2-unit square.

- Adjoin a 3-unit square.

- Adjoin a 5-unit square.

- Use a compass to draw quarter-circle arcs inside each square. Locate the arcs so that they form a continuous spiral.

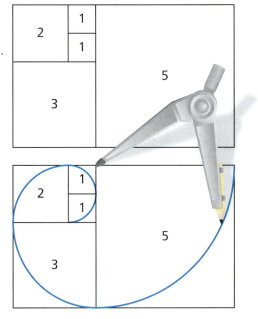

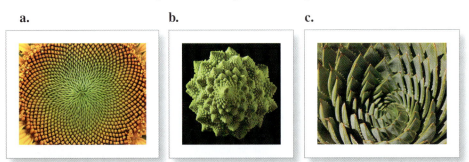

You can use sticky notes and thumbtacks to create a Fibonacci spiral.

✓ **Checkpoint** Help at *Math*.and**YOU**.com

Use the Internet to identify the following Fibonacci spirals in nature.

a. **b.** **c.**

	A	B
1	**Term**	**Ratio**
2	0	
3	1	
4	1	1
5	2	2
6	3	1.5
7	5	1.6666667
8	8	1.6
9	13	1.625
10	21	1.6153846
11	34	1.6190476
12	55	1.6176471
13	89	1.6181818
14	144	1.6179775
15	233	1.6180556
16	377	1.6180258
17	610	1.6180371
18	987	1.6180328
19	1597	1.6180344
20	2584	1.6180338
21	4181	1.6180341
22		

EXAMPLE 4   **Using the Golden Ratio**

The spreadsheet shows that when you find the ratio of any two successive terms in the Fibonacci sequence (divide the larger by the smaller), you approach the limit of 1.6180339887. . .. This is called the *golden ratio*.

In art, a rectangle whose side lengths are in this ratio is considered aesthetically pleasing. Identify some uses of this "golden rectangle" in art and architecture.

Golden Rectangle

SOLUTION

The dimensions of the front of the Parthenon in Athens are roughly that of a golden rectangle.

Switch plate

Mona Lisa's face

✓ **Checkpoint**   Help at *Math*.and**Y❂U**.com

Use the Internet to find other examples of the use of the golden ratio in art or architecture.

Other Patterns in Mathematics

You have learned about linear patterns (7.1), exponential patterns (7.2), quadratic patterns (7.3), and Fibonacci patterns (7.4). There are many other types of mathematical patterns. Two of them are shown in Examples 5 and 6.

Kepler's laws and his assertion that the planets orbit the Sun in elliptical orbits with varying speeds disagreed with the accepted models of Aristotle, Ptolemy, and Copernicus.

EXAMPLE 5   Analyzing Kepler's Third Law

In the heart of the Scientific Revolution in Europe, Johannes Kepler analyzed the astronomical observations of Tycho Brahe and, in 1609, published his first 2 laws of planetary motion.

1. The orbit of every planet is an ellipse, with the Sun at one of the two foci.

2. A line joining a planet and the Sun sweeps out equal areas during equal intervals of time.

His third law of planetary motion was not published until 9 years later. It concerns the pattern in the following table. The period is the time (in years) it takes a planet to make one orbit around the Sun. The mean distance is the average distance (in astronomical units) between a planet and the Sun as the planet passes through its elliptical orbit.

Planet	Mercury	Venus	Earth	Mars	Jupiter	Saturn
Period	0.241	0.615	1.000	1.881	11.862	29.457
Mean distance	0.387	0.723	1.000	1.524	5.203	9.537

Can you see the pattern?

SOLUTION

After many years, Kepler noticed that the square of the period is the cube of the mean distance. This relationship is summarized in Kepler's Third Law of Planetary Motion.

3. The square of the period of a planet is directly proportional to the cube of its mean distance from the Sun.

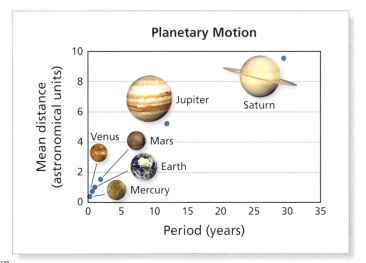

✓ **Checkpoint**   Help at

Verify Kepler's Third Law by completing the following table.

Planet	Mercury	Venus	Earth	Mars	Jupiter	Saturn
(Period)^2						
(Mean distance)^3						

In the United States, the 41st parallel forms the border between Wyoming and Utah, Wyoming and Colorado, and Colorado and Nebraska.

EXAMPLE 6   **Analyzing Hours of Daylight**

The graph shows how the hours of daylight vary at any location on the 41st parallel in the northern hemisphere. Describe this pattern.

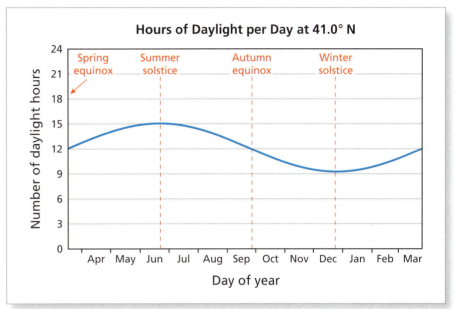

SOLUTION

There are four critical points on the graph.

- **Summer solstice:** This day corresponds to the greatest number of hours of daylight at any location in the northern hemisphere.
- **Winter solstice:** This day corresponds to the least number of hours of daylight at any location in the northern hemisphere.
- **Spring & autumn equinox:** On these 2 days, every location in the northern hemisphere receives equal amounts of daylight and darkness—12 hours of daylight and 12 hours of darkness.

This pattern is called a *sine wave* or a *sinusoid*. It continuously oscillates above and below a mean value.

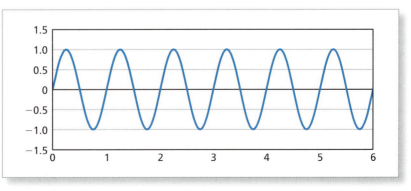

✓ **Checkpoint**   Help at *Math*.and*YOU*.com

Describe other occurrences in nature that can be modeled by a sine wave. Explain your reasoning.

7.4 Exercises

Fibonacci In Exercises 1–6, use the Internet to describe how the Fibonacci sequence is related to the object shown. *(See Examples 1 and 2.)*

1. Daisies

2. Phyllotaxis

3. Pineapple

4. Pinecone

5. Sneezewort

6. Human body

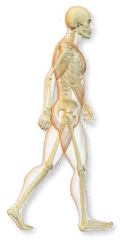

Golden Rectangle In Exercises 7–10, determine whether the golden ratio applies to the object. *(See Example 4.)*

7.

8.

9.

10.

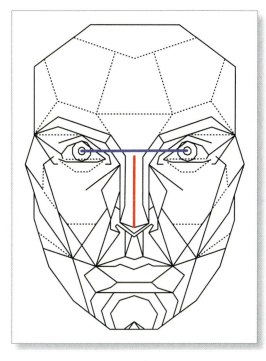

Beauty In Exercises 11–14, use the information below. *(See Example 4.)*

Since ancient Greece, symmetry and the golden ratio often have been thought to embody ideal beauty. Dr. Stephen Marquardt has studied human beauty for years across both genders and for all races, cultures, and eras. Through his research, he developed and patented the Repose Frontal (RF) Mask, which he claims is the most beautiful shape for a human face.

11. Use the Internet to describe why Dr. Marquardt believes that the RF Mask models the perfect human face.

12. Compare the distance between the pupils to the length of the nose on the RF Mask. What do you notice?

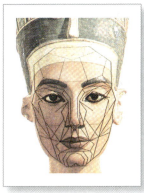

13. The RF Mask is applied to the face of the Egyptian queen Nefertiti (14th century B.C.). According to the RF Mask, is Nefertiti beautiful? Explain.

14. It is common for humans to find or see patterns in everything they do. Do you think the golden ratio is a law of nature or just a coincidental pattern detected by humans? Explain your reasoning.

Triangular Numbers In Exercises 15–17, use the information below. *(See Examples 5 and 6.)*

The sequence of triangular numbers is

1, 3, 6, 10, 15, 21, 28, 36, 45, ...

15. Describe the pattern.

16. Describe how the sequence of triangular numbers is related to the following diagram.

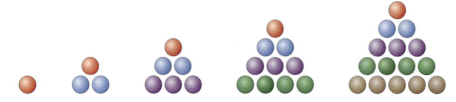

17. Is 115 a triangular number? Explain your reasoning.

Lucas Sequence In Exercises 18–20, use the information below. *(See Examples 5 and 6.)*

The Lucas sequence is named in honor of mathematician François Édouard Anatole Lucas.

The Lucas sequence is

1, 3, 4, 7, 11, 18, 29, 47, 76, ...

18. Describe the pattern.

19. Describe how the Lucas sequence is related to the cactus.

20. Use the spreadsheet to find the ratio of successive Lucas numbers. What number do you approach as the numbers get larger?

	A	B
1	**Term**	**Ratio**
2	1	
3	3	
4	4	1.333333333
5	7	1.75
6	11	
7	18	
8	29	
9	47	
10	76	
11	123	
12	199	
13	322	
14	521	
15	843	
16	1364	
17	2207	
18	3571	
19	5778	
20	9349	
21	15127	

▶ Extending Concepts

Golden Angle In Exercises 21 and 22, use the Internet and the information below.

In the figure, the golden angle is the angle subtended by the smaller red arc.

21. What is the measure of the golden angle in degrees?
Explain why this angle is called the golden angle.

22. Describe how the golden angle is related to phyllotaxis.

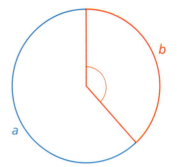

Golden Triangle In Exercises 23–26, use the Internet and the information below.

A golden triangle is an isosceles triangle in which the ratio of the length of one of the longer sides to the length of the shorter side is the golden ratio. The base angles are 72° each, and the smaller angle is 36°.

23. Does a triangle with the following side lengths approximate a golden triangle? Explain.

    a. 8 ft, 8 ft, 5 ft   **b.** 21 cm, 13 cm, 18 cm

    c. 55 m, 34 m, 55 m   **d.** 10 in., 14 in., 14 in.

24. When a base angle of a golden triangle is bisected, the angle bisector divides the opposite side in a golden ratio and forms two smaller isosceles triangles. The blue triangle (shown below) that is created from the bisection is a golden triangle. This process can be continued indefinitely, creating smaller and smaller golden triangles. Use the bisection process of a golden triangle, a compass, and a ruler to draw the Fibonacci spiral.

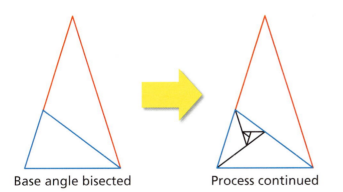

Base angle bisected   Process continued

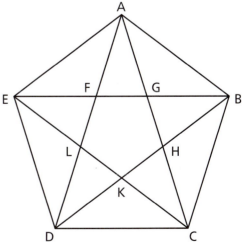

25. The figure at the right is a pentagon with all its diagonals drawn. How are the diagonals and the sides related to the golden ratio?

26. How many golden triangles can you find in the figure at the right?

7.3–7.4 Quiz

Heart Rate Recovery In Exercises 1–3, use the information below.

Heart rate recovery is the reduction in heart rate from the rate at peak exercise to the rate 1 minute after the exercise has stopped. It can be used as a predictor of mortality. Heart rate is measured in beats per minute (bpm). The table shows the relative risk of death for various heart rate recoveries.

Heart rate recovery (bpm)	6	8	10	12	14	16	18	20
Relative risk of death	6.38	5.18	4.14	3.26	2.54	1.98	1.58	1.34

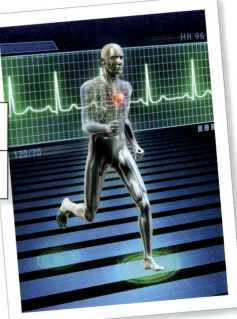

1. Describe the pattern of relative risk of death.

2. Extend the pattern to find the relative risk of death when the heart rate recovery is 22 beats per minute.

3. Use a spreadsheet to graph the data. Describe the graph.

Golden Heartbeat In Exercises 4–6, use the information below.

The main components of an electrocardiogram (EKG) are the P wave, the electrical activity in the atria; the QRS complex, the electrical activity in the ventricles; and the T wave, the electrical recovery of the ventricles. The electrocardiograms of human heartbeats vary considerably depending on a variety of factors.

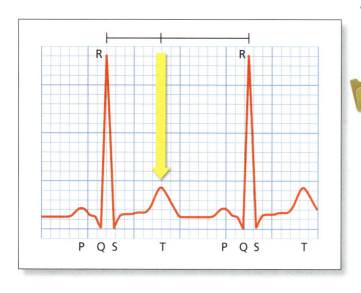

4. Some people believe that a peaceful heartbeat produces a rhythm related to the golden ratio. Use the EKG to describe how a peaceful heartbeat is related to the golden ratio.

5. The table shows the approximate ratio of each successive pair of Fibonacci numbers. Graph the ratios. Identify any similarities to the EKG.

1	/	0	→	∞
1	/	1	=	1
2	/	1	=	2
3	/	2	=	1.5
5	/	3	=	1.667
8	/	5	=	1.6
13	/	8	=	1.625
21	/	13	=	1.615
34	/	21	=	1.619

6. Use the Internet to describe how blood pressure is related to the golden ratio.

Chapter 7   Summary

	Section Objectives	How does it apply to you?

Section 1

Recognize and describe a linear pattern. → A sequence of numbers has a linear pattern when each successive number increases (or decreases) by the same amount. *(See Examples 1 and 2.)*

Use a linear pattern to predict a future event. → You can extend a linear pattern to predict a value. *(See Example 3.)*

Recognize a proportional pattern. → A pattern with two variables is proportional when one of the variables is a constant multiple of the other variable. *(See Examples 5 and 6.)*

Section 2

Recognize and describe an exponential pattern. → A sequence of numbers has an exponential pattern when each successive number increases (or decreases) by the same percent. *(See Examples 1 and 2.)*

Use an exponential pattern to predict a future event. → You can extend an exponential pattern to predict a value. *(See Examples 3 and 4.)*

Compare exponential and logistic growth. → Logistic growth accounts for the physical boundaries that limit exponential growth in nature. *(See Examples 5 and 6.)*

Section 3

Recognize and describe a quadratic pattern. → A sequence of numbers has a quadratic pattern when its sequence of second differences is constant. *(See Examples 1 and 2.)*

Use a quadratic pattern to predict a future event. → You can extend a quadratic pattern to predict a value. *(See Examples 3 and 4.)*

Compare linear, quadratic, and exponential growth. → You can determine whether a sequence follows a linear, exponential, or quadratic pattern, or none of these patterns. *(See Examples 5 and 6.)*

Section 4

Recognize and describe the Fibonacci pattern. → In the Fibonacci sequence of numbers, each number is the sum of the 2 previous numbers, starting with 0 and 1. *(See Examples 1 and 2.)*

Analyze geometric Fibonacci patterns. → You can identify and describe the Fibonacci sequence in art and nature. *(See Examples 3 and 4.)*

Recognize and describe other patterns in mathematics. → There are many other types of mathematical patterns. You can analyze them to better understand the world around you. *(See Examples 5 and 6.)*

Chapter 7 Review Exercises

Section 7.1

Temperature and Resistance The table shows the resistances of a coil of copper wire at various temperatures. **In Exercises 1–4, use the table.**

Temperature (°C)	Resistance (ohms)
20	100.00
21	100.38
22	100.76
23	101.14
24	101.52
25	101.90
26	102.28

1. Does the table relating temperature and resistance represent a linear pattern? Explain your reasoning.

2. Use a spreadsheet to graph the data. Is the graph linear?

3. Find the resistance of the coil when the temperature is 30°C.

4. Find the resistance of the coil when the temperature is 48°C.

Length (meters)	Resistance (ohms)
0	0
4	0.068
8	0.136
12	0.204
16	0.272
20	0.340
24	0.408

Length and Resistance The table shows the resistances of a coil of copper wire for various lengths. **In Exercises 5 and 6, use the table.**

5. Is the length of the wire proportional to its resistance? Make a scatter plot of the data to verify your answer.

6. Extend the pattern in the table to find the resistance for each length of the copper wire.

   a. 26 meters

   b. 28 meters

   c. 30 meters

Voltage and Current **In Exercises 7 and 8, use the information below.**

Electric current is proportional to voltage.

7. Suppose a wire connected to a 3-volt battery has a current of 15 amperes. What is the current when the wire is connected to a 9-volt battery?

8. Suppose a wire connected to a 1.5-volt battery has a current of 20 amperes. What is the current when the wire is connected to a 4.5-volt battery?

Section 7.2

 Flour Beetles The graph shows the growth of a flour beetle population in a natural environment. In Exercises 9–12, use the graph.

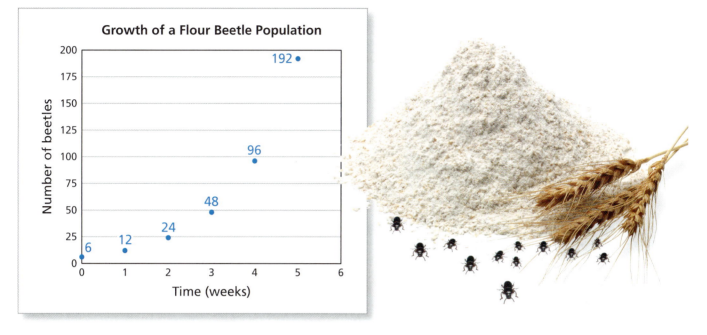

Growth of a Flour Beetle Population

Number of beetles vs *Time (weeks)*

192
96
48
24
12
6

9. At what rate is the flour beetle population increasing?

10. Predict the number of flour beetles in week 6.

11. When does the population exceed 4000 beetles?

12. How many weeks does it take the population to quadruple?

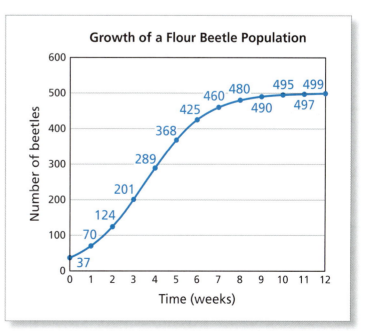

 Flour Beetles In a laboratory, the growth of a flour beetle population in a container of flour is monitored for 12 weeks. The graph shows the growth of the population. In Exercises 13–16, use the graph.

13. What type of growth does the graph display?

14. What is the maximum sustainable population?

15. Make a table that shows the change in the number of flour beetles for each week. Discuss any trends.

16. Make a table that shows the percent change in the number of flour beetles for each week. Discuss any trends.

Growth of a Flour Beetle Population

Number of beetles vs *Time (weeks)*

37, 70, 124, 201, 289, 368, 425, 460, 480, 490, 495, 497, 499

Section 7.3

Golden Gate Bridge **In Exercises 17–22, use the information below.**

The Golden Gate Bridge is a suspension bridge that spans the opening of the San Francisco Bay into the Pacific Ocean. It has two main suspension cables that pass over the tops of two main towers at 500 feet above the roadway. Each of the main cables has a diameter of about 3 feet. The table shows the heights of the main cables above the roadway relative to the distance from the middle of the bridge.

Distance (ft)	−2000	−1500	−1000	−500	0	500	1000	1500	2000
Height (ft)	456	260	120	36	8	36	120	260	456

17. Describe the pattern in the table.

18. Use a spreadsheet to graph the data in the table. Does the graph appear to be linear or quadratic? Explain your reasoning.

19. Use the graph from Exercise 18 to predict the height of the cables 2500 feet from the middle of the bridge. Is your prediction possible? Explain your reasoning.

20. A maintenance worker is painting a main cable when he accidently drops a paintbrush. The table shows the heights of the paintbrush at various times. Is the pattern in the table linear, exponential, quadratic, or none of these? Explain your reasoning.

Time (sec)	Height (ft)
0	576
0.5	572
1	560
1.5	540
2	512
2.5	476
3	432

© Golden Gate Bridge District, *www.goldengate.org*

21. Use a spreadsheet to graph the data in Exercise 20. Describe the graph.

22. Use the graph from Exercise 21 to predict how long it takes the paintbrush to reach the water.

Section 7.4

DNA Deoxyribonucleic acid (DNA) is the genetic material in all known living organisms and some viruses. DNA contains two strands wrapped around each other in a double helix. In Exercises 23 and 24, use the Internet.

23. Describe how the Fibonacci sequence is related to each full cycle of a DNA double helix.

24. Determine whether the golden ratio applies to a DNA double helix.

B-Form DNA In B-form DNA, the intertwined strands make two grooves of different widths, referred to as the major groove and the minor groove. In Exercises 25 and 26, use the B-form DNA shown.

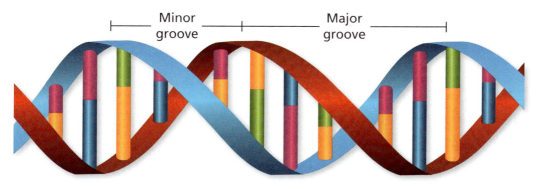

Minor groove   Major groove

25. Use the Internet to describe how the Fibonacci sequence is related to the major and minor grooves of B-form DNA.

26. Determine whether the golden ratio applies to B-form DNA.

DNA Cross-section
In Exercises 27 and 28, use the cross-sectional view of a DNA double helix shown.

27. Show how two pentagons can be used to construct the cross-sectional view of a DNA double helix.

28. How is the cross-sectional view of a DNA double helix related to the golden ratio?

8 The Mathematics of Likelihood

8.1 Assigning a Measure to Likelihood

▶ Use probability to describe the likelihood of an event.

▶ Analyze the likelihood of a risk.

▶ Use likelihood to describe actuarial data.

8.2 Estimating Likelihood

▶ Find a theoretical probability.

▶ Find an experimental probability.

▶ Estimate a probability using historical results.

8.3 Expected Value

▶ Find an expected value involving two events.

▶ Find an expected value involving multiple events.

▶ Use expected value to make investment decisions.

8.4 Expecting the Unexpected

▶ Find the probability of independent events.

▶ Find the probability that an event does not occur.

▶ Find counterintuitive probabilities.

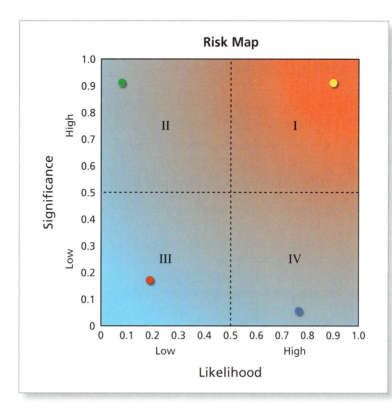

Risk Map

Example 4 on page 355 shows a risk map. Which quadrant on this map might represent the risk of catching a cold during the winter?

8.1 Assigning a Measure to Likelihood

▶ Use probability to describe the likelihood of an event.
▶ Analyze the likelihood of a risk.
▶ Use likelihood to describe actuarial data.

Describing the Likelihood of an Event

The **probability** of an event is a number that measures the likelihood that the event will occur. Probabilities are between 0 and 1, including 0 and 1. The diagram relates likelihoods (described in words) and probabilities (numbers from 0 to 1).

Words	Impossible	Unlikely	Equally likely to happen or not happen	Likely	Certain
Fraction	0	$\frac{1}{4}$	$\frac{1}{2}$	$\frac{3}{4}$	1
Decimal	0	0.25	0.5	0.75	1
Percent	0%	25%	50%	75%	100%

EXAMPLE 1 Describing Likelihoods

Describe the likelihood of each event in words.

NASA says there is no chance of the 886-foot asteroid Apophis smashing into Earth in its first flyby in 2029, and only a 1-in-250,000 chance of a collision in 2036.

Probability of an Asteroid or a Meteoroid Hitting Earth			
Asteroid	**Diameter**	**Probability of impact**	**Date**
• Meteoroid	6 in.	0.75	Any day
• Apophis	886 ft	0	2029
• 2000 SG344	121 ft	1/435	2068–2110
• 2008 TC3	4 m	1	2008 (occurred)

SOLUTION

• On any given day, it is *likely* that a meteoroid of this size will enter Earth's atmosphere. If you have ever seen a "shooting star," you have seen one.

• A probability of zero means this event is ***impossible***.

• With a probability of 1/435 ≈ 0.23%, this event is *unlikely*.

• With a probability of 1, this event is ***certain***. It occurred in 2008.

✓ **Checkpoint**

Describe each event as unlikely, equally likely to happen or not happen, or likely. Explain your reasoning.

a. The oldest child in a family is a girl.

b. The two oldest children in a family are both girls.

EXAMPLE 2   **Putting Probability into Words**

The map is a hurricane strike probability map for Hurricane Charley from August 2004. It maps the probability that the center of the storm passes within 75 miles of a location during a 72-hour time interval.

Hurricane Strike Probability Map

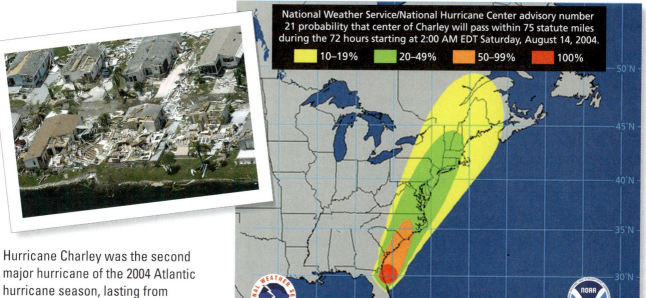

Hurricane Charley was the second major hurricane of the 2004 Atlantic hurricane season, lasting from August 9 to August 15. It was a strong Category 4 hurricane with maximum winds near 150 miles per hour. It was the strongest hurricane to have hit Florida since 1992.

Describe the likelihood that Hurricane Charley passed within 75 miles of each of the 4 colored regions on the map.

SOLUTION

- ● **100%:** Hurricane Charley was *certain* to pass within 75 miles of the red region on the map.

- ● **50–99%:** Hurricane Charley was *very likely* to pass within 75 miles of the orange region on the map.

- ● **20–49%:** Hurricane Charley was *not very likely* to pass within 75 miles of the green region on the map.

- ● **10–19%:** Hurricane Charley was *unlikely* to pass within 75 miles of the yellow region on the map.

✓ **Checkpoint**

Help at *Math*.and*YOU*.com

Describe the likelihood that the next U.S. presidential election has the pattern in electoral college votes. If it does, who wins, red or blue?

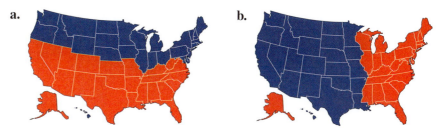

a.

b.

Analyzing the Likelihood of a Risk

 **EXAMPLE 3   Analyzing the Likelihood of a Risk**

Evaluate the following statement.

> "The annual risk of being killed in a plane crash for the average American is about 1 in 11 million."   "How Risky Is Flying?," David Ropeik

SOLUTION

Here is David Ropeik's own evaluation of the statement.

> "On that basis, the risk looks pretty small. Compare that, for example, to the annual risk of being killed in a motor vehicle crash for the average American, which is about 1 in 5,000. But if you think about those numbers, problems crop up right away. First of all, you are not the average American. Nobody is. Some people fly more and some fly less and some don't fly at all. So if you take the total number of people killed in commercial plane crashes and divide that into the total population, the result, the risk for the average American, may be a good general guide to whether the risk is big or small, but it's not specific to your personal risk."

In his article, Ropeik goes on to say the following.

> ". . . numbers are a great way to put risk in general perspective, and there is no question that by most metrics, flying is a less risky way to travel than most others. But wait: Just when you thought it was safe to use numbers to put risk in perspective . . . Numbers are not the only way—not even the most important way—we judge what to be afraid of. Risk perception is not just a matter of the facts. It's also a matter of the other things we know (e.g., airline companies are in financial trouble) and our experiences (maybe you took a really scary, turbulent flight once) and our life circumstances (my wife was more afraid of flying when our kids were little). And on top of all that, several general characteristics make some risks feel scarier than others."

David Ropeik taught risk communication at the Harvard School of Public Health. He is a coauthor of *Risk: A Practical Guide for Deciding What's Really Safe and What's Really Dangerous in the World Around You.*

✓ **Checkpoint**   Help at *Math*.and**Y⊙U**.com

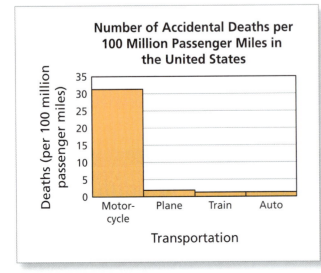

Number of Accidental Deaths per 100 Million Passenger Miles in the United States

Use the bar graph to compare the risk of using the different means of transportation.

There are about 2 accidental deaths for every 100 million airplane miles flown. What does this say about the risk of taking a 2000-mile plane flight? Is it true that the more you fly, the more you increase the likelihood of an accident?

EXAMPLE 4   **Comparing Significance and Risk**

Give an example of each of the four possible risks.

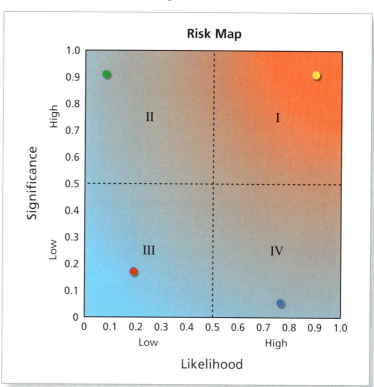

Risk Map

SOLUTION

- An example of a risk that is significant but unlikely is contracting a serious illness when you are young.

- An example of a risk that is both significant and likely is contracting a serious illness if you are extremely overweight.

- An example of a risk that is insignificant but fairly likely is catching a cold during the year.

- An example of a risk that is insignificant and unlikely is being stung by a bee (unless, of course, you are allergic to bee stings).

✓ **Checkpoint**   Help at Math.andYOU.com

Give an example of each of the following types of risks. Then identify where you think the risk is located on the risk map. Explain your reasoning.

a. Prevent-at-source risks: Threaten the achievement of company objectives

b. Detect-and-monitor risks: Should be monitored on a rotational basis to ensure that they are detected before they occur

c. Monitor risks: Should be monitored to ensure that they are being appropriately managed and that their significance has not changed due to changing business conditions

d. Low-control risks: Require minimal monitoring and control unless subsequent risk assessments show a substantial change, prompting a move to another risk category

RISK ASSESSMENT SURVEY

Business owners use a risk assessment survey to determine the significance and likelihood of business risks. A risk map is then used to plot the significance and likelihood of the business risk occurring.

Using Likelihood to Describe Actuarial Data

One profession that makes great use of probability is the actuarial profession. An **actuary** is a person who assesses the risk of an event occurring and calculates the cost of that risk. Most actuaries work in the insurance industry.

EXAMPLE 5   Using Actuarial Data

Explain how an actuary for an insurance company might use the following information.

> A life insurance policy is a wager on when you will die. Insurance companies put a lot of work into determining how long you will live. How do insurance companies determine the probability of when you will die? Actuaries analyze data to determine factors that influence a person's probability of dying. A person's age is a very significant factor.
>
> For instance, in 2005, the probability of dying for a Californian in the 25–34 age group was 84 in 100,000. The probability nearly doubled to 161 in 100,000 for a Californian in the 35–44 age group. This doubling pattern generally continued for consecutive 10-year age groups. The probability of dying for a Californian in the 75–84 age group was abut 60 times greater than that of a Californian in the 25–34 age group.
>
> Actuaries also consider a person's gender when determining the probability of dying. Generally, for a man and woman of the same age, the man has a greater risk of dying.

SOLUTION

The paragraph is basically saying that the older you are, the more likely you are to die. It claims that through age 74 the probability of dying roughly doubles each time your age increases by 10 years. This is shown in the graph. An actuary uses this information by creating questionnaires for people who are applying for life insurance and by setting the insurance rates based on the answers to the questions in the questionnaires.

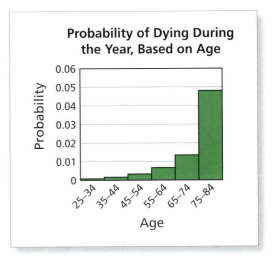

Probability of Dying During the Year, Based on Age

 **Checkpoint**   Help at *Math*.and**Y☺U**.com

You have taken a job at a large corporation that provides life insurance for its employees. As part of the employment process, you are asked to supply your age, gender, height, weight, blood pressure, cholesterol level, race, and marital status. You are also asked whether you smoke, drink alcohol, exercise regularly, or participate in dangerous activities, such as skydiving. What is your opinion about this?

EXAMPLE 6   **Analyzing an Actuarial Graph**

Discuss the actuarial graph. Does it seem reasonable for life expectancy in America today?

World Life Expectancies

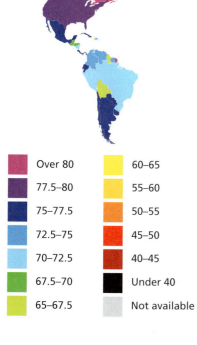

Over 80	60–65
77.5–80	55–60
75–77.5	50–55
72.5–75	45–50
70–72.5	40–45
67.5–70	Under 40
65–67.5	Not available

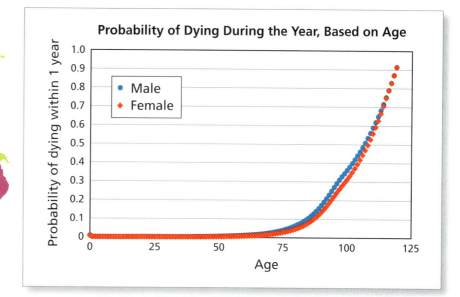

Probability of Dying During the Year, Based on Age

SOLUTION

The graph shows several things.

- As a person gets older, the probability that the person will die during the year increases.

- Males have a slightly greater probability of dying than females. For instance, at age 90, a male has a 17% chance of dying during the year, while a female has only a 14% chance of dying during the year.

The graph seems reasonable for life expectancy in America today.

✓ **Checkpoint**

Help at *Math*.and**YOU**.com

a. Discuss the actuarial graph.

b. How would an actuary use the data?

c. Does the graph seem reasonable? Explain your reasoning.

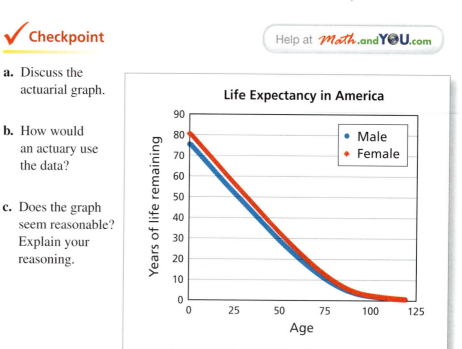

Life Expectancy in America

8.1 Exercises

Describing Likelihood In Exercises 1 and 2, complete the table. Then describe the likelihood of each event in words. *(See Example 1.)*

1.

Probability of Fears			
	Fraction	Decimal	Percent
Dark	$\frac{1}{20}$		
Heights		0.36	
Identity theft	$\frac{25}{38}$		
Thunder or lightning			11%

2.

Probability of Health Issues			
	Fraction	Decimal	Percent
Diagnosed with cancer			41%
Die from flu	$\frac{1}{345,100}$		
Has health insurance		0.847	
Obese			26.9%

Snowfall In Exercises 3–6, describe the likelihood that at least 1 inch of snow will accumulate in the city on the morning of December 25. *(See Example 2.)*

WHITE CHRISTMAS

5% 10% 25% 40% 50% 60% 75% >90% HISTORIC PROBABILITY

The Weather Channel weather.com

06 Aug 2011 06:00 GMT / 06 Aug 2011 02:00 AM EDT

3. Richmond, Virginia

4. Madison, Wisconsin

5. Denver, Colorado

6. Philadelphia, Pennsylvania

7. **Fatal Occupational Injuries** The graph shows the rates of fatal occupational injuries per 100,000 full-time workers for several industries. *(See Example 3.)*

Fatal Occupational Injuries

a. Compare the rates of fatal occupational injuries for the different industries.

b. Why do different industries have different rates?

8. **Fatal Occupational Injuries** The graph shows the numbers of fatal occupational injuries for the same industries as in Exercise 7. *(See Example 3.)*

a. Compare the numbers of fatal occupational injuries for the different industries.

b. Explain why financial activities and mining have the same number of fatal occupational injuries, but their rates in Exercise 7 are different.

Comparing Significance and Risk In Exercises 9–14, determine the likelihood and significance of the event. *(See Example 4.)*

9. A nuclear plant meltdown

10. Contracting a foodborne illness during your lifetime

11. A flood damaging property on a hilltop

12. A person with high blood pressure contracting heart disease

13. A deployed air bag causing a severe injury

14. A heavy smoker contracting cancer

Life Expectancy The graphs show the life expectancies of females and males from 1980 through 2007 for 5 countries. In Exercises 15–20, use the graphs. *(See Examples 5 and 6.)*

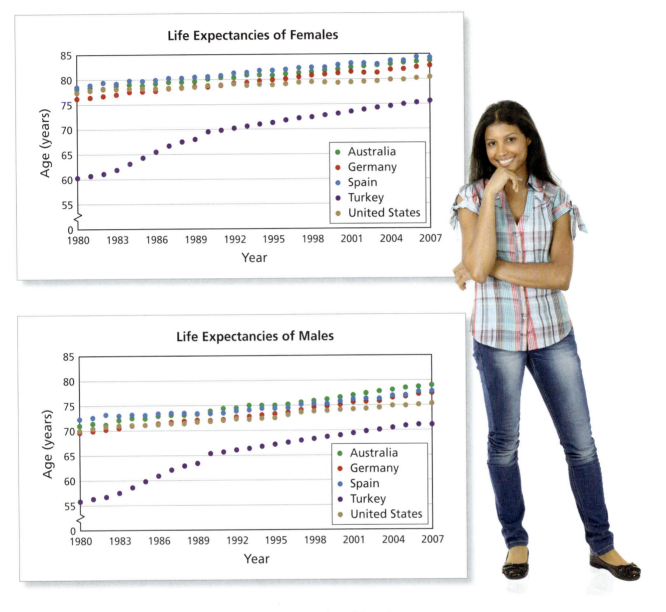

15. Discuss any trends in the graph of the life expectancies of females.

16. Discuss any trends in the graph of the life expectancies of males.

17. Order the countries by female life expectancy from least to greatest for 1980 and 2007. Compare the orders.

18. Order the countries by male life expectancy from least to greatest for 1980 and 2007. Compare the orders.

19. Compare the life expectancies of females and males for each country. What do you notice?

20. Which country had the greatest increase in life expectancy for females? Which country had the greatest increase in life expectancy for males?

▶ Extending Concepts

Comparing Events **In Exercises 21–23, use the spinners.**

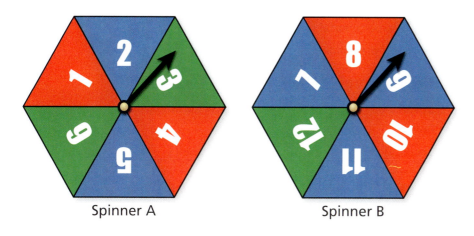

Spinner A   Spinner B

21. You spin spinner A. Which event has a greater likelihood? Explain your reasoning.

   a. Event 1: Spinning a number less than 1

      Event 2: Spinning a number
           greater than 1

   b. Event 1: Spinning a multiple of 3

      Event 2: Spinning a number
           greater than 3

22. You spin spinner B. Which event
has a greater likelihood? Explain
your reasoning.

   a. Event 1: Spinning green

      Event 2: Spinning red

   b. Event 1: Spinning an even number

      Event 2: Spinning a prime number

23. You spin both spinners. Which event
has a greater likelihood? Explain
your reasoning.

   a. Event 1: Spinning blue on spinner A

      Event 2: Spinning blue on spinner B

   b. Event 1: Spinning a multiple of 2 on spinner A

      Event 2: Spinning a multiple of 6 on spinner B

24. Impossible and Certain Events Describe two events that are impossible.
Describe two events that are certain. Explain your reasoning.

8.2 Estimating Likelihood

▶ Find a theoretical probability.

▶ Find an experimental probability.

▶ Estimate a probability using historical results.

Finding a Theoretical Probability

Here are three ways to estimate (or find exactly) the probability that an event will occur.

1. **Theoretical probability** is used in cases where all possible outcomes of an event are known and can be counted. Gambling is a common example.
2. **Experimental probability** is used in cases where a representative sample can be taken and counted. Quality control is a common example.
3. **Historical probability** is used in cases where past occurrences are assumed to be representative of future occurrences. Weather prediction is a common example.

Theoretical Probability

Suppose an event can occur n ways out of a total number T of equally likely outcomes. The **probability** that the event will occur is

$$\text{Probability} = \frac{\text{number of favorable outcomes}}{\text{total number of possible outcomes}}$$

$$= \frac{n}{T}.$$

n favorable outcomes

T possible outcomes

EXAMPLE 1   Finding a Theoretical Probability

The Daily Number is a common type of state lottery. To play, you choose a 3-digit number. Find the probability of winning for the following cases.

a. You pick one 3-digit number.   **b.** You pick ten 3-digit numbers.

SOLUTION

There are 1000 possible numbers (000 to 999) that players can select.

a. $\text{Probability} = \dfrac{1}{1000}$ **b.** $\text{Probability} = \dfrac{10}{1000} = \dfrac{1}{100}$

✓ **Checkpoint**   Help at *Math*.and**Y●U**.com

Most states have some form of state lottery. Altogether, these lotteries bring in about $52 billion annually. Of this amount, about $32 billion is paid out in prize money. The remainder is retained by the states.

The Daily Number pays $500 for a $1 winning ticket. Suppose you buy ten $1 tickets each day for 100 days. How many times do you expect to win? During the 100 days, how much do you spend? How much do you win?

EXAMPLE 2   **Finding a Theoretical Probability**

The NBA Draft Lottery is an annual event held by the National Basketball Association. The 14 teams that missed the playoffs in the previous season participate in a lottery to determine the order for drafting players. The lottery is weighted so that the team with the worst record has the best chance of obtaining a higher draft pick. The lottery determines the first three picks of the draft.

1. 250 combinations	
2. 199 combinations	**The NBA Draft Lottery**
3. 156 combinations	The 14 teams are ordered (from worst to best) and randomly given 4-digit combinations, as shown at the left.
4. 119 combinations	
5. 88 combinations	
6. 63 combinations	To conduct the lottery, 14 balls numbered 1 through 14 are put into a lottery machine. Four of the numbers are drawn. The team that has that combination of numbers (order does not matter) wins the first pick. There are 1001 possible combinations of 4 numbers (from 1 through 14). However, the combination 11-12-13-14 does not qualify, leaving only 1000 valid combinations.
7. 43 combinations	
8. 28 combinations	
9. 17 combinations	
10. 11 combinations	
11. 8 combinations	
12. 7 combinations	
13. 6 combinations	
14. 5 combinations	

a. What is the probability that the team with the worst record wins the first pick?

b. What is the probability that team #14 wins the first pick?

SOLUTION

a. Probability $= \dfrac{250}{1000} = \dfrac{1}{4} = 25\%$

b. Probability $= \dfrac{5}{1000} = \dfrac{1}{200} = 0.5\%$

The National Basketball Association (NBA) is the leading professional basketball league in North America. It has 30 teams. It was formed in 1949 when the National Basketball League merged with the Basketball Association of America.

✓ **Checkpoint**   Help at *Math*.and**Y☺U**.com

Enter the number of combinations for each of the 14 teams in the NBA Draft Lottery into a spreadsheet.

c. Use the spreadsheet to find the probability that each team wins the first pick.

d. Find the total of the probabilities column. What can you conclude?

	A	B	C
	Team	**Combinations**	**Probability of 1st Pick**
1			
2	1	250	25.0%
3	2	199	
4	3	156	
5	4	119	
6	5	88	
7	6	63	
8	7	43	
9	8	28	
10			

DATA

Finding an Experimental Probability

In Examples 1 and 2, you could find an exact probability because you could determine the exact number of favorable outcomes and the exact number of possible outcomes. In real-life situations, it is sometimes difficult or impossible to determine this information. In these cases, you can try to find a sample that is representative.

EXAMPLE 3 **Finding an Experimental Probability**

To form a theory about the inheritance of eye color, a geneticist records the eye color of 2400 sets of parents and their children, as shown below.

a. From this sample, what is the probability that a blue-eyed parent and a brown-eyed parent have a blue-eyed child?

b. What can you conclude about the eye color of the children of a blue-eyed parent and a brown-eyed parent?

Percent of Light-Colored Eyes in Europe

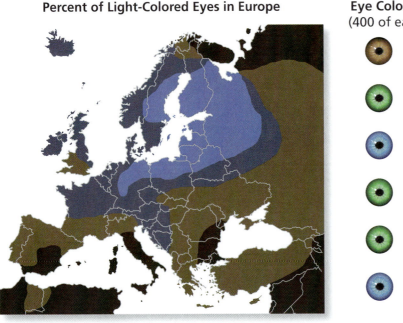

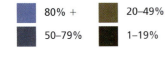

80% + 20–49%
50–79% 1–19%

© Eupedia.com

Eye Color of Parents
(400 of each pattern)

Number of Children with Given Eye Color

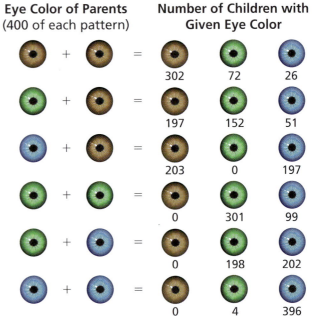

SOLUTION

a. Of the 400 children with a blue-eyed parent and a brown-eyed parent, there are 197 blue-eyed children. So, you can estimate the probability to be

$$\text{Probability} = \frac{197}{400} = 49.25\%.$$

b. From this sample, it appears that the children of a blue-eyed parent and a brown-eyed parent are equally likely to have blue eyes or brown eyes.

✓ **Checkpoint**

Help at ***Math*.andYOU.com**

From the above sample, what is the probability that a child of two brown-eyed parents will not have brown eyes?

EXAMPLE 4   **Finding an Experimental Probability**

A fast-food restaurant performs the following market research before adding a broiled chicken sandwich to its menu.

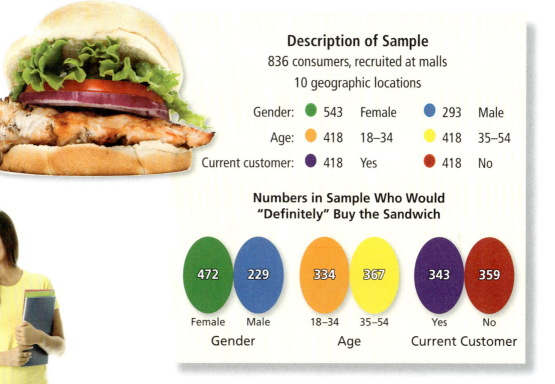

Description of Sample

836 consumers, recruited at malls

10 geographic locations

Gender:	● 543 Female	● 293 Male
Age:	● 418 18–34	● 418 35–54
Current customer:	● 418 Yes	● 418 No

Numbers in Sample Who Would "Definitely" Buy the Sandwich

472	229	334	367	343	359
Female	Male	18–34	35–54	Yes	No
Gender		Age		Current Customer	

Assume that the sample of 836 consumers is representative of the general population of people who are between the ages of 18 and 54. Describe the probability that a female buys the sandwich. Describe the probability that a male buys the sandwich.

SOLUTION

Female: 472 of the 543 women sampled said they would buy the sandwich.

$$\text{Probability} = \frac{472}{543} \approx 0.87 = 87\%$$

So, a female is *very likely* to buy the sandwich.

Male: 229 of the 293 men sampled said they would buy the sandwich.

$$\text{Probability} = \frac{229}{293} \approx 0.78 = 78\%$$

So, although the sandwich appears to be somewhat less appealing to men than to women, you can still conclude that a male is *likely* to buy the sandwich.

✓ **Checkpoint**

Describe the probability of each person buying the sandwich. Explain your reasoning.

a. A person in the 18–34 age group

b. A person who is not a current customer

Finding samples of people who are truly representative of the general population is the major challenge to market research and polling companies. The saying, "Will it play in Peoria?" is often used to ask whether a product will appeal to a broad demographic.

Estimating a Probability Using Historical Results

One of the most common ways to estimate the probability of an event is to look at how often it occurred in the past.

EXAMPLE 5   **Interpreting a Historical Probability**

Apiphobia is a fear of bees. Cyclophobia is a fear of bicycles. According to the bubble graph, how many times more likely is it for a person to die in a bicycling accident than by a bee sting?

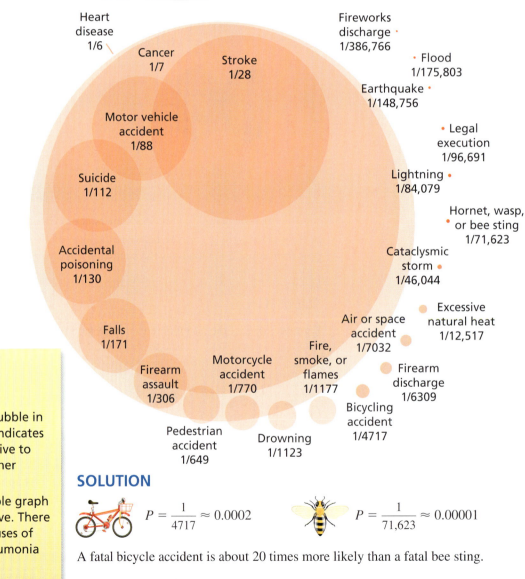

Heart disease 1/6

Cancer 1/7

Stroke 1/28

Fireworks discharge 1/386,766

Flood 1/175,803

Earthquake 1/148,756

Motor vehicle accident 1/88

Legal execution 1/96,691

Lightning 1/84,079

Suicide 1/112

Hornet, wasp, or bee sting 1/71,623

Accidental poisoning 1/130

Cataclysmic storm 1/46,044

Excessive natural heat 1/12,517

Air or space accident 1/7032

Falls 1/171

Fire, smoke, or flames 1/1177

Firearm discharge 1/6309

Firearm assault 1/306

Motorcycle accident 1/770

Bicycling accident 1/4717

Pedestrian accident 1/649

Drowning 1/1123

Study Tip

The area of each bubble in the bubble graph indicates its probability relative to the value of the other probabilities.

Note that the bubble graph is not comprehensive. There are many other causes of death, such as pneumonia and diabetes.

SOLUTION

$$P = \frac{1}{4717} \approx 0.0002 \qquad P = \frac{1}{71,623} \approx 0.00001$$

A fatal bicycle accident is about 20 times more likely than a fatal bee sting.

✓ **Checkpoint**   Help at *Math*.and**YOU**.com

What is the probability that a person dies from heart disease, cancer, or a stroke? Explain your reasoning.

EXAMPLE 6   **Analyzing a Historical Probability**

The graph shows the probability that there will be at least 1 inch of snow on the ground on December 25. The estimates are based on data over a 30-year period. How do these data correlate with the latitude of each location?

Snow on December 25

City	Probability
Marquette, MI	100%
Montpelier, VT	93%
Wausau, WI	93%
Bismarck, ND	87%
Portland, ME	83%
Lander, WY	77%
Blue Canyon, CA	74%
Minneapolis, MN	73%
Madison, WI	67%
Great Falls, MT	57%
Flagstaff, AZ	56%
Salt Lake City, UT	53%
Denver, CO	50%
Chicago, IL	40%
Boise, ID	30%
Reno, NV	20%
Winslow, AZ	20%
Washington, DC	13%
New York, NY	10%
Prescott, AZ	10%
Albuquerque, NM	3%

SOLUTION

Enter the latitude and probability of snow for each city into a spreadsheet. Then use the spreadsheet to make a scatter plot as shown. The scatter plot indicates only a slight correlation between latitude and the probability of snow.

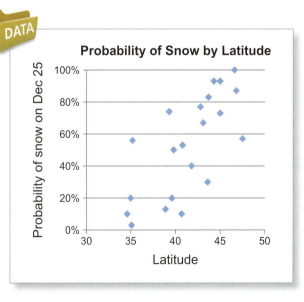

Probability of Snow by Latitude

✓ **Checkpoint**   Help at *Math*.and**YOU**.com

Latitude seems to be a weak predictor of snow on December 25. What other factors can you use to improve the prediction?

8.2 Exercises

Dodecahedron Die A dodecahedron die has 12 sides numbered 1 through 12. You roll a dodecahedron. In Exercises 1–6, find the probability of the event. *(See Example 1.)*

1. Rolling a 6

2. Rolling an 11

3. Rolling a number less than 9

4. Rolling a multiple of 4

5. Rolling an odd number

6. Rolling a prime number

7. Raffle A charity sells 1000 tickets for a raffle. There is a grand prize of $200 and 4 other prizes of $50. You buy one ticket. *(See Example 1.)*

a. What is the probability that you win the grand prize?

b. What is the probability that you win a prize?

8. Lottery The table shows the payouts for the 600,000 people who played the lottery yesterday. You randomly choose one person who played the lottery yesterday. Find the probability that the person is in each payout group. *(See Example 2.)*

Payout	People
$0	568,375
$3	18,245
$4	9820
$7	3417
$100	136
$10,000	6
$200,000	1
$16,000,000	0

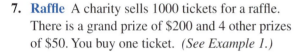

Internet Usage The tables show the results of a survey that asked whether the Internet is a main news source. In Exercises 9–12, find the probability of the event. *(See Examples 3 and 4.)*

Age group	Yes	No
18–29	259	141
30–49	179	193
50–64	128	245
65 and over	50	305

Region	Yes	No
Northeast	179	269
Midwest	128	208
South	194	279
West	115	128

9. A person in the 18–29 age group says "yes"

10. A person in the 65 and over age group says "yes"

11. A person in any age group says "no"

12. A person in any region says "no"

13. **Data Analysis** Compare your answers to Exercises 11 and 12. What do you notice? Explain.

New Year's Resolutions The circle graphs show the results of a survey that asked men and women in different body mass index (BMI) categories whether losing weight is one of their New Year's resolutions. In Exercises 14–16, use the graphs. *(See Examples 3 and 4.)*

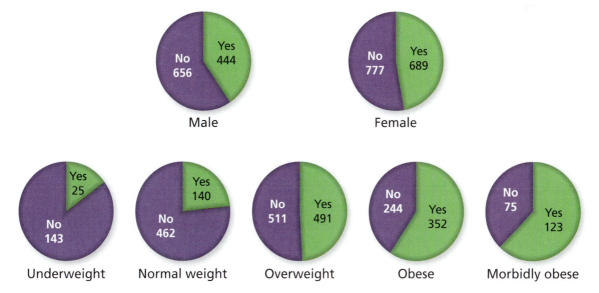

14. Find the probability that losing weight is one of the New Year's resolutions for each gender. Then describe the likelihood.

15. Find the probability that losing weight is one of the New Year's resolutions for a person in each BMI category. Then describe the likelihood.

16. Describe any trends you see in the graphs.

Football The graph shows the positions of the most valuable players (MVPs) for the first 45 Super Bowls. (*Note:* There are 46 players in the graph because there were co-MVPs in Super Bowl XII.) In Exercises 17–22, use the graph. (*See Examples 5 and 6.*)

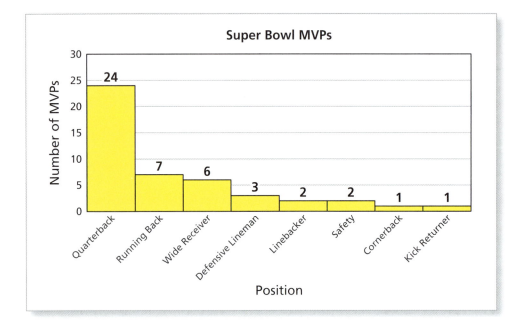

17. Find the probability that a quarterback is a Super Bowl MVP.

18. Find the probability that an offensive player (quarterback, running back, or wide receiver) is a Super Bowl MVP.

19. Find the probability that a defensive or special teams player (defensive lineman, linebacker, safety, cornerback, or kick returner) is a Super Bowl MVP.

20. How many times more likely is it for a Super Bowl MVP to be a quarterback than a wide receiver?

21. How many times more likely is it for a Super Bowl MVP to be a running back than a linebacker?

22. How many times more likely is it for a Super Bowl MVP to be a quarterback than a defensive or special teams player?

**Super Bowl MVPs
(2002–2011)**

2011	Aaron Rodgers, quarterback
2010	Drew Brees, quarterback
2009	Santonio Holmes, wide receiver
2008	Eli Manning, quarterback
2007	Peyton Manning, quarterback
2006	Hines Ward, wide receiver
2005	Deion Branch, wide receiver
2004	Tom Brady, quarterback
2003	Dexter Jackson, safety
2002	Tom Brady, quarterback

▶ **Extending Concepts**

Comparing Probabilities **In Exercises 23–25, you have a standard deck of cards.**

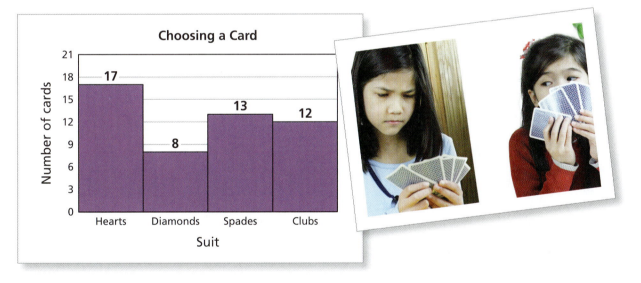

23. You randomly choose a card from the deck. Find the theoretical probability of choosing a card of each suit.

24. The bar graph shows the results of randomly choosing 1 card, recording its suit, and placing it back in the deck for 50 trials. Find the experimental probability of choosing a card of each suit.

25. Compare the probabilities you found in Exercises 23 and 24.

Addition Rule The probability that one of two events occurs is

$$\text{Probability that either event occurs} = \begin{pmatrix}\text{probability}\\\text{of event 1}\end{pmatrix} + \begin{pmatrix}\text{probability}\\\text{of event 2}\end{pmatrix} - \begin{pmatrix}\text{probability of}\\\text{both events}\end{pmatrix}.$$

In Exercises 26–28, you randomly choose a card from a standard deck of cards. Find the probability.

26. Choosing a heart or a 6

27. Choosing a black suit or a 2

28. Choosing a face card or a diamond

8.1–8.2   Quiz

1. **Winter Storm** The table shows the probabilities of a winter storm in New York for several months. Complete the table. Then describe the likelihood of each event in words.

Month	Fraction	Decimal	Percent
November			4.4%
December	$\frac{25}{74}$		
January		0.081	
February			30.1%
March	$\frac{5}{34}$		

Snowfall in Montana In Exercises 2–5, describe the likelihood that at least 1 inch of snow will accumulate in the city.

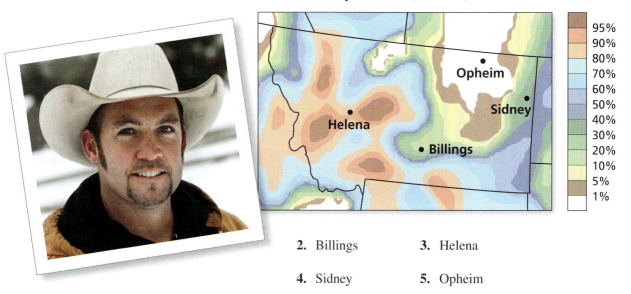

Probability of 1 Inch (or more) of Snow

95%	
90%	
80%	
70%	
60%	
50%	
40%	
30%	
20%	
10%	
5%	
1%	

2. Billings

3. Helena

4. Sidney

5. Opheim

Randomly Selected Digits In Exercises 6–8, consider a computer that randomly selects a 4-digit number. Each digit can be any number from 0 to 9.

6. How many possible numbers are there?

7. Find the probability that the number is greater than or equal to 3000.

8. Find the probability that the number is divisible by 1000.

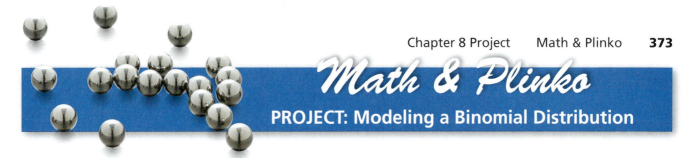

Math & Plinko

PROJECT: Modeling a Binomial Distribution

1. Use the *Plinko Simulator* at *Math.andYou.com.*

- Set the number of rows to 25.
- Set the probability of falling to the right at 0.5.
- Drop a ball by pressing "Start."
- Drop 99 more balls. You do not have to wait for one drop to complete before starting the next drop.

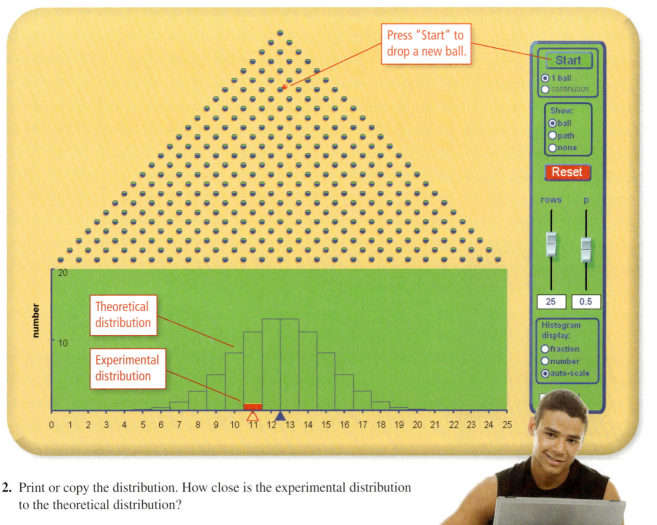

2. Print or copy the distribution. How close is the experimental distribution to the theoretical distribution?

3. This type of distribution is called a *binomial distribution*. The name comes from the fact that as the ball hits a peg, it can fall in two ways (right or left). Estimate the probability that the ball lands in each of the positions from 0 through 25. Explain your reasoning.

4. Is it possible that a ball lands in position 25? What is the likelihood of this? Explain your reasoning.

5. Describe an event in real life that can be modeled by a binomial distribution. Adjust the simulator to fit the event. Run a simulation and summarize your results.

8.3 Expected Value

▶ Find an expected value involving two events.

▶ Find an expected value involving multiple events.

▶ Use expected value to make investment decisions.

Finding an Expected Value Involving Two Events

The **expected value** of an "experiment" is the long-run average—if the experiment could be repeated many times, the expected value is the average of all the results.

Expected Value

Consider an experiment that has only two possible events. The expected value of the experiment is

$$\text{Expected value} = \left(\begin{array}{c}\text{probability}\\\text{of event 1}\end{array}\right)\left(\begin{array}{c}\text{payoff for}\\\text{event 1}\end{array}\right) + \left(\begin{array}{c}\text{probability}\\\text{of event 2}\end{array}\right)\left(\begin{array}{c}\text{payoff for}\\\text{event 2}\end{array}\right).$$

EXAMPLE 1 **Finding an Expected Value**

In a state lottery, a single digit is drawn from each of four containers. Each container has 10 balls numbered 0 through 9. To play, you choose a 4-digit number and pay $1. If your number is drawn, you win $5000. If your number is not drawn, you lose your dollar. What is the expected value for this game?

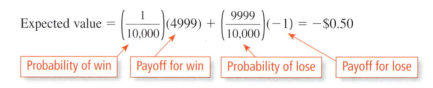

SOLUTION

The probability of winning is 1/10,000. The probability of losing is 9999/10,000.

$$\text{Expected value} = \left(\frac{1}{10,000}\right)(4999) + \left(\frac{9999}{10,000}\right)(-1) = -\$0.50$$

Probability of win ‖ Payoff for win ‖ Probability of lose ‖ Payoff for lose

So, on average, you should expect to lose $0.50 each time you play the game.

✓ **Checkpoint**  Help at *Math*.andYOU.com

Play the *Lottery Simulator* at *Math.andYou.com*. Set the number of games to 100,000. Discuss your results in the context of expected value. States are often criticized for falsely raising people's expectations of winning and for encouraging a form of regressive tax on the poor. What is your opinion of this?

The total annual cost of fire in the United States is about 2.5% of the domestic gross product.

Find an Expected Value

You take out a fire insurance policy on your home. The annual premium is $300. In case of fire, the insurance company will pay you $200,000. The probability of a house fire in your area is 0.0002.

a. What is the expected value?

b. What is the insurance company's expected value?

c. Suppose the insurance company sells 100,000 of these policies. What can the company expect to earn?

SOLUTION

$$200,000 - 300$$

a. Expected value $= (0.0002)(199,700) + (0.9998)(-300) = -\$260.00$

           Fire   No Fire

   The expected value over many years is $-\$260$ per year. Of course, your hope is that you will never have to collect on fire insurance for your home.

b. The expected value for the insurance company is the same, except the perspective is switched. Instead of $-\$260$ per year, it is $+\$260$ per year. Of this, the company must pay a large percent for salaries and overhead.

c. The insurance company can expect to gross $30,000,000 in premiums on 100,000 such policies. With a probability of 0.0002 for fire, the company can expect to pay on about 20 fires. This leaves a gross profit of $26,000,000.

✓ **Checkpoint**   Help at *Math*.and**Y🌐U**.com

In the circle graph, why is the percent for property damage greater than the percent for fire insurance premiums?

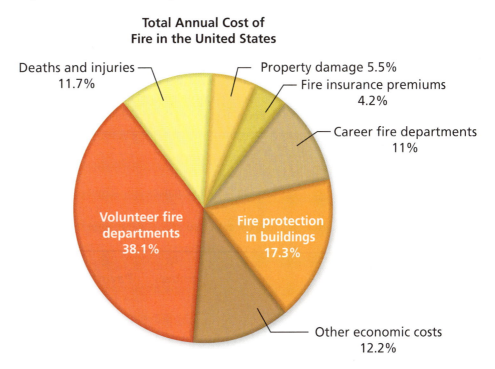

Total Annual Cost of Fire in the United States

Deaths and injuries 11.7%
Property damage 5.5%
Fire insurance premiums 4.2%
Career fire departments 11%
Volunteer fire departments 38.1%
Fire protection in buildings 17.3%
Other economic costs 12.2%

Finding an Expected Value Involving Multiple Events

EXAMPLE 3   **Comparing Two Expected Values**

A child asks his parents for some money. The parents make the following offers.

Father's offer: The child flips a coin. If the coin lands heads up, the father will give the child $20. If the coin lands tails up, the father will give the child nothing.

Mother's offer: The child rolls a 6-sided die. The mother will give the child $3 for each dot on the up side of the die.

Which offer has the greater expected value?

SOLUTION

Father's offer:

$$\text{Expected value} = \left(\frac{1}{2}\right)(20) + \left(\frac{1}{2}\right)(0) = \$10$$

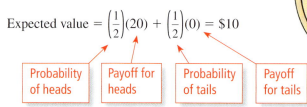

| Probability of heads | Payoff for heads | Probability of tails | Payoff for tails |

Mother's offer: There are six possible outcomes.

DATA

	A	B	C	D
1	**Number**	**Payoff**	**Probability**	**Expected Value**
2	1	$3.00	16.67%	$0.50
3	2	$6.00	16.67%	$1.00
4	3	$9.00	16.67%	$1.50
5	4	$12.00	16.67%	$2.00
6	5	$15.00	16.67%	$2.50
7	6	$18.00	16.67%	$3.00
8	**Total**			**$10.50**

Even though the mother's offer has a slightly higher expected value, the best the child can do with the mother's offer is $18, whereas the child has a 50% chance of receiving $20 with the father's offer.

✓ **Checkpoint**   Help at

The child's uncle makes a different offer. The child rolls a 12-sided die. The uncle will give the child $2 for each dot on the up side of the die. Use a spreadsheet to find the expected value of this offer. Which offer would you take? Explain.

	A	B	C	D
1	**Number**	**Payoff**	**Probability**	**Expected Value**
2	1	$2.00	8.33%	$0.17
3	2	$4.00	8.33%	$0.33
4	3	$6.00	8.33%	$0.50
5	4	$8.00	8.33%	$0.67

EXAMPLE 4   **Using a Decision Tree**

Your company is considering developing one of two cell phones. Your development and market research teams provide you with the following projections.

🟡 **Cell phone A:**

Cost of development: $2,500,000

Projected sales: 50% chance of net sales of $5,000,000
30% chance of net sales of $3,000,000
20% chance of net sales of $1,500,000

🟢 **Cell phone B:**

Cost of development: $1,500,000

Projected sales: 30% chance of net sales of $4,000,000
60% chance of net sales of $2,000,000
10% chance of net sales of $500,000

Which model should your company develop? Explain.

SOLUTION

A *decision tree* can help organize your thinking.

Probability	Profit	Expected Value
50%	$2.5 million	0.5(2.5)
30%	$0.5 million	0.3(0.5)
A 20%	−$1 million	+ 0.2(−1)
		$1.2 million
30%	$2.5 million	0.3(2.5)
60%	$0.5 million	0.6(0.5)
B 10%	−$1 million	+ 0.1(−1)
		$0.95 million
Neither 100%	$0	1.0(0) = **$0 million**

Although cell phone A has twice the risk of losing $1 million, it has the greater expected value. So, using expected value as a decision guideline, your company should develop cell phone A.

✓ **Checkpoint**   Help at *Math*.and**YOU**.com

Which of the following should your company develop? Explain.

🔴 **Cell phone C:** Cost of development: $2,000,000

Projected sales: 40% chance of net sales of $5,000,000
40% chance of net sales of $3,000,000
20% chance of net sales of $1,500,000

🔵 **Cell phone D:** Cost of development: $1,500,000

Projected sales: 15% chance of net sales of $4,000,000
75% chance of net sales of $2,000,000
10% chance of net sales of $500,000

As of 2010, it was estimated that there were over 5 billion cell phone subscriptions worldwide. With this massive market, the enticement to invest in the development of new and innovative products is strong.

Using Expected Value to Make Investment Decisions

EXAMPLE 5 **Using Expected Value**

Analyze the mathematics in the following description of Daniel Kahneman and Amos Tversky's "Prospect Theory: An Analysis of Decision under Risk."

Daniel Kahneman, a professor at Princeton University, became the first psychologist to win the Nobel Prize in Economic Sciences. The prize was awarded for his "prospect theory" about investors' "illusion of control."

"A problem is positively framed when the options at hand generally have a perceived probability to result in a positive outcome. Negative framing occurs when the perceived probability weighs over into a negative outcome scenario. In one of Kahneman and Tversky's (1979) experiments, the participants were to choose one of two scenarios, an 80% possibility to win $4,000 and the 20% risk of not winning anything as opposed to a 100% possibility of winning $3,000. Although the riskier choice had a higher expected value ($4,000 × 0.8 = $3,200), 80% of the participants chose the safe $3,000. When participants had to choose between an 80% possibility to lose $4,000 and the 20% risk of not losing anything as one scenario, and a 100% possibility of losing $3,000 as the other scenario, 92% of the participants picked the gambling scenario. This framing effect, as described in . . . Prospect Theory, occurs because individuals over-weigh losses when they are described as definitive, as opposed to situations where they are described as possible. This is done even though a rational economical evaluation of the two situations lead to identical expected value. People tend to fear losses more than they value gains. A $1 loss is more painful than the pleasure of a $1 gain."

Johan Ginyard

SOLUTION

Here are the first two options the participants were given.

Expected Value

> Greater expected value →

Option 1: 80% chance of gaining $4000 $(0.8)(4000) + (0.2)(0) = \$3200$
         20% chance of gaining $0

Option 2: 100% chance of gaining $3000 $(1.0)(3000) = \$3000$

> Preferred by participants →

Here are the second two options the participants were given.

Expected Value

> Preferred by participants →

Option 1: 80% chance of losing $4000 $(0.8)(-4000) + (0.2)(0) = -\$3200$
         20% chance of losing $0

Option 2: 100% chance of losing $3000 $(1.0)(-3000) = -\$3000$

> Greater expected value →

What Kahneman and Tversky found surprising was that in neither case did the participants intuitively choose the option with the greater expected value.

✓ **Checkpoint**   Help at *Math*.and**YOU**.com

Describe other situations in which people fear losses more than they value gains.

EXAMPLE 6   Comparing Two Expected Values

A *speculative investment* is one in which there is a high risk of loss. What is the expected value for each of the following for a $1000 investment?

a. Speculative investment

- Complete loss: 40% chance
- No gain or loss: 15% chance
- 100% gain: 15% chance
- 400% gain: 15% chance
- 900% gain: 15% chance

b. Conservative investment

- Complete loss: 1% chance
- No gain or loss: 35% chance
- 10% gain: 59% chance
- 20% gain: 5% chance

SOLUTION

From 1973 through 2010, the Standard and Poor 500 Index had an average annual gain of 11.5%. During this time, its greatest annual gain was 37.4% in 1995, and its greatest annual loss was 37% in 2008.

a. Speculative investment

	A	B	C	D
1	**Result**	**Payoff**	**Probability**	**Expected Value**
2	Complete loss	-$1,000	40%	-$400
3	No gain or loss	$0	15%	$0
4	100% gain	$1,000	15%	$150
5	400% gain	$4,000	15%	$600
6	900% gain	$9,000	15%	$1,350
7	**Total**		**100%**	**$1,700**

b. Conservative investment

	A	B	C	D
1	**Result**	**Payoff**	**Probability**	**Expected Value**
2	Complete loss	-$1,000	1%	-$10
3	No gain or loss	$0	35%	$0
4	10% gain	$100	59%	$59
5	20% gain	$200	5%	$10
6	**Total**		**100%**	**$59**

This example points out the potential gain and the risk of investment. The speculative investment has an expected value of $1700, which is a high return on investment. If you had the opportunity to make 100 such investments, you would have a high likelihood of making a profit. But, when making only 1 such investment, you have a 40% chance of losing everything.

✓ **Checkpoint**   Help at *Math*.and**Y⊕U**.com

Which of the following investments is better? Explain your reasoning.

c. Speculative investment

- Complete loss: 20% chance
- No gain or loss: 35% chance
- 100% gain: 35% chance
- 400% gain: 5% chance
- 2000% gain: 5% chance

d. Conservative investment

- Complete loss: 2% chance
- No gain or loss: 38% chance
- 20% gain: 55% chance
- 30% gain: 5% chance

8.3 Exercises

Life Insurance The table shows the probabilities of dying during the year for various ages. In Exercises 1–6, use the table. *(See Examples 1 and 2.)*

Probability of Dying During the Year		
Age	Male	Female
21	0.001420	0.000472
22	0.001488	0.000487
23	0.001502	0.000496
24	0.001474	0.000503
25	0.001430	0.000509
26	0.001393	0.000519
27	0.001366	0.000535
28	0.001362	0.000561
29	0.001379	0.000595
30	0.001406	0.000637

1. A 23-year-old male pays $275 for a 1-year $150,000 life insurance policy. What is the expected value of the policy for the policyholder?

2. A 28-year-old female pays $163 for a 1-year $200,000 life insurance policy. What is the expected value of the policy for the policyholder?

3. A 27-year-old male pays $310 for a 1-year $175,000 life insurance policy. What is the expected value of the policy for the insurance company?

4. A 25-year-old female pays $128 for a 1-year $120,000 life insurance policy. What is the expected value of the policy for the insurance company?

5. A 26-year-old male pays $351 for a 1-year $180,000 life insurance policy.

   a. What is the expected value of the policy for the policyholder?

   b. What is the expected value of the policy for the insurance company?

   c. Suppose the insurance company sells 10,000 of these policies. What is the expected value of the policies for the insurance company?

6. A 30-year-old female pays $259 for a 1-year $250,000 life insurance policy.

   a. What is the expected value of the policy for the policyholder?

   b. What is the expected value of the policy for the insurance company?

   c. Suppose the insurance company sells 10,000 of these policies. What is the expected value of the policies for the insurance company?

Consumer Electronics Company A consumer electronics company is considering developing one of two products. In Exercises 7–10, use a decision tree to decide which model the company should develop. *(See Examples 3 and 4.)*

7. Laptop A: Cost of development: $8 million

Projected Sales	
Probability	Net sales (in millions)
10%	$16
70%	$12
20%	$6

Laptop B: Cost of development: $10 million

Projected Sales	
Probability	Net sales (in millions)
30%	$18
50%	$14
20%	$8

8. MP3 Player A:
Cost of development: $5 million

Projected Sales	
Probability	Net sales (in millions)
20%	$10
60%	$8
20%	$2

MP3 Player B:
Cost of development: $3 million

Projected Sales	
Probability	Net sales (in millions)
40%	$6
40%	$5
20%	$1

9. E-reader A: Cost of development: $3 million

Projected Sales	
Probability	Net sales (in millions)
20%	$10
45%	$6
25%	$5
10%	$0.5

E-reader B: Cost of development: $4 million

Projected Sales	
Probability	Net sales (in millions)
10%	$12
40%	$10
30%	$4
20%	$1

10. Camera A:
Cost of development: $5 million

Projected Sales	
Probability	Net sales (in millions)
10%	$13
30%	$10
20%	$8
25%	$6
15%	$2

Camera B:
Cost of development: $3.5 million

Projected Sales	
Probability	Net sales (in millions)
20%	$10
35%	$7.5
25%	$5.5
10%	$3.5
10%	$0.5

Option Comparison In Exercises 11–14, compare the two options. *(See Example 5.)*

11.

	Probability	Gain
Option 1	100%	$1000
	0%	$0
Option 2	60%	$2000
	40%	$0

12.

	Probability	Gain
Option 1	100%	−$1000
	0%	$0
Option 2	70%	−$2000
	30%	$1000

13.

	Probability	Gain
Option 1	80%	$1000
	20%	$3000
Option 2	90%	$2000
	10%	$0

14.

	Probability	Gain
Option 1	75%	$500
	25%	$1500
Option 2	50%	−$500
	50%	$2000

Investment Comparison You want to invest $1000. In Exercises 15 and 16, find the expected values for the two investments. *(See Example 6.)*

15.

Speculative investment	Conservative investment
• Complete loss: 10% chance	• Complete loss: 1% chance
• No gain or loss: 20% chance	• No gain or loss: 39% chance
• 150% gain: 40% chance	• 50% gain: 40% chance
• 200% gain: 20% chance	• 100% gain: 20% chance
• 700% gain: 10% chance	

16.

Speculative investment	Conservative investment
• Complete loss: 30% chance	• Complete loss: 5% chance
• No gain or loss: 25% chance	• No gain or loss: 15% chance
• 100% gain: 20% chance	• 30% gain: 60% chance
• 500% gain: 15% chance	• 60% gain: 20% chance
• 1000% gain: 10% chance	

▶ Extending Concepts

 Investment Portfolio The table shows the rates of return of two stocks for different economic states. In Exercises 17 and 18, use the table.

Economic State	Probability	Rate of return	
		Stock V	Stock W
Boom	20%	28%	−5%
Normal	65%	12%	7%
Recession	15%	−16%	23%

17. Compare the expected rates of return of the two stocks.

18. You invest 50% of your money in stock V and 50% of your money in stock W. What is the expected rate of return?

 Investment Portfolio The table shows the rates of return of three stocks for different economic states. In Exercises 19–22, find the expected rate of return for the portfolio.

Economic state	Probability	Rate of return		
		Stock X	Stock Y	Stock Z
Boom	20%	16%	27%	3%
Normal	65%	9%	13%	7%
Recession	15%	−5%	−23%	21%

19.

Portfolio mix:

50% invested in stock X

20% invested in stock Y

30% invested in stock Z

20.

Portfolio mix:

70% invested in stock X

10% invested in stock Y

20% invested in stock Z

21.

Portfolio mix:

50% invested in stock X

25% invested in stock Y

25% invested in stock Z

22.

Portfolio mix:

60% invested in stock X

30% invested in stock Y

10% invested in stock Z

8.4 Expecting the Unexpected

▶ Find the probability of independent events.

▶ Find the probability that an event does not occur.

▶ Find counterintuitive probabilities.

Finding the Probability of Independent Events

Two events are **independent** if the occurrence of one does not affect the occurrence of the other. For instance, the event "consumer likes chocolate" should be independent of the event "consumer has more than one cell phone."

Probability of Independent Events

The probability that two independent events occur is the product of their individual probabilities.

$$\text{Probability that both events occur} = \left(\begin{array}{c} \text{probability} \\ \text{of event 1} \end{array}\right)\left(\begin{array}{c} \text{probability} \\ \text{of event 2} \end{array}\right)$$

This principle can be applied to three or more independent events.

EXAMPLE 1 **Finding the Probability of Independent Events**

In the game of *Yahtzee*, rolling five dice with the same number is called a YAHTZEE. What is the probability of rolling five 6s?

SOLUTION

Rolling one die has no effect on any of the other dice. So, the events are independent. For each die, the probability of rolling a 6 is one-sixth. So, the probability of rolling five 6s is

$$\text{Probability of rolling five 6s} = \left(\frac{1}{6}\right)\left(\frac{1}{6}\right)\left(\frac{1}{6}\right)\left(\frac{1}{6}\right)\left(\frac{1}{6}\right)$$

$$= \left(\frac{1}{6}\right)^5$$

$$= \frac{1}{7776}.$$

So, you have a 1 in 7776 chance of rolling five 6s on any given roll.

Yahtzee was first marketed by Edwin Lowe in 1956.

✓ **Checkpoint**  Help at *Math.andYOU.com*

In the game of *Yahtzee*, you are allowed three rolls of the dice, and you are allowed to keep any of the dice you want. With 3 rolls, it can be shown that the probability of rolling any YAHTZEE is about 1 out of 22. Visit *Math.andYOU.com* to play a dice game simulator. How many times do you play the game before you roll a YAHTZEE?

EXAMPLE 2   **Using Independent Events to Detect Fraud**

Benford's law states that in real-life data, the leading digit is 1 almost one-third of the time. Digits greater than 1 occur as the leading digit with decreasing frequencies, as shown in the table.

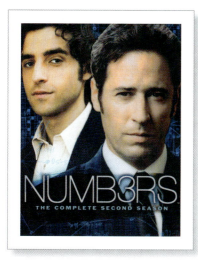

Benford's law was used by the character Charlie Eppes to help solve a case in the "Running Man" episode of the television crime drama *NUMB3RS*.

Leading digit	Probability
1	30.1%
2	17.6%
3	12.5%
4	9.7%
5	7.9%
6	6.7%
7	5.8%
8	5.1%
9	4.6%

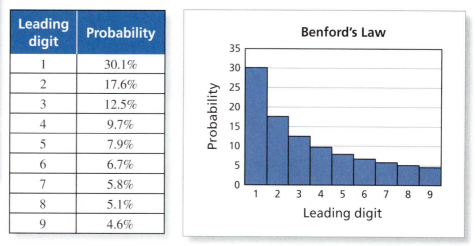

You work for the IRS. As a means of randomly selecting tax returns to audit, you propose the following. A computer randomly selects one of the amounts from the first page and a second amount from the second page. If both amounts have 9 as a leading digit (as in $93.28 and $901.92), the return will be audited.

a. Using Benford's law, what percent of the returns should be audited?

b. Suppose each digit is equally likely to occur as a leading digit. What percent should be audited?

SOLUTION

a. According to Benford's law, you should expect to select 9 as the leading digit only 4.6% of the time. The probability of selecting two 9s is

Probability of selecting two 9s = $(0.046)(0.046) \approx 0.00212$.

So, about 0.2% of the returns (about 2 out of 1000) should be audited.

b. If each digit is equally likely to occur as a leading digit, then a 9 should be selected as the leading digit one-ninth of the time. The probability of selecting two 9s is

Probability of selecting two 9s = $\left(\frac{1}{9}\right)\left(\frac{1}{9}\right) \approx 0.012$.

So, about 1.2% of the returns (about 12 out of 1000) should be audited.

✓ **Checkpoint**   Help at *Math*.and**Y☉U**.com

The following excerpt from *The New York Times* indicates that Benford's law can be used to spot fraudulent tax returns. Explain how this could happen.

> "Dr. Hill is one of a growing number of statisticians, accountants and mathematicians who are convinced that an astonishing mathematical theorem known as Benford's Law is a powerful . . . tool for pointing suspicion at frauds, embezzlers, tax evaders, sloppy accountants . . .."

Of the 70 (as of 2010) women to win the Academy Award for Best Actress, there are 4 pairs who share the same birthday: Jane Wyman and Diane Keaton (Jan 5), Joanne Woodward and Elizabeth Taylor (Feb 27), Barbra Streisand and Shirley MacLaine (Apr 24), and Helen Mirren and Sandra Bullock (Jul 26).

Finding the Probability That an Event Does Not Occur

Probability That an Event Does Not Occur

If the probability that an event occurs is p, then the probability that the event does not occur is

$$\text{Probability that event does not occur} = 1 - p.$$

EXAMPLE 3 Finding the Probability of an Event

A classroom has 35 students. What is the probability that at least two of them have the same birthday?

SOLUTION

To answer this question, you can use a technique that is frequently used in probability. That is, it is often easier to find the probability that an event *does not* occur, and then subtract the result from 1 to find the probability that it *does* occur.

$$\overbrace{}^{\text{35 factors}}$$

$$\begin{array}{l}\text{Probability that all} \\ \text{35 students have} \\ \textit{different} \text{ birthdays}\end{array} = \left(\frac{366}{366}\right)\left(\frac{365}{366}\right)\left(\frac{364}{366}\right)\left(\frac{363}{366}\right)\cdots\left(\frac{333}{366}\right)\left(\frac{332}{366}\right) \approx 0.187$$

$$\begin{array}{l}\text{Probability that at} \\ \text{least 2 students have} \\ \text{the } \textit{same} \text{ birthday}\end{array} = 1 - 0.187 = 0.813$$

So, the probability that at least 2 of the students have the same birthday is about 81.3%. Surprising, isn't it?

✓ Checkpoint

Help at *Math*.and**Y⊕U**.com

Use a spreadsheet to extend the result of the above example to 40 students.

	A	B	C
1	**Number of Students**	**Unused Birthdays**	**Probability of Different Birthdays**
2	1	366	100.00%
3	2	365	99.73%
4	3	364	99.18%
5	4	363	98.37%
36	35	332	18.65%
37	36	331	
38	37	330	
39	38	329	
40	39	328	
41	40	327	

EXAMPLE 4   **Finding the Probability of an Event**

You are an actuary for a retirement pension company. When a person retires at age 65, you use the following table to help determine the amount the person can withdraw monthly from his or her account.

Probability of Person Age 65 Living to Certain Age		
Age	**Male**	**Female**
80	59%	70%
85	38%	51%
90	19%	30%
95	6%	12%

For some retirement accounts, a person can take the option of having monthly withdrawals (at a lesser amount) as long as the person *or* the person's spouse survives. Consider a man and a woman who are each 65 years old.

a. What is the probability that at least 1 of them lives to age 80?

b. What is the probability that at least 1 of them lives to age 90?

SOLUTION

When analyzing questions like these, be sure that you do not start multiplying probabilities without thinking carefully. For instance, in these questions it is easier to work with the probability that both people *do not* survive.

a. Probability that both *do not* survive to age 80
$$= (1 - 0.59)(1 - 0.70) = 0.123$$ (Man Woman; Age 80)

Probability that at least one *does* survive to age 80
$$= 1 - 0.123 = 0.877$$

So, there is an 87.7% chance that at least 1 will survive to age 80.

b. Probability that both *do not* survive to age 90
$$= (1 - 0.19)(1 - 0.30) = 0.567$$ (Man Woman; Age 90)

Probability that at least one *does* survive to age 90
$$= 1 - 0.567 = 0.433$$

So, there is a 43.3% chance that at least 1 will survive to age 90.

✓ **Checkpoint**   Help at *Math*.and**YOU**.com

Rework the probabilities in the above example in the case that (c) both people are women and (d) both people are men.

Counterintuitive Probabilities

EXAMPLE 5   **Finding a Counterintuitive Probability**

On the game show *Let's Make a Deal*, the contestant is given the choice of three doors. Behind two of the doors are goats. Behind the other door is a new car.

Step 1: The contestant randomly chooses one of the doors.

Step 2: The game show host knows the location of the car. After the contestant chooses a door, the host reveals the goat behind one of the remaining doors. Then the host asks the contestant, "Do you want to switch doors?"

Step 3: The contestant either switches or stays with his or her first choice.

Based on probability, what should the contestant do?

Monty Hall hosted the game show *Let's Make a Deal* for many years. In 1990, Marilyn vos Savant, a columnist at *Parade* magazine, published the solution of the "Monty Hall Problem." The magazine received about 10,000 responses, most of which said that Marilyn was wrong. She was, however, correct.

SOLUTION

On the show, many contestants stay with their original choice. However, consider the table. Because it is irrelevant, assume the contestant chooses door 1.

Door 1	Door 2	Door 3	Result if switching	Result if staying
Car	Goat	Goat	Goat	**Car**
Goat	Car	Goat	**Car**	Goat
Goat	Goat	Car	**Car**	Goat
			Win: 2 out of 3	**Win: 1 out of 3**

So, by switching, the contestant doubles the likelihood that he or she wins the car. For most people, this result seems counterintuitive.

✓ **Checkpoint**   Help at *Math*.and**YOU**.com

Play the *Monty Hall Game* at *Math.andYou.com* 20 times by staying and 20 times by switching. Do your outcomes agree with the probabilities in the example?

EXAMPLE 6   **Winning a Lottery Twice: Fraud?**

The likelihood of winning a major state lottery is very small. Yet some people have won state lotteries more than once. Does this indicate fraud? Explain your reasoning.

Lottery Luck Strikes Twice in Three Months

ST. LOUIS - You are as likely to win the lottery as you are to be struck by lightning. Ernest Pullen, a Missouri resident, has won the lottery twice in a matter of three months. In June, he won his first cash prize for $1 million. In September, he won his second cash prize for $2 million. One lottery official called it an amazing coincidence.

SOLUTION

This problem is difficult because there is not enough information to analyze it completely. Even so, consider a simplified example.

- There is a drawing every day with one winner.

- The same million people play each day.

- The probability of winning with one ticket is one-millionth.

With these assumptions, the mathematical strategy for the solution is the same as that in the "Birthday Problem" in Example 3 on page 386.

$$\underbrace{\phantom{\left(\frac{1{,}000{,}000}{1{,}000{,}000}\right)\left(\frac{999{,}999}{1{,}000{,}000}\right)\left(\frac{999{,}998}{1{,}000{,}000}\right) \cdots \left(\frac{999{,}636}{1{,}000{,}000}\right)}}_{\text{365 factors}}$$

Probability that all 365 tickets have *different* owners $= \left(\dfrac{1{,}000{,}000}{1{,}000{,}000}\right)\left(\dfrac{999{,}999}{1{,}000{,}000}\right)\left(\dfrac{999{,}998}{1{,}000{,}000}\right) \cdots \left(\dfrac{999{,}636}{1{,}000{,}000}\right)$

≈ 0.936

Probability that at least 2 tickets have the *same* owner $= 1 - 0.936 = 0.064$

So, there is about a 6.4% chance that during 1 year, someone will win the lottery twice. At the end of 2 years, this probability increases to 23.4%. By the end of 3 years, it increases to 45.1%.

Be sure you see that this is not the probability that *you* (or any other given person) will win the lottery twice. It is simply the probability that *someone* will win it twice. The point is this—having someone win twice is not an indication of fraud.

✓ **Checkpoint**

Help at *Math*.and**YOU**.com

You win a state lottery for $1 million and you have the choice of receiving $50,000 a year for 20 years or $500,000 in a lump-sum payment now.
(a) Suppose you live in California and take the lump-sum payment. How much state and federal tax do you have to pay on your winnings? (See Section 5.2.)
(b) Canadian citizens do not have to pay income tax on lottery winnings. Do you agree that Americans should have to pay state income tax on a lottery sponsored by the state? Explain.

8.4 Exercises

Computer-Generated Sequence A computer randomly generates a sequence of symbols. The circle graph shows the probability that the computer generates each of the symbols. In Exercises 1–7, find the probability that the sequence occurs. *(See Examples 1 and 2.)*

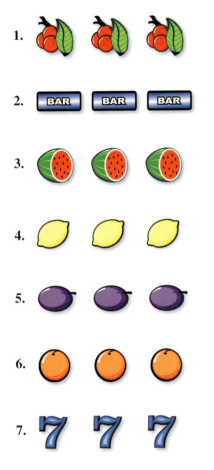

1.

2. BAR BAR BAR

3.

4.

5.

6.

7. 7 7 7

Probabilities of Symbols

12%
BAR 10%
8%
7%
6%
5%
Blank 50%
7 2%

Around 1900, Charles Fey invented a slot machine called the Liberty Bell. It had three spinning reels with a total of five symbols: horseshoes, diamonds, spades, hearts, and a Liberty Bell. Three bells produced the biggest payoff, which was 10 nickels.

8. **Slot Machine** Complete the spreadsheet to find the expected value of each of the events in Exercises 1–7. Then find the probability that a given spin is a "no win." Suppose these symbols represent a slot machine game. What is the expected value?

	A	B	C	D
1	**Event**	**Probability**	**Payoff**	**Expected Value**
2	3 cherries		150	
3	3 bars		200	
4	3 watermelons		250	
5	3 lemons		400	
6	3 plums		1,000	
7	3 oranges		7,500	
8	3 sevens		12,000	
9	No win		-1	
10				

Dice In Exercises 9–12, find the probability of the event. *(See Examples 3 and 4.)*

9. You roll three dice. What is the probability of rolling at least two of the same number?

10. You roll five dice. What is the probability of rolling at least two of the same number?

11. You roll six dice. What is the probability of rolling at least one odd number?

12. You roll seven dice. What is the probability of rolling at least two of the same number?

Marbles There are 26 marbles in a bag. The graph shows the color distribution of the marbles. In Exercises 13–16, use the graph. Assume that you randomly draw one marble from the bag and you put the marble back before drawing another. *(See Examples 3 and 4.)*

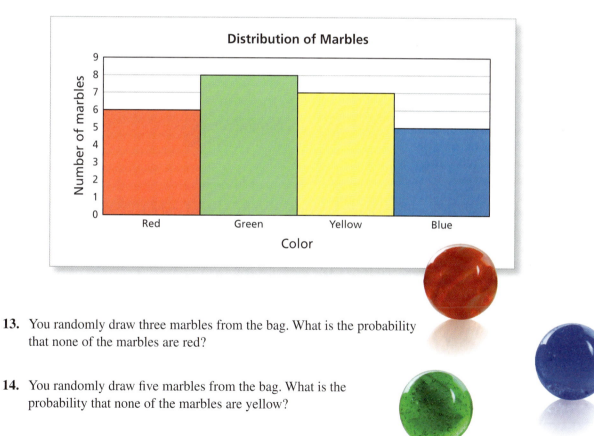

13. You randomly draw three marbles from the bag. What is the probability that none of the marbles are red?

14. You randomly draw five marbles from the bag. What is the probability that none of the marbles are yellow?

15. You randomly draw four marbles from the bag. What is the probability that at least one marble is green?

16. You randomly draw six marbles from the bag. What is the probability that at least one marble is blue?

17. **Coin Conundrum** Your friend flips two coins and tells you that at least one coin landed heads up. What is the probability that both coins landed heads up? *(See Example 5.)*

18. **Boy or Girl Paradox** Your friend has two children. At least one of the children is a boy. Considering the order that the children were born, what is the probability that the other child is a girl? *(See Example 5.)*

19. **Treasure Beyond Measure** You are a contestant on the fictional game show *Treasure Beyond Measure*. The host presents you with five unopened treasure chests. One chest contains $1,000,000. The other four contain nothing. You randomly choose two of the chests. The host, knowing which chest contains the money, opens two of the remaining chests and shows that they contain nothing. The host then asks you, "Do you want to switch your two chests for the one chest that I didn't open?" Based on probability, what should you do? *(See Example 5.)*

20. **Ultimate Monty Hall** You are a contestant on the fictional game show *Ultimate Monty Hall*. The host presents you with 101 unopened doors. Behind one door is a car. Behind the other 100 doors are goats. You randomly choose 50 of the doors. The host, knowing the location of the car, reveals the goats behind 50 of the remaining doors. The host then asks you, "Do you want to switch your 50 doors for the 1 door that I didn't open?" Based on probability, what should you do? *(See Example 5.)*

Card Puzzle In Exercises 21 and 22, use the information below. *(See Examples 5 and 6.)*

You deal two of the cards shown to your friend.

21. Your friend says, "I have at least one ace." What is the probability that your friend's other card is an ace?

22. Your friend says, "I have the ace of spades." What is the probability that your friend's other card is an ace?

▶ Extending Concepts

Bayes' Theorem In Exercises 23–26, use the information below.

For any two events with probabilities greater than 0,

$$\text{Probability of event 1 given event 2} = \frac{\left(\begin{array}{c}\text{probability of event 2}\\\text{given event 1}\end{array}\right)\left(\begin{array}{c}\text{probability}\\\text{of event 1}\end{array}\right)}{\text{probability of event 2}}.$$

23. You have the following information about students at a college.

    ● 49% of the students are male.

    ● 11% of the students are nursing majors.

    ● 9% of the nursing majors are male.

    What is the probability that a student is a nursing major given that the student is male?

24. You have the following information about students at a college.

    ● 51% of the students are female.

    ● 10% of the students are history majors.

    ● 60% of the history majors are female.

    What is the probability that a student is a history major given that the student is female?

25. You have the following information about voters in a local mayoral election.

    ● 61% of voters were registered Republican.

    ● 53% of voters voted Republican.

    ● 86% of voters who voted Republican were registered Republican.

    What is the probability that a voter voted Republican given that the voter was registered Republican?

26. You have the following information about voters in a local congressional election.

    ● 74% of voters were registered Democrat.

    ● 62% of voters voted Democrat.

    ● 79% of voters who voted Democrat were registered Democrat.

    What is the probability that a voter voted Democrat given that the voter was registered Democrat?

8.3–8.4   Quiz

Multiple-Choice Quiz You take a 10-question multiple-choice quiz. Each question has four choices. In Exercises 1–7, suppose you guess when answering the questions.

1. Suppose you gain 1 point when you answer a question correctly and 0 points when you answer a question incorrectly. What is the expected value of the quiz?

2. Suppose you gain 1 point when you answer a question correctly and lose ¼ point when you answer a question incorrectly. What is the expected value of the quiz?

3. What is the probability that you answer every question incorrectly?

4. What is the probability that you answer every question correctly?

5. What is the probability that you answer only the first question correctly?

6. What is the probability that you answer only the first five questions correctly?

7. What is the probability that you answer at least one question correctly?

Name _____   Date _____

1. Who was the first president of the United States?

Ⓐ George Washington   Ⓑ Thomas Jefferson   Ⓒ John Adams   Ⓓ Gerald Ford

2. Who wrote the *Declaration of Independance*?

Ⓐ Thomas Jefferson   Ⓑ Benjamin Franklin   Ⓒ Samuel Adams   Ⓓ John Hancock

3. Who wrote the *The Wealth of Nations*?

Ⓐ John Maynard Keynes Ⓑ Adam Smith   Ⓒ Thomas Paine   Ⓓ Thomas Johnson

4. Who was the president of the Confederate States?

Ⓐ Robert E. Lee   Ⓑ Jefferson Davis   Ⓒ John C. Calhoun   Ⓓ Edwin Stanton

5. Who was Thomas Jefferson's first vice president?

Ⓐ George Clinton   Ⓑ Aaron Burr   Ⓒ James Madison   Ⓓ Spiro Agnew

6. In what year did Alaska officially become a state?

Ⓐ 1859   Ⓑ 1911   Ⓒ 1945   Ⓓ 1959

7. Which president was never married?

Ⓐ Grover Cleveland   Ⓑ Andrew Jackson   Ⓒ James Buchanan   Ⓓ Ulysses S. Grant

8. Which president was called "the Great Communicator"?

Ⓐ Franklin Roosevelt   Ⓑ John Kennedy   Ⓒ Jimmy Carter   Ⓓ Ronald Reagan

9. Which president's foreign policy was to "speak softly and carry a big stick"?

Ⓐ William McKinley   Ⓑ Theodore Roosevelt Ⓒ William Taft   Ⓓ James Polk

10. Which president's 1972 visit to China helped improve the relationship between the U.S. and China?

Ⓐ Lyndon Johnson   Ⓑ Harry Truman   Ⓒ Richard Nixon   Ⓓ Dwight Eisenhower

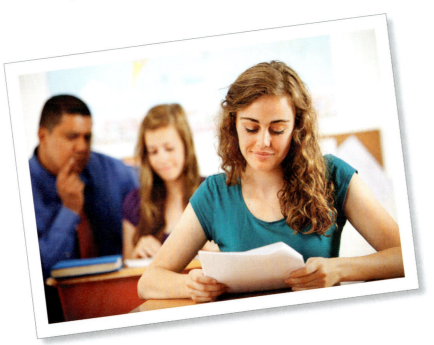

Chapter 8   Summary

Section Objectives		How does it apply to you?

Section 1

Use probability to describe the likelihood of an event.	→	Probabilities are numbers between 0 and 1, including 0 and 1. An event with a probability of 0 is impossible. An event with a probability of 1 is certain. *(See Examples 1 and 2.)*
Analyze the likelihood of a risk.	→	You can assess the risk involved in a given situation. *(See Examples 3 and 4.)*
Use likelihood to describe actuarial data.	→	Actuaries use probabilities to calculate the costs of risks. *(See Examples 5 and 6.)*

Section 2

Find a theoretical probability.	→	You can find the probability of an event in cases where all possible outcomes of an event are known and can be counted. *(See Examples 1 and 2.)*
Find an experimental probability.	→	You can find the probability of an event in cases where a representative sample can be taken and counted. *(See Examples 3 and 4.)*
Estimate a probability using historical results.	→	You can find the probability of an event in cases where past occurrences are assumed to be representative of future occurrences. *(See Examples 5 and 6.)*

Section 3

Find an expected value involving two events.	→	You can determine the long-run average of an experiment. *(See Examples 1 and 2.)*
Find an expected value involving multiple events.	→	You can extend the concept of expected value to situations in which there are more than two events. *(See Examples 3 and 4.)*
Use expected value to make investment decisions.	→	You can use expected value to compare different investments. *(See Examples 5 and 6.)*

Section 4

Find the probability of independent events.	→	You can find the probability that two events occur when the occurrence of one does not affect the occurrence of the other. *(See Examples 1 and 2.)*
Find the probability that an event does not occur.	→	If you know the probability that an event occurs, then you can calculate the probability that the event does not occur. *(See Examples 3 and 4.)*
Find counterintuitive probabilities.	→	You can solve problems in which the probability of an event is counterintuitive. *(See Examples 5 and 6.)*

Chapter 8   Review Exercises

Section 8.1

1. **Describing Likelihood** The table shows the probabilities of several events. Complete the table. Then describe the likelihood of each event in words.

Event	Fraction	Decimal	Percent
Being an organ donor	$\frac{4}{15}$		
Eats breakfast		0.61	
Having a dream that comes true			42.9%
Household with television			98.2%

Snowfall in Washington In Exercises 2–5, describe the likelihood that at least 1 inch of snow will accumulate in the city.

2. Olympia

3. Packwood

4. Richland

5. Seattle

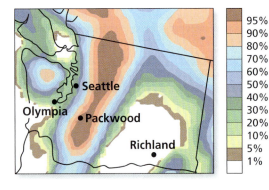

Probability of 1 Inch (or more) of Snow

6. **Deaths** The graph shows the death rate per 100,000 people in the United States for several causes. Compare the rates for the causes of death.

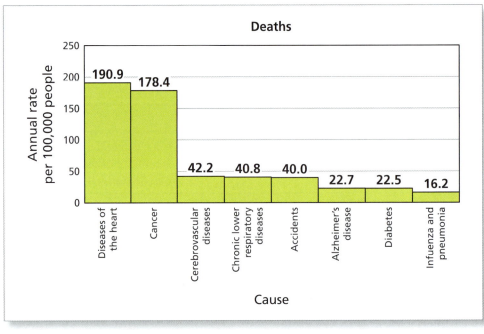

Deaths

Section 8.2

Platonic Solids In Exercises 7–10, use the information below.

There are five Platonic solids: tetrahedron, cube, octahedron, dodecahedron, and icosahedron.

Tetrahedron   Cube   Octahedron   Dodecahedron   Icosahedron

7. Why do these solids make natural shapes for dice?

8. Which shape is the best for dice? Why? Which is the worst? Why?

9. A die has the shape of an icosahedron, with consecutively numbered sides starting at 1. What is the probability of rolling a number that is greater than 5?

10. There are only five possible Platonic solids. What is the definition of a Platonic solid?

MP3 Player The tables show the results of a survey that asked adults whether they own an MP3 player. In Exercises 11 and 12, use the tables.

Age group	Yes	No
18–29	630	212
30–49	479	343
50–64	250	553
65 and over	52	482

Geographic location	Yes	No
Rural	120	270
Suburban	700	750
Urban	591	615

11. Find the probability that an adult in each age group owns an MP3 player. Then describe the likelihood.

12. Find the probability that an adult in each geographic location owns an MP3 player. Then describe the likelihood.

Section 8.3

Home Appliance Company In Exercises 13–16, assume that you are the owner of a home appliance company.

13. You take out a $500,000 fire insurance policy on your factory. The annual premium is $2750. The probability of a fire is 0.005. What is the expected value?

14. You take out a $250,000 flood insurance policy on your factory. The annual premium is $3770. The probability of a flood is 0.014. What is the expected value?

15. Your company is considering developing one of two toaster models. Use a decision tree to decide which model your company should develop.

Toaster A: Cost of development: $500,000   **Toaster B:** Cost of development: $750,000

Projected Sales	
Probability	Net sales (in thousands)
25%	$1000
65%	$600
10%	$250

Projected Sales	
Probability	Net sales (in thousands)
30%	$1200
55%	$800
15%	$500

16. Your company is considering developing one of two microwave models. Use a decision tree to decide which model your company should develop.

Microwave A: Cost of development: $1,000,000   **Microwave B:** Cost of development: $900,000

Projected Sales	
Probability	Net sales (in thousands)
20%	$1600
65%	$1400
15%	$800

Projected Sales	
Probability	Net sales (in thousands)
35%	$1500
45%	$1300
20%	$750

17. **Option Comparison** Compare the two options.

	Probability	Gain
Option 1	100%	$1000
	0%	$0
Option 2	50%	$500
	50%	$2000

18. **Investment Comparison** You want to invest $1000. Find the expected values for the two investments.

Speculative investment	Conservative investment
● Complete loss: 30% chance	● Complete loss: 5% chance
● No gain or loss: 25% chance	● No gain or loss: 30% chance
● 100% gain: 25% chance	● 25% gain: 60% chance
● 200% gain: 15% chance	● 50% gain: 5% chance
● 500% gain: 5% chance	

Section 8.4

Summer Weather The table shows the probability of sunshine or rain on any given day in July for a town. In Exercises 19–22, use the table.

Weather	Probability
☀️	75%
🌧️	25%

19. What is the probability of rain three days in a row?

20. What is the probability of sunshine five days in a row?

21. What is the probability of rain at least once during a week?

22. What is the probability of sunshine at least once during a week?

Winter Weather The probability of snow on any given day in December for a town is 30%. In Exercises 23–26, use this information.

23. What is the probability of snow four days in a row?

24. What is the probability of no snow six days in a row?

25. What is the probability of snow at least once during a week?

26. What is the probability of no snow at least once during a week?

9 The Mathematics of Description

9.1 Information Design

▶ Use stacked area graphs to represent the changing parts of a whole.

▶ Use a radar graph and an area graph to represent data.

▶ Graphically represent data sets that have several variables.

9.2 Describing "Average"

▶ Use mean, median, and mode to describe the average value of a data set.

▶ Read and understand box-and-whisker plots and histograms.

▶ Understand the effect of outliers on averages.

9.3 Describing Dispersion

▶ Use standard deviation to describe the dispersion of a data set.

▶ Use standard deviation to describe a data set that is normally distributed.

▶ Compare different types of distributions.

9.4 Describing by Sampling

▶ Use a randomly chosen sample to describe a population.

▶ Determine whether a sample is representative of a population.

▶ Determine a sample size to obtain valid inferences.

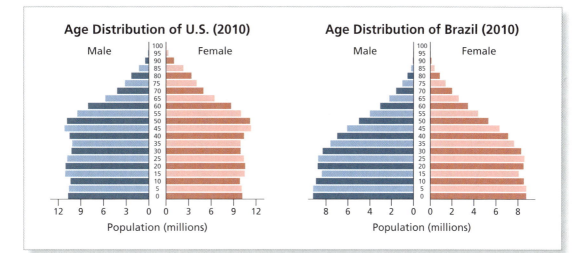

Age Distribution of U.S. (2010)

Male   Female

Population (millions)

Age Distribution of Brazil (2010)

Male   Female

Population (millions)

Example 1 on page 412 uses population pyramids to describe
the distribution of ages in the United States and in Brazil.
How do the two distributions differ?

9.1 Information Design

▶ Use stacked area graphs to represent the changing parts of a whole.

▶ Use a radar graph and an area graph to represent data.

▶ Graphically represent data sets that have several variables.

Stacked Area Graphs

Information design is the presentation of data and information so that people can understand and use it. Throughout this text, you have seen many types of information design: bar graphs, circle graphs, scatter plots, line graphs, and bubble graphs. In this section, you will look at several more graphical ways to organize and present data.

> **EXAMPLE 1** **Reading a Stacked Area Graph**

Describe the information presented in the *stacked area graph*.

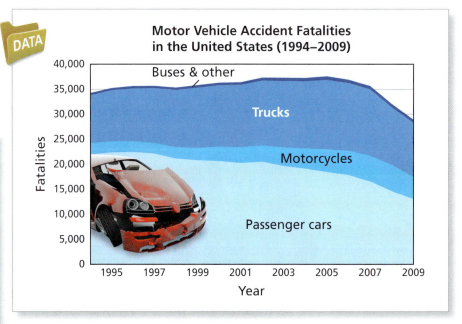

DATA

Motor Vehicle Accident Fatalities in the United States (1994–2009)

Buses & other

Trucks

Motorcycles

Passenger cars

The picture of data and information that you show people has everything to do with their understanding of the data and information.

SOLUTION

Here are some observations.

- The total number of fatalities was relatively constant from 1994 to 2007.
- The number of passenger car fatalities has decreased.
- In 1994, passenger cars accounted for about two-thirds of all motor vehicle fatalities. By 2009, they accounted for only about one-half of all motor vehicle fatalities.
- The number of motorcycle fatalities has increased.

✓ **Checkpoint**   Help at *Math*.and**YOU**.com

The data for Example 1 is available at *Math.andYou.com*. Use the data to compare motorcycle fatalities in 1994 and 2009.

EXAMPLE 2   **Interpreting a Stacked Area Graph**

A household began using the Internet in 1995. The stacked area graph shows the composition of Internet traffic for the household from 1995 to 2010. Suppose you see this graph online. What comments can you make about the graph?

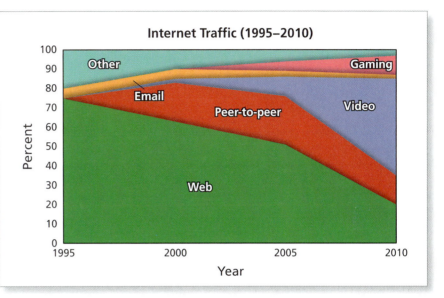

According to a report by Sandvine Network Demographics, streaming videos from Netflix represent 29.7% of peak downsteam traffic for North American fixed access networks. This makes Netflix the single largest source of peak downstream traffic for these networks.

SOLUTION

Here are some comments.

> COMMENT:
>
> What does the vertical axis represent? Time? Bits? Users?

This is a good question. The answer is "bits."

Here are some other comments.

> COMMENT:
>
> In the graph, the term Web does not mean Internet. It means HTML data. Someone looking at the graph may be misled and assume that the term Web is referring to the Internet.

> COMMENT:
>
> One minute of streaming video is about 2 MB. This is about 100 times the size of a typical e-mail and about 10 times the size of a typical web page. So, you would have to send 100 e-mails or visit 10 web pages to use the same amount of bandwidth as a single minute of streaming video.

This graph illustrates that you must be careful with how you present data. Information design is a powerful tool. When it is used incorrectly, the results can be very misleading.

✓ **Checkpoint**

Help at *Math*.and**YOU**.com

From the graph, can you conclude that the household's Web usage, in total bits downloaded, declined from 1995 to 2010? Explain.

Radar Graphs and Area Graphs

| EXAMPLE 3 | **Reading a Radar Graph** |

Describe the information presented in the *radar graph*.

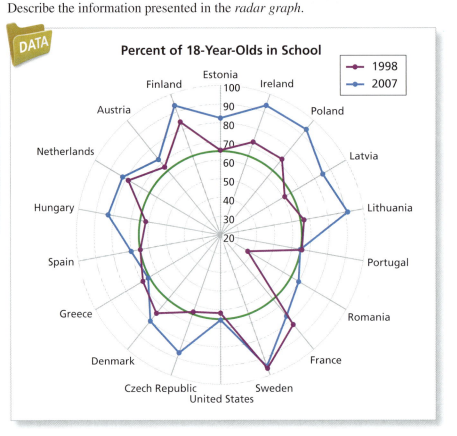

Percent of 18-Year-Olds in School

During the 1989 revolution against Communist dictator Nicolae Ceausescu, a fire destroyed about 500,000 books in Bucharest's Central University Library. After the fall of communism, Romania moved into a period of economic reform. The library was repaired and reopened in 2001. Romania joined the European Union in 2007.

SOLUTION

Here are some observations.

- The percent increased in all countries (the **blue graph** is farther away from the center than the **purple graph**) except for France, Greece, Portugal, and Sweden.

- In France and Greece, the percent decreased.

- In Portugal and Sweden, the percent stayed about the same.

- By drawing a **circle** at 65%, you can see that in 2007, the United States, Portugal, and Greece had the least percent of 18-year-olds in school.

- The country that made the greatest increase from 1998 to 2007 was Romania. You might assume that Romania had a great cultural change during these 10 years.

✓ **Checkpoint** Help at

What other observations can you make about the radar graph?

EXAMPLE 4   **Drawing an Area Graph**

The following area graph is taken from *Information is Beautiful* by David McCandless. Describe the information and patterns presented in the graph.

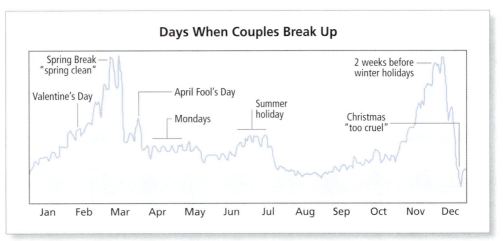

Days When Couples Break Up

David McCandless's book on information design is "a visual guide to the way the world really works."

SOLUTION

The graph shows the results of tracking 10,000 Facebook status updates to determine when people in relationships broke up.

There are several patterns that you can observe in the graph.

- The two peak times during the year when people break up are in the month leading up to Spring Break and the month leading up to Christmas.

- The spike in the number of breakups on April Fool's Day may be a result of people changing their relationship statuses as a joke.

- The frequency of breakups is relatively low from late July through October.

- The day of the year when people are least likely to break up is Christmas.

✓ **Checkpoint**   Help at *Math*.and*YOU*.com

What observations can you make about the area graph below?

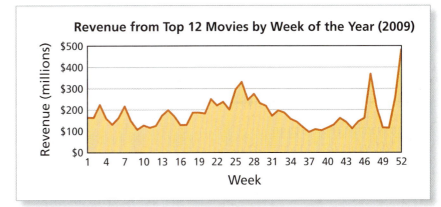

Revenue from Top 12 Movies by Week of the Year (2009)

Information Design with Several Variables

EXAMPLE 5 **Animating Information**

At *Gapminder.org/world*, you can watch "Wealth & Health of Nations" to see how countries have changed from 1800 through the present. Describe some of the variables displayed in the graph.

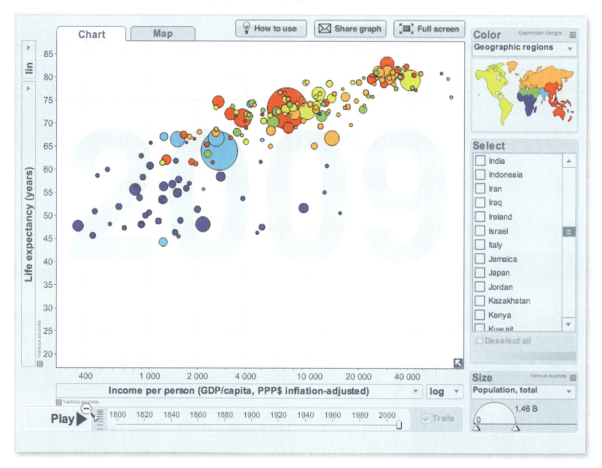

SOLUTION

Here are some of the variables.

You can watch a short lecture called "200 years that changed the world" at *Gapminder.org.*

- The vertical axis measures *life expectancy*. The higher the bubble, the greater the life expectancy.

- The horizontal axis measures *income per person*. The farther to the right the bubble, the greater the income per person.

- The area of the bubble measures *population size*. The bigger the bubble, the larger the population.

- The color of the bubble indicates the *geographic region*.

- The date on the coordinate plane indicates the *year*.

✓ **Checkpoint**  Help at *Math.andYOU.com*

 Watch the position of China in the animation from 1800 through the present. Describe the changes during the 200+ years.

EXAMPLE 6   **Reading a Design with Several Variables**

The stream graph shows trends in the top 25 movies at the box office for each weekend in 2010. Describe the variables displayed in the graph.

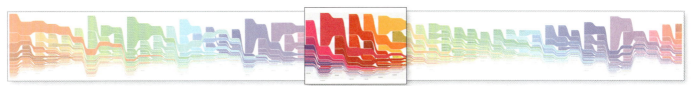

Revenue for Top 25 Movies in 2010

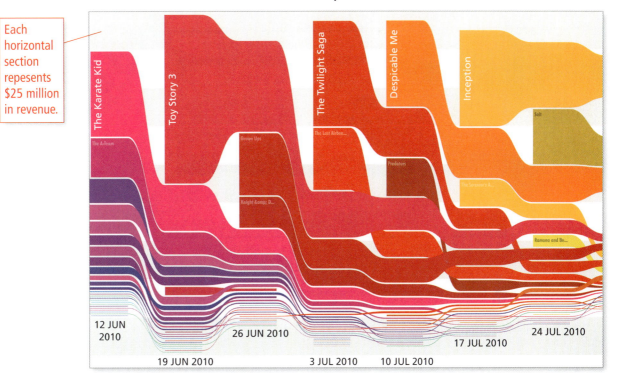

Each horizontal section repesents $25 million in revenue.

12 JUN 2010   19 JUN 2010   26 JUN 2010   3 JUL 2010   10 JUL 2010   17 JUL 2010   24 JUL 2010

Toy Story 3 is a 3D computer-animated film by Disney/Pixar. During 2010, it had the greatest box office receipts for the weekends of June 19 and June 26.

SOLUTION

Here are the variables.

- The color identifies the *title* of the movie.

- The horizontal axis shows the *weekend* during the year.

- The height of the color shows the *revenue* for the weekend.

- The position of the color for each weekend indicates the *order* from highest box office receipts to lowest box office receipts.

✓ **Checkpoint**   Help at *Math*.and**YOU**.com

Describe how the colors in the graph can help you identify movies that have particularly long runs at the box office.

9.1 Exercises

USAF Fighter Force The stacked area graph shows the composition of the aircraft in the United States Air Force (USAF) fighter force. In Exercises 1–6, use the graph. *(See Examples 1 and 2.)*

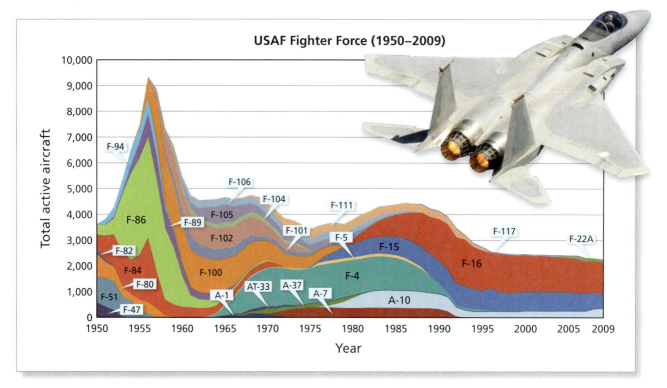

1. Describe the information presented in the stacked area graph.

2. Which aircraft was the largest component of the USAF fighter force in the 1970s?

3. Find the size of the USAF fighter force in 2009 as a percent of the peak size in the 1950s.

4. In which decade did the USAF fighter force have the least variety of aircraft? Explain your reasoning.

5. Do you agree with the comment shown? Explain your reasoning.

COMMENT:

The F-16 is losing favor as the most popular fighter jet in the USAF fighter force.

6. What other comments can you make about the stacked area graph?

Competitor Profiles In Exercises 7–10, use the radar graph. *(See Example 3.)*

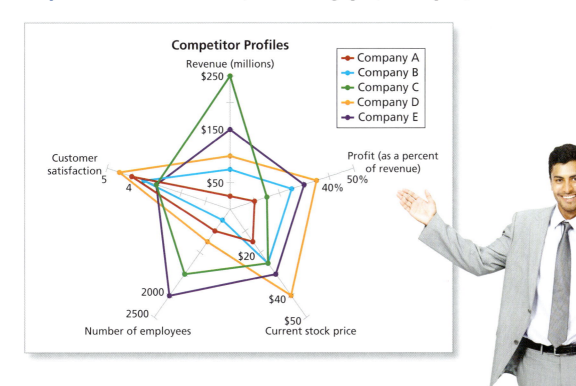

7. Which competitor has the greatest customer satisfaction?

8. Which competitor has the most employees?

9. Which competitor has the greatest profit percent? Which competitor makes the most profit? Explain your reasoning.

10. When do you think it would be beneficial to use a radar graph in information design?

Stock Price The area graph shows the stock price of company C during 2010. In Exercises 11–14, use the area graph. *(See Example 4.)*

11. Describe the information and patterns presented in the graph.

12. What was the stock price in week 31?

13. Suppose you bought 100 shares of stock in week 15 and sold them in week 51. Did you lose money or earn a profit?

14. What was the highest percent return an investor could have earned during 2010? Explain your reasoning.

Candlestick Chart A candlestick chart can be used to monitor the movements of a stock price. In Exercises 15–22, use the chart. *(See Examples 5 and 6.)*

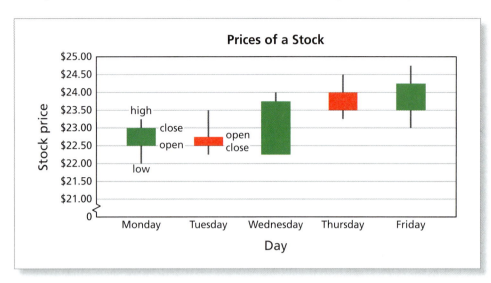

15. Describe the variables displayed in the graph.

16. What does it mean when the candlestick is green? red?

17. What is the opening price on Monday? Tuesday?

18. What is the closing price on Wednesday? Thursday?

19. On which day did the stock have the greatest gain? loss?

20. Explain why the candlestick for Wednesday does not have a lower shadow.

21. What is the highest percent return you could have earned during this week? Explain your reasoning.

22. Use the Internet to find other ways to display stock prices. Which type of design do you prefer?

The techniques used in candlestick charting originated in 17th-century Japan. Homma, a rice trader from Japan, is credited with early versions of candlestick charting, which have evolved over many years.

▶ Extending Concepts

Life Chart In Exercises 23–28, use the following information.

Your habits, behaviors, and activities indicate who you are as a person. A life chart gives a visual representation of your daily information over time. An example of a life chart is shown.

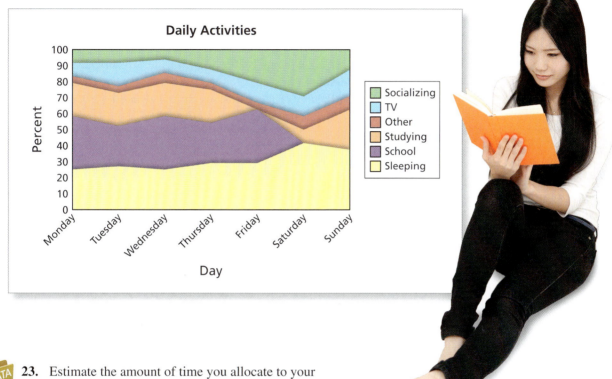

23. Estimate the amount of time you allocate to your daily activities for a typical week. Use the data to create a life chart like the one shown.

24. How might your life chart for a year differ from your life chart for a week?

25. Explain why a chart like the one in Example 2 is better suited for the information displayed in a life chart than a chart like the one in Example 1.

26. Many people try to make lifestyle changes such as getting more sleep, saving more money, and losing weight. Do you think a life chart can help you make lifestyle changes over time? Explain your reasoning.

27. Other than time allocation, what are some other personal data that you can track to help manage your life?

28. Organize some data from one of the topics you listed in Exercise 27. Create an information design to present the data in a way that you have not seen in this text.

9.2 Describing "Average"

▶ Use mean, median, and mode to describe the average value of a data set.

▶ Read and understand box-and-whisker plots and histograms.

▶ Understand the effect of outliers on averages.

Mean, Median, and Mode

Some data sets have typical values that are representative of the entire set. For instance, a typical adult thumb is about 1 inch wide. (Historically, this is how inches were measured.) In such data sets, there are three basic ways to describe the "average" of the data set. These measures are called *measures of central tendency.* This is part of a field called *descriptive statistics.*

Study Tip

Here is an example of mean, median, and mode.

Data: 1, 2, 2, 2, 3, 4, 6, 7, 9

Mean: $\frac{36}{9} = 4$

Median:
middle number = 3

Mode: most frequent = 2

Mean, Median, and Mode

To find the **mean,** add all the values in the data set and divide by the number of values in the set.

To find the **median,** arrange the values in order. The number in the middle or the mean of the two middle values is the median.

To find the **mode,** look for the value that occurs most often in the data set.

EXAMPLE 1 **Estimating the Mean, Median, and Mode**

The population pyramid shows the age distributions of males and females in the United States. Estimate the mean, median, and mode for males and females.

SOLUTION

You can use a spreadsheet and the actual data to determine the mean, median, and mode.

Mean: The mean age for males is about 36, and the mean age for females is about 39.

Median: The median age for both males and females is between 35 and 39.

Mode: The mode age for both males and females is between 45 and 49.

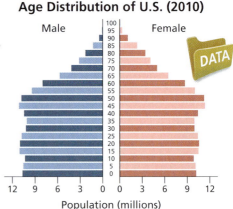

Age Distribution of U.S. (2010)

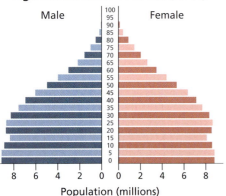

Age Distribution of Brazil (2010)

✓ **Checkpoint**   Help at *Math*.and**YOU**.com

How does the population pyramid of the United States differ from the population pyramid of Brazil? Explain your reasoning.

EXAMPLE 2   **Comparing Population Pyramids**

Compare the mean, median, and mode of the three types of population pyramids. Describe the economic repercussions of each type.

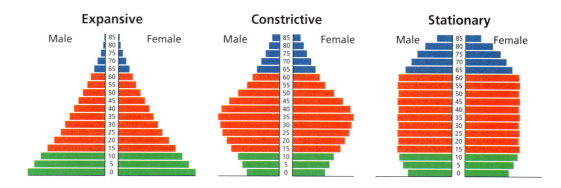

SOLUTION

Expansive: A large percent of the population is in the **young age group** and **working age group**. The mean, median, and mode lie toward the beginning of the age spectrum. The working age group must support a relatively large population of youths.

Constrictive: A large percent of the population is in the **working age group**. The mean, median, and mode are similar to those of the United States in Example 1. This group must provide for the young and the old.

Stationary: There is a larger percent of the population in the **old age group** than in the other two population pyramids. The mean and the median lie toward the middle of the age spectrum. The working age group must support a relatively large population of old people.

 **Checkpoint**   Help at *Math*.and**YOU**.com

Match the population pyramid with Afghanistan, Canada, and Mexico. Explain your reasoning.

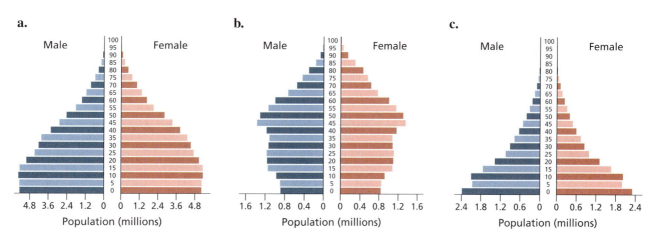

Box-and-Whisker Plots and Histograms

EXAMPLE 3 Reading a Box-and-Whisker Plot

The box-and-whisker plot shows the weights of the 59 players on the 2010 Chicago Bears football team. Use the graph to analyze the players' weights.

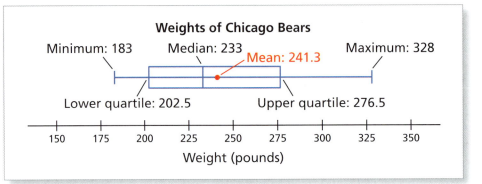

Weights of Chicago Bears

Minimum: 183   Median: 233   Mean: 241.3   Maximum: 328

Lower quartile: 202.5   Upper quartile: 276.5

Weight (pounds)

The mean weight of National Football League (NFL) quarterbacks is about 24 pounds less than the mean weight of the players in general.

SOLUTION

A box-and-whisker plot conveys a wealth of information. Here are some observations you can make about the weights of the players.

- About 15 players weighed between 183 and 202.5 pounds.

- About 15 players weighed between 202.5 and 233 pounds.

- About 15 players weighed between 233 and 276.5 pounds.

- About 15 players weighed between 276.5 and 328 pounds.

- There was a greater range of weights for the heaviest 25% of the players (about 50 pounds) than there was for the lightest 25% of the players (about 20 pounds).

You can make your own box-and-whisker plots at *Math.andYou.com*.

✔ **Checkpoint**   Help at *Math.andYOU.com*

The box-and-whisker plot shows the weights of the 40-man roster of the 2010 Los Angeles Angels of Anaheim baseball team. Use the graph to analyze the players' weights.

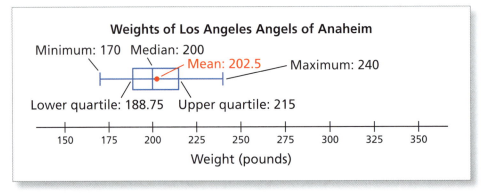

Weights of Los Angeles Angels of Anaheim

Minimum: 170   Median: 200   Mean: 202.5   Maximum: 240

Lower quartile: 188.75   Upper quartile: 215

Weight (pounds)

EXAMPLE 4   Reading Histograms

The color-coded histograms show the data from Example 3 in a different way. Compare the histograms to the box-and-whisker plots. What new information do the histograms provide?

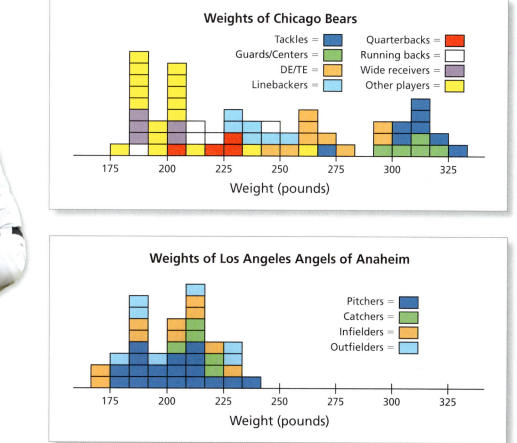

The mean weight of Major League Baseball (MLB) pitchers is not very different from the mean weight of the players in general.

SOLUTION

The histograms do not show the medians or quartiles of the data sets.

On the other hand, the histograms show how the data are distributed. You can make the following observations that are not evident in the box-and-whisker plots.

- For *football players*, there is a correlation between weight and position. Tackles, guards, centers, tight ends, and defensive ends tend to weigh considerably more than the other players.

- For *baseball players*, there is not a strong correlation between weight and position.

 **Checkpoint**   Help at *Math*.and**Y🌐U**.com

Make histograms showing the weights of players on another NFL team and another MLB team. Are the histograms similar to the two above? Explain your reasoning.

The Effect of Outliers on Averages

An **outlier** in a data set is a value that lies outside (is much smaller or larger than) most of the other values in the data set.

EXAMPLE 5   Analyzing Outliers and Averages

The 2009 U.S. Census report on income in the United States uses median income to describe the average income. Why does the report use median income instead of mean income to represent the average income?

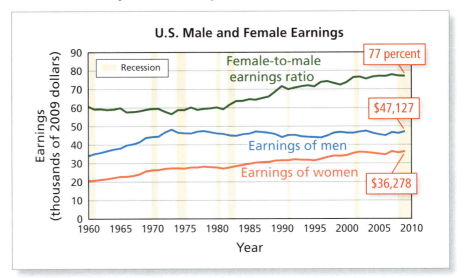

U.S. Male and Female Earnings

SOLUTION

The mean annual household income in the United States was about $67,976 in 2009. The median household income was $49,777. The circle graph below helps explain the discrepancy between these two averages. The very high incomes in the top 20% caused the mean to overestimate the typical household income. Because the median is not as strongly affected by outliers as the mean, it is used to measure average income.

Each year, the U.S. Census Bureau issues a report called *Income, Poverty, and Health Insurance Coverage in the United States.* This one was issued in September 2010.

Mean Household Income

Bottom 20%: Mean = $11,552 **3.4%**

4th 20%: Mean = $29,257   **8.6%**

3rd 20%: Mean = $49,534   **14.6%**

Top 20%: Mean = $170,844   **50.3%**

2nd 20%: Mean = $78,694   **23.2%**

✔ **Checkpoint**   Help at *Math*.and**YOU**.com

Discuss other instances where the median is a better measure of the average value than the mean.

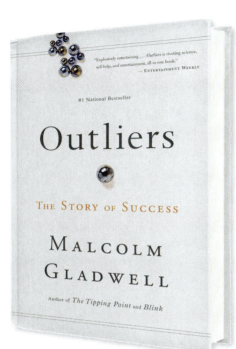

In Malcolm Gladwell's book *Outliers: The Story of Success*, the position of being an "outlier" is enviable. When using the mean as an average, however, outliers can cause undesired effects.

EXAMPLE 6   **Analyzing Data**

The following gives the estimated IQs of 12 recent presidents of the United States. It also gives their placements in the 2010 Siena College Research Institute Presidential Ranking Survey (with 43 being the lowest and 1 being the highest).

President	IQ	Poll Rating
Franklin Roosevelt	140	1
Harry Truman	128	9
Dwight Eisenhower	132	10
John Kennedy	151	11
Lyndon Johnson	128	16
Richard Nixon	131	30
Gerald Ford	127	28
Jimmy Carter	145	32
Ronald Reagan	130	18
George H. Bush	130	22
William Clinton	149	13
George W. Bush	125	39

Assuming these data are valid, which of the following statements are valid?

a. On average, U.S. presidents have above average IQs.

b. As president, the higher your IQ, the more popular you will be.

SOLUTION

a. This statement is certainly valid. Standardized IQ scores follow a distribution with an average IQ of 100. The median of the above IQs is about 131. The mean is about 135.

b. The scatter plot compares estimated IQs with poll ratings. If the statement were true, the scatter plot would show a pattern that moved from the lower left to the upper right. So, according to the above list, this conclusion is not valid.

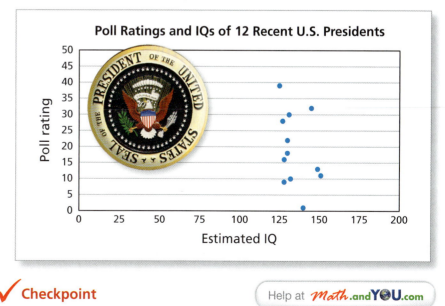

✓ **Checkpoint**   Help at *Math*.and**YOU**.com

Suppose a president with an IQ of 200 is elected. How does this affect the mean and the median of the data?

9.2 Exercises

Height The double stem-and-leaf plot shows the heights of players on a women's basketball team and players on a men's basketball team. In Exercises 1–8, use the double stem-and-leaf plot. *(See Examples 1 and 2.)*

Player Height

Women's Team	Stem	Men's Team
9 8 5	6	
7 6 5 5 4 4 3 2	7	2 4 4 6 7 7 8 9 9
	8	0 1 2 3 4 4 5

Key:

Women's: 5|6| = 65-inch-tall woman

Men's: |7|2 = 72-inch-tall man

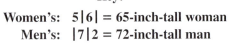

1. Compare the mean of each data set.

2. What is the mode of each data set?

3. Compare the median of each data set.

4. What percent of the players on the men's team are at least 7 feet tall?

5. Which measure of central tendency best describes the average height of the players on the women's basketball team?

6. Which measure of central tendency best describes the average height of the players on the men's basketball team?

7. Suppose each player's height on the women's team is 2 inches greater. How does this affect the mean and the median of the data?

8. The 5 starting players on the men's team have a mean height of 80.2 inches. Find the heights of 5 possible starters.

Annual Cost of College The box-and-whisker plot shows the annual cost for undergraduate tuition, room, and board at 21 public and private colleges. In Exercises 9–12, use the box-and-whisker plot. *(See Example 3.)*

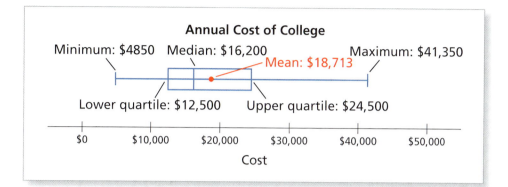

Annual Cost of College

Minimum: $4850   Median: $16,200   Maximum: $41,350

Mean: $18,713

Lower quartile: $12,500   Upper quartile: $24,500

$0 $10,000 $20,000 $30,000 $40,000 $50,000

Cost

9. How many colleges have an annual cost between $12,500 and $24,500?

10. Are there more colleges above or below the mean annual cost? Explain your reasoning.

11. The annual costs are divided into four quartiles, each of which contains about 25% of the data. In which quartile are the data most spread out? least spread out? Explain your reasoning.

12. The annual cost to attend college A is $10,780. In which quartile does this college belong?

Annual Cost of College The color-coded histogram shows the data from the box-and-whisker plot above in a different way. In Exercises 13–16, use the histogram. *(See Example 4.)*

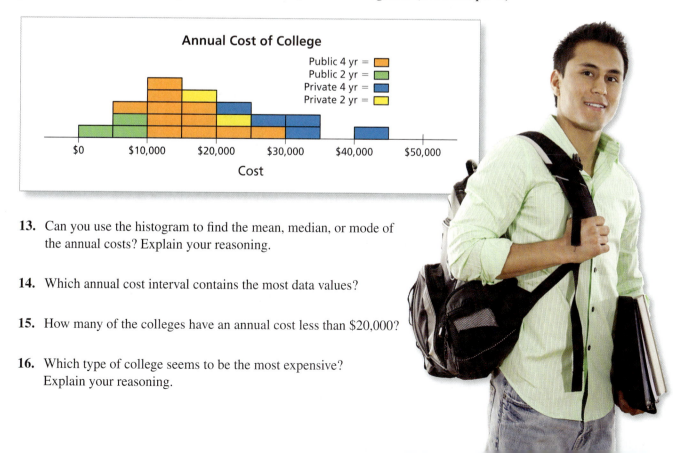

Annual Cost of College

Public 4 yr =
Public 2 yr =
Private 4 yr =
Private 2 yr =

$0 $10,000 $20,000 $30,000 $40,000 $50,000

Cost

13. Can you use the histogram to find the mean, median, or mode of the annual costs? Explain your reasoning.

14. Which annual cost interval contains the most data values?

15. How many of the colleges have an annual cost less than $20,000?

16. Which type of college seems to be the most expensive? Explain your reasoning.

Deadliest Earthquakes The double line graph shows the world's deadliest earthquake for each year from 1990 through 2010. In Exercises 17–22, use the double line graph. *(See Examples 5 and 6.)*

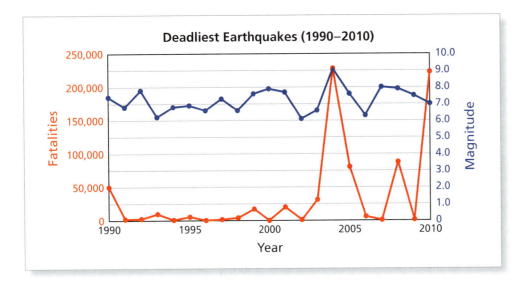

17. Which years appear to be outliers when considering the numbers of fatalities?

18. Which years appear to be outliers when considering the magnitudes of the earthquakes?

19. Find the effect of outliers on averages.

   a. Use a spreadsheet to find the mean and the median number of fatalities.

   b. Use a spreadsheet to find the mean and the median number of fatalities without the two most significant outliers.

   c. Do these outliers have a greater effect on the mean or the median? Explain your reasoning.

20. How do outliers affect the mode of a data set?

21. Does the mean or the median number of fatalities better describe the average number of fatalities each year? Explain your reasoning.

22. Determine whether each statement is valid.

   a. On average, the most fatal earthquakes have a magnitude around 6.0.

   b. The number of fatalities from an earthquake depends on the earthquake's magnitude.

▶ Extending Concepts

Printer Speeds A consumer testing service is determining the printing speed, in pages per minute (ppm), of three printers. The table shows the results of five print jobs for each printer. In Exercises 23 and 24, use the table.

	Job 1	Job 2	Job 3	Job 4	Job 5
Printer A (ppm)	28	32	30	27	32
Printer B (ppm)	30	28	34	28	32
Printer C (ppm)	28	31	32	27	31

23. Each printer has a different manufacturer, and each manufacturer advertises that its printer has a faster average printing speed than its competitors' printers. Which measure of central tendency (mean, median, or mode) would each of the following manufacturers use to support its claim? Explain your reasoning.

   a. the manufacturer of printer A

   b. the manufacturer of printer B

   c. the manufacturer of printer C

24. The midrange is $\dfrac{\text{(maximum data value)} + \text{(minimum data value)}}{2}$.

   a. Which printer has the fastest average speed according to the midrange?

   b. Why do you think the midrange is rarely used as a measure of central tendency?

Printer Prices The table below shows the prices of all the printers in stock at an electronics store. In Exercises 25–28, use the table.

Prices of Printers in Stock					
$45	$89	$72	$118	$68	$89
$105	$76	$95	$182	$120	$48
$68	$89	$128	$52	$108	$99

25. Make a box-and-whisker plot of the printer prices.

26. Describe the shape of the distribution.

27. Make a histogram of the printer prices.

28. What new information does the histogram give you about the distribution?

9.1–9.2   Quiz

World Education Rankings In Exercises 1–8, use the chart and the following information.

The chart shows the world education rankings from the Organisation for Economic Co-operation and Development (OECD).

1. Describe the information presented in the rankings chart.

2. Are the countries listed in order of greatest total score? Explain your reasoning.

3. Do you agree with the following comment? Explain your reasoning.

   COMMENT:

   > Overall, the United States scored higher than Germany.

4. Did Austria or Poland score higher in math? Explain your reasoning.

5. Do the following for each column in the chart.

   a. Use the chart to find the median and mode.

   b. Use a spreadsheet to find the mean.

6. Which measure of central tendency best describes the average of the math scores? Explain your reasoning.

7. Describe the distribution of math scores.

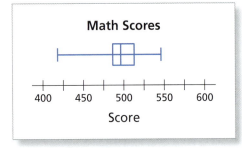

Math Scores

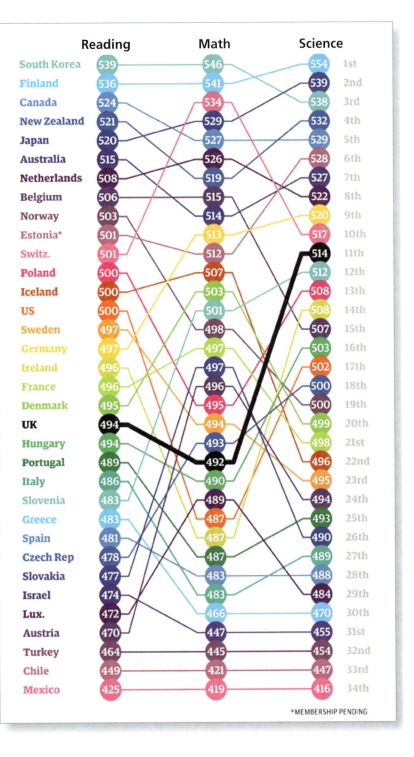

8. How might a politician or policymaker use this chart?

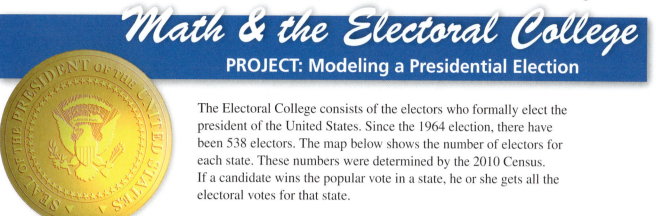

Math & the Electoral College

PROJECT: Modeling a Presidential Election

The Electoral College consists of the electors who formally elect the president of the United States. Since the 1964 election, there have been 538 electors. The map below shows the number of electors for each state. These numbers were determined by the 2010 Census. If a candidate wins the popular vote in a state, he or she gets all the electoral votes for that state.

1. Use the *Electoral College Simulator* at *Math.andYou.com*. It is possible to win the presidential election and not win the popular vote. This has happened four times in the United States (see page 114). The simulator below shows an imaginary election in which the winner receives 52.8% of the electoral votes, but only 28.8% of the popular vote. Use the simulator to find the minimum popular vote that a candidate can receive and still win.

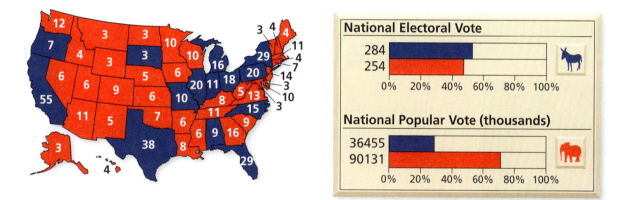

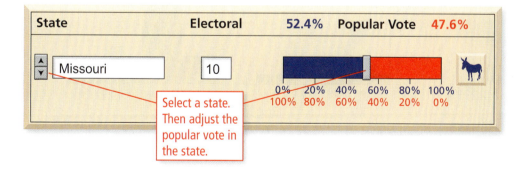

2. Is it possible to become president and not have a single person west of the Mississippi River vote for you? Use the simulator to verify your answer.

3. Is it possible to win the election without carrying one of the "big" states (California, Texas, Florida, New York, Pennsylvania, Illinois, Ohio, Michigan, Georgia)? Use the simulator to verify your answer.

4. Why was the Electoral College created? Do you agree with this system? If not, describe a different system that you think would be better. For instance, do you think the system used in Canada is better? Explain your reasoning.

9.3 Describing Dispersion

▶ Use standard deviation to describe the dispersion of a data set.

▶ Use standard deviation to describe a data set that is normally distributed.

▶ Compare different types of distributions.

Standard Deviation

Study Tip

In statistics, the rule at the right is called Chebyshev's inequality. Its estimates are conservative. On page 426, you will see a stronger result for data sets that follow a normal distribution.

Standard deviation is a measurement that shows how much variation or dispersion there is from the mean. A small standard deviation indicates that the data are clustered tightly around the mean. A large standard deviation indicates that the data are spread out over a large range of values.

Standard Deviation and Dispersion

In any data set, at least 75% of the values lie within 2 standard deviations of the mean. At least 89% lie within 3 standard deviations, and at least 94% lie within 4 standard deviations.

EXAMPLE 1 Describing Dispersion

The histogram shows the distribution of the lengths of 45 cuckoo eggs.

a. Describe the dispersion.

b. What percent of the lengths lie within 2 standard deviations of the mean?

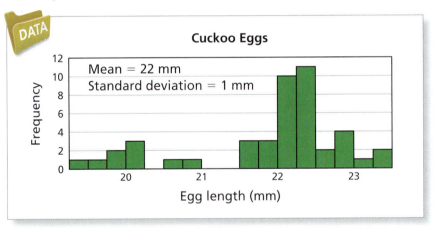

Cuckoo Eggs

Mean = 22 mm
Standard deviation = 1 mm

Rather than building its own nest, the common cuckoo lays its eggs in the nests of other birds. When it hatches, the cuckoo chick eventually pushes the other eggs and chicks from the nest.

SOLUTION

a. 27 of the lengths lie within $\frac{1}{2}$ of a standard deviation of the mean.

b. 41 of the lengths, or about 91%, lie within 2 standard deviations of the mean.

✓ **Checkpoint**  Help at *Math*.and**Y⊕U**.com

The data set for Example 1 is available at *Math.andYou.com*. Use the *Histogram Generator* at *Math.andYou.com* to display the data. Experiment with different interval widths. The histogram above uses an interval width of 0.25. Try using an interval width of 0.5.

EXAMPLE 2   **Describing Dispersion**

The histogram shows the distribution of the mean daily temperatures at McGuire Air Force Base from 1955 through 2010. Describe the dispersion.

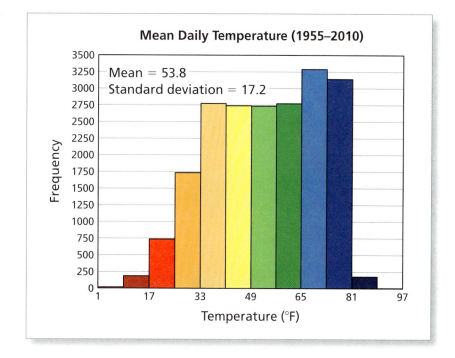

Mean Daily Temperature (1955–2010)

Mean = 53.8
Standard deviation = 17.2

SOLUTION

The vast majority of the temperatures lie within 1 standard deviation of the mean. That is, they range from 36.6°F to 71°F.

 **Checkpoint**   Help at *Math*.and**YOU**.com

The line graph shows 101-day moving averages of the data represented in the histogram. The temperatures are plotted from 1955 through 2010. Suppose you are doing a study on global warming. Which of the two graphic displays would be more helpful? Explain your reasoning.

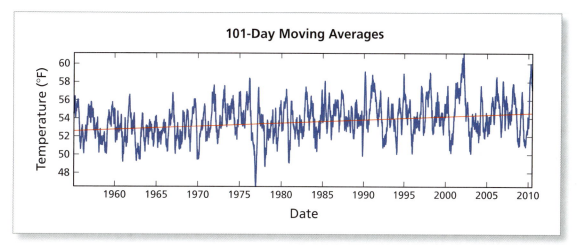

101-Day Moving Averages

Standard Deviation and Normal Distribution

In many naturally occurring data sets, a histogram of the data is often bell shaped. In statistics, such data sets are said to have a **normal distribution.**

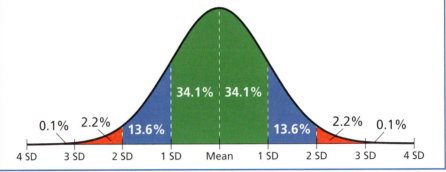

Standard Deviation and a Normal Distribution

A normal distribution is symmetrical about its mean. 68.2% (about two-thirds) of the data lie within 1 standard deviation of the mean. 95.4% of the data lie within 2 standard deviations of the mean.

EXAMPLE 3 Analyzing a Famous Normal Distribution

A famous data set was collected in Scotland in the mid-1800s. It contains the chest sizes (in inches) of 5738 men in the Scottish Militia. What percent of the chest sizes lie within 1 standard deviation of the mean?

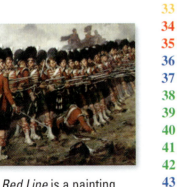

The Thin Red Line is a painting by Robert Gibb. It was painted in 1881. Only the left portion of the painting is shown above.

Chest Size	Number of Men
33	3
34	18
35	81
36	185
37	420
38	749
39	1073
40	1079
41	934
42	658
43	370
44	92
45	50
46	21
47	4
48	1

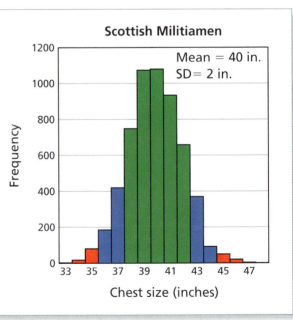

SOLUTION

The number of chest sizes within 1 standard deviation of the mean is 749 + 1073 + 1079 + 934 + 658 = 4493. This is about 78.3% of the total, which is somewhat more than the percent predicted by a normal distribution.

✓ **Checkpoint**

Help at *Math*.and**YOU**.com

What percent of the chest sizes lie within 2 standard deviations of the mean?

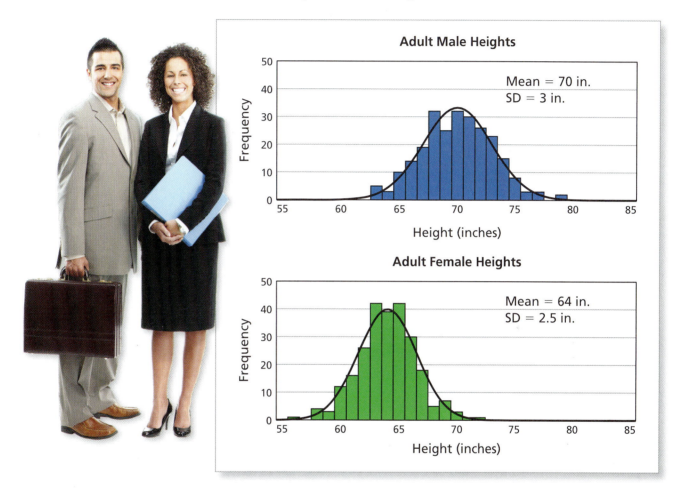

> **EXAMPLE 4**   **Comparing Distributions**

The graphs show the distributions of samples of heights of adult American males and females. There are 250 people in each sample.

a. What is the significance of the smaller standard deviation for females?

b. Estimate the percent of male heights between 67 inches and 73 inches.

SOLUTION

a. Standard deviation is a measure of dispersion. The larger the standard deviation, the more the data are spread out. So, if these two samples are representative of male and female heights in the United States, you can conclude that male heights have a greater variation than female heights.

b. The data for male heights appear to be normally distributed. If this is true, then you can conclude that about 68% of adult male heights are between 67 inches (5' 7") and 73 inches (6' 1").

 **Checkpoint**

 Help at *Math*.and**YOU**.com

In a study, Timothy Judge and Daniel Cable found that each additional inch of height is worth an extra $789 per year in salary. According to Judge, "Height matters for career success." Do you agree with Judge's claim? Explain.

Comparing Different Types of Distributions

While normal distributions (bell-shaped) do occur commonly in real life, there are other types of distributions. Here are three of them.

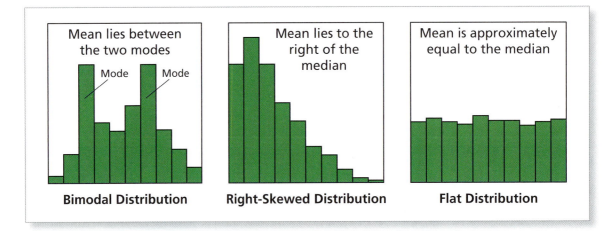

EXAMPLE 5   **Comparing Distributions**

The histograms show the distributions of game lengths (in number of turns) when there are two players. Describe the differences in the distributions.

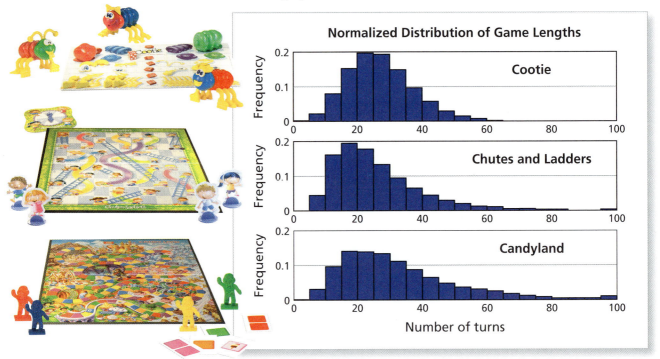

SOLUTION

Candyland tends to take the most turns to finish. *Chutes and Ladders* tends to take the least.

 **Checkpoint**   Help at *Math*.andY⊙U.com

Suppose you work in the marketing department for a game company. How would you use the information in Example 5 to target different age groups? Explain.

EXAMPLE 6   Analyzing Bimodal Distribution

The graph shows the distribution of the full-time salaries of 22,665 people who graduated from law school in 2006. How would you explain the bimodal distribution?

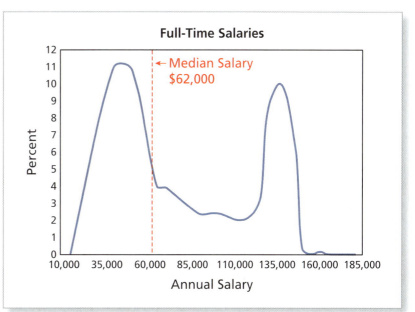

Full-Time Salaries

← Median Salary
$62,000

Employed graduates from the law school class of 2006 took the following types of jobs: private practice (55.8%), business (14.2%), other government (10.6%), judicial clerk (9.8%), public interest (5.4%), academic (1.7%), military (1.1%), and unknown (1.3%).

SOLUTION

According to the National Association for Law Placement (NALP), 71% of employed graduates from the class of 2006 took jobs at small firms (50 or fewer lawyers) or in nonfirm settings. Only 20% took jobs at large firms (more than 100 lawyers).

The bimodal distribution represents a cluster of lawyers at small and midsize firms earning between $40,000 and $50,000 and a cluster of lawyers at large firms earning between $135,000 and $145,000.

✓ Checkpoint

Help at *Math*.and**Y☺U**.com

Suppose that you combine the data from the two data sets in Example 4 into one data set. What will the histogram look like? Explain how you can discover two (or more) subpopulations within a given larger population.

Adult Male Heights

Mean = 70 in.
SD = 3 in.

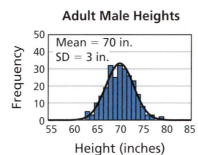

Adult Female Heights

Mean = 64 in.
SD = 2.5 in.

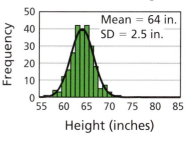

Law schools are ranked each year. One of the recent rankings by *U.S. News and World Report* listed the following as the top five law schools in the United States (with full-time tuition).

1. Yale University ($50,750)
2. Harvard University ($45,450)
3. Stanford University ($46,581)
4. Columbia University ($50,428)
5. University of Chicago ($45,405)

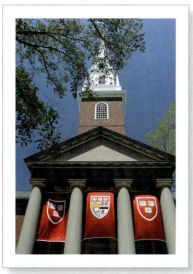

9.3 Exercises

Rainbow Trout Eggs The histogram shows the distribution of the diameters of 100 rainbow trout eggs. In Exercises 1–3, use the histogram. *(See Examples 1 and 2.)*

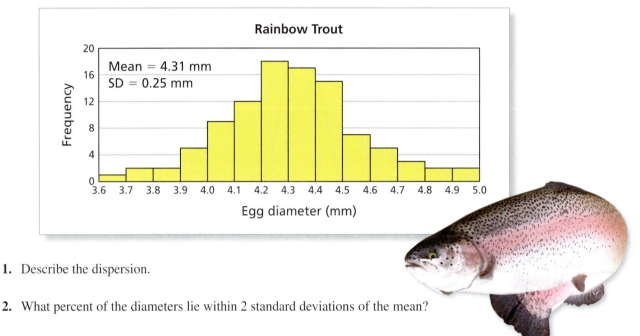

Rainbow Trout

Mean = 4.31 mm
SD = 0.25 mm

1. Describe the dispersion.

2. What percent of the diameters lie within 2 standard deviations of the mean?

3. What percent of the diameters lie within 3 standard deviations of the mean?

Temperatures The histogram shows the distribution of the daily high temperatures in Pittsburgh for 2010. In Exercises 4–6, use the histogram. *(See Examples 1 and 2.)*

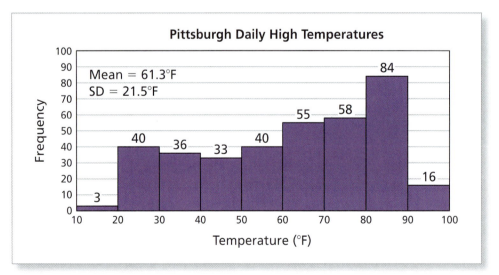

Pittsburgh Daily High Temperatures

Mean = 61.3°F
SD = 21.5°F

4. Describe the dispersion.

5. What percent of the temperatures lie within 1 standard deviation of the mean?

6. What percent of the temperatures lie within 2 standard deviations of the mean?

Factory Employees The graph shows the distribution of the years worked by the
1820 employees of a cereal factory. In Exercises 7–10, use the histogram. *(See Example 3.)*

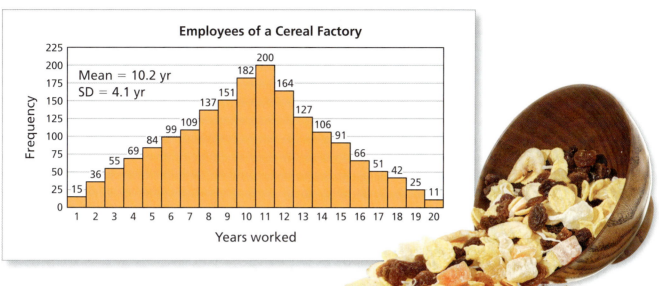

Employees of a Cereal Factory

7. Does this data set have a normal distribution?
    Explain your reasoning.

8. What percent of the employees lie within 1 standard
    deviation of the mean?

9. What percent of the employees lie within 2 standard deviations of the mean?

10. Compare the percents in Exercises 8 and 9 with the percents given by the
    normal distribution.

Cereal The graphs show the distributions of samples of weights of boxes of cereal
filled by two machines. In Exercises 11 and 12, use the histograms. *(See Example 4.)*

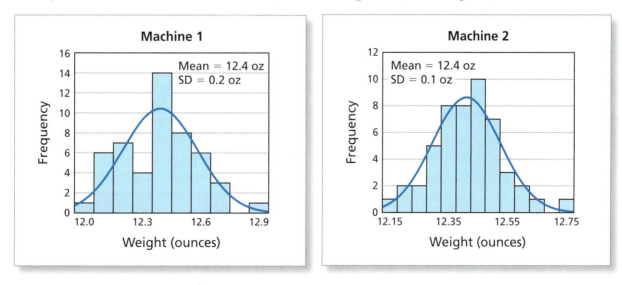

11. What is the significance of the smaller standard deviation for machine 2?

12. The boxes must weigh at least 12.2 ounces. Estimate the percent of boxes from each
    machine that pass the weight requirement.

SAT Scores The histograms show the distributions of SAT scores for mathematics and writing for males and females in a recent year. In Exercises 13 and 14, use the histograms. *(See Example 5.)*

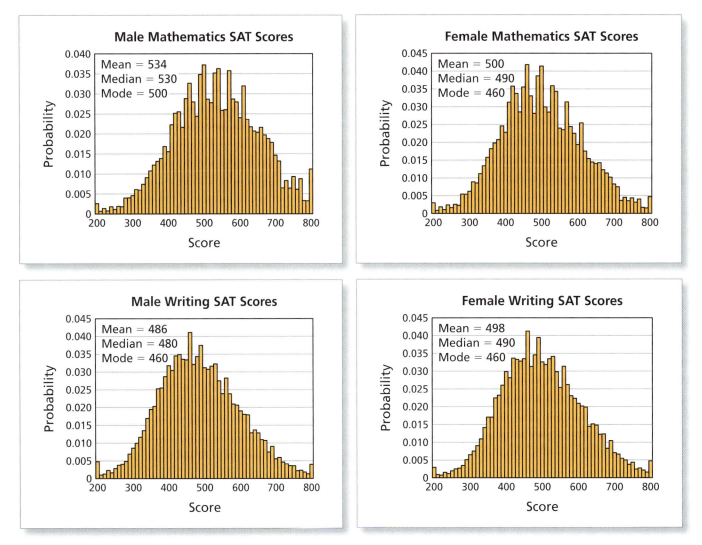

13. Describe the differences in the distributions of mathematics SAT scores for males and females. Describe the differences in the distributions of writing SAT scores for males and females.

14. Describe the differences in the distributions of mathematics SAT scores and writing SAT scores for males. Describe the differences in the distributions of mathematics SAT scores and writing SAT scores for females.

Test Preparation The histogram shows the distribution of the total numbers of hours students at a school study for the SAT over 12 months. In Exercises 15 and 16, use the histogram. *(See Example 6.)*

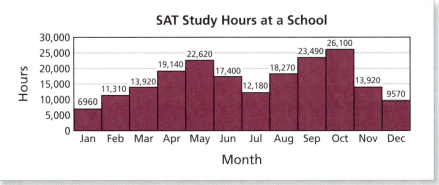

15. What are the two modes of this bimodal distribution? How would you explain the bimodal distribution?

16. What percent of the study hours are in October?

▶ Extending Concepts

Hodgkin Lymphoma The histogram shows the distribution of the ages at which people were diagnosed with Hodgkin lymphoma in a recent year. The table shows the population of the United States for each age group in the same year. In Exercises 17 and 18, use the histogram and the table.

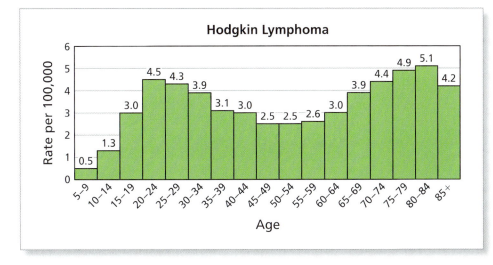

Age	Population
5–9	20,313,416
10–14	20,103,526
15–19	21,628,099
20–24	21,322,253
25–29	21,441,834
30–34	19,515,671
35–39	20,846,774
40–44	21,394,190
45–49	22,802,020
50–54	21,431,624
55–59	18,541,377
60–64	15,081,608
65–69	11,332,535
70–74	8,806,448
75–79	7,385,163
80–84	5,825,975
85+	5,449,770

17. The distribution of the ages in the graph is bimodal. How would you explain this?

18. Use the rates and the populations to create a histogram that estimates the number of people diagnosed with Hodgkin lymphoma for each age group. Is the histogram bimodal? Explain.

Standard Deviation The formula for the sample standard deviation of a data set is given by

$$s = \sqrt{\dfrac{\sum x^2 - \dfrac{(\sum x)^2}{n}}{n-1}}$$

where x represents each value in the data set and n is the number of values in the data set. The symbol Σ indicates a sum of values. So, $\sum x$ is the sum of the data values, and $\sum x^2$ is the sum of the squared data values. In Exercises 19 and 20, complete the table and then use the formula to find the sample standard deviation of the data set.

19.

x	x^2
16	
12	
23	
20	
18	
15	
18	
19	
$\sum x =$	$\sum x^2 =$

20.

x	x^2
33	
24	
27	
26	
30	
29	
26	
31	
34	
20	
$\sum x =$	$\sum x^2 =$

9.4 Describing by Sampling

▶ Use a randomly chosen sample to describe a population.

▶ Determine whether a sample is representative of a population.

▶ Determine a sample size to obtain valid inferences.

Inferring from a Sample

You can best describe any population when you have data for the entire population. Every 10 years, the U.S. Census Bureau attempts to do this. It is costly. The 2010 Census cost the United States $13.1 billion!

A complete census is often unpractical. So, governments, researchers, and businesses attempt to describe populations by taking a **representative sample.** You can be assured that a sample is representative if it is *randomly chosen* and *large enough.*

Study Tip

To obtain smaller margins of error, increase the size of the sample.

Inferring from a Sample

If a sample is randomly taken from a population, then the sample mean can be used to estimate the population mean, given the following limitations.

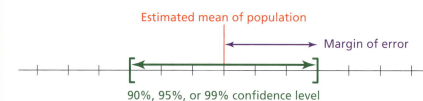

Estimated mean of population

Margin of error

90%, 95%, or 99% confidence level

EXAMPLE 1   **Estimating a Population Mean by Sampling**

The histogram shows the distribution of the shoulder heights of a sample of 40 male black bears in Great Smoky Mountains National Park. Use a 90% confidence level and a 95% confidence level to estimate the population mean shoulder height.

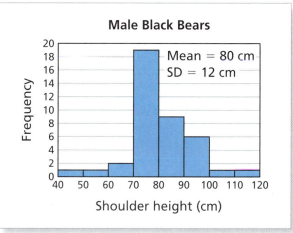

Male Black Bears

Mean = 80 cm
SD = 12 cm

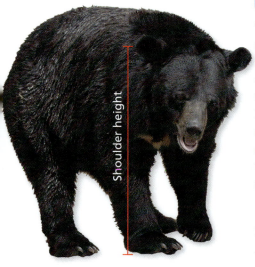

Shoulder height

SOLUTION

Using a confidence interval calculator, you can obtain the following.

- **90% Confidence level:** The population mean is 80 cm ± 3.1 cm.
- **95% Confidence level:** The population mean is 80 cm ± 3.7 cm.

✓ **Checkpoint**   Help at *Math.andYOU.com*

Use the *Confidence Interval Calculator* at *Math.andYou.com* and a 99% confidence level to estimate the population mean shoulder height.

Here is the basic idea of statistical sampling.

1. Take a representative sample of the population.
2. Find the mean and the standard deviation of the sample.
3. Determine how confident you want to be of your *inference* (90%, 95%, 99%).
4. Use a confidence interval calculator to determine the margin of error.
5. *Infer* that the mean of the population is equal to the mean of the sample (± the margin of error).

This process is called **statistical inference.**

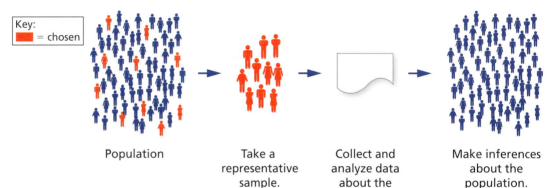

Key:
⬛ = chosen

Population → Take a representative sample. → Collect and analyze data about the sample. → Make inferences about the population.

EXAMPLE 2   Estimating by Sampling

In 1879, Albert Michelson conducted an experiment to measure the speed of light. He conducted the experiment 100 times and obtained the data represented in the histogram. Use a 95% confidence level to estimate the speed of light.

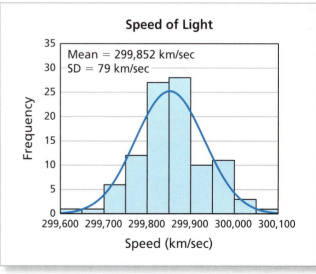

Speed of Light

Mean = 299,852 km/sec
SD = 79 km/sec

SOLUTION

Using a confidence interval calculator and a 95% confidence level, you can infer that the speed of light is

299,852 kilometers per second ± 15 kilometers per second.

In 1926, Michelson improved his estimate to 299,796 ± 4 kilometers per second. The accepted value today is 299,792.458 kilometers per second.

Albert Michelson and Albert Einstein met at Mount Wilson Observatory in 1931, shortly before Michelson's death that same year.

 **Checkpoint**    Help at *Math*.and**Y⚙U**.com

A researcher records the temperature of 130 people. The mean of the sample is 98.25°F and the standard deviation is 0.73°F. Does this study allow for the accepted human body temperature of 98.6°F? Explain your reasoning.

Determining Whether a Sample is Representative

Statistical inference only applies when the sample is randomly chosen from the population. Even then, there may be other factors that make the sample unrepresentative. For a sample to be random, each member of the population must have an equal chance of being chosen.

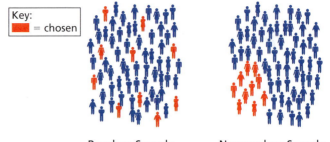

Random Sample   Nonrandom Sample

Obtaining a sample that is truly representative is a major problem faced by researchers, polling organizations, and market research departments.

EXAMPLE 3   **Analyzing How a Poll Can Go Wrong**

Comment on the following article from NBC News.

"When there's plentiful public polling and stable national trends to base their assumptions upon, their prognostications are usually more or less close to the mark. But politics is unpredictable, and pollsters use different methods to determine who's up and who's down. Every once in a while, they get it wrong.

The famous erroneous 1948 'Dewey defeats Truman' banner headline in the first edition of the *Chicago Tribune* was the result of polls and conventional wisdom that turned out to be dramatically off base. After Los Angeles Mayor Tom Bradley's unanticipated loss in the 1982 California governors' race, many blamed racial bias undetected in public opinion polls. After Barack Obama's dramatic win in the Iowa caucuses in 2008, the media's overwhelming assumption that he would continue his march to victory days later in New Hampshire was due to polling. Those assumptions were shattered when Hillary Clinton wound up winning—prompting weeks of media navel-gazing and questions of 'how did we get it so wrong?'"

SOLUTION

There are many reasons a poll can go wrong. Here are a few.

- The people being polled do not represent a random selection from the population.

- The questions in the poll can be leading or confusing, as in "Do you plan to vote for Proposition 4, which is against ownership of pit bulls?"

- The people being polled may not know how they will react to a product or how they plan to vote.

On election night in 1948, the press deadline for the *Chicago Tribune* required that the first post-election issue go to press before the states had reported many of their results. Conventional wisdom, supported by polls, predicted that Dewey would win. So, the *Tribune* went to press with the incorrect headline "DEWEY DEFEATS TRUMAN."

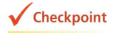 **Checkpoint**

 Help at *Math*.and**YOU**.com

What other reasons might cause a poll to go wrong?

When the population consists of people, obtaining a representative sample can be difficult. For various reasons, the sample can be **biased,** which means that it is not representative of the entire population.

> ### EXAMPLE 4   Analyzing Biased Samples

Why might the following samples be biased?

a. In 1936, *The Literary Digest* used the results of more than two million responses to a survey to predict that the Republican presidential candidate, Alfred Landon, would defeat the incumbent, Franklin Roosevelt, by a large margin.

b. In 2010, the Pew Research Center conducted a pre-election survey of likely voters and found Republican candidates leading by 12 points, 51% to 39%.

c. In the 1940s, the Metropolitan Life Insurance Company introduced its standard height-weight tables for men and women. To create the tables, the company used the weights of people who were insured with the company and lived the longest.

SOLUTION

Alfred Mossman "Alf" Landon (1887–1987) was the 26th governor of Kansas. He was the Republican nominee in the 1936 presidential election.

a. *The Literary Digest* primarily collected its sample from people who, during the depression, could afford cars, phones, and magazine subscriptions. This sample overrepresented higher income people, who were more likely to vote for the Republican candidate. Also, Landon supporters may have been more likely to return the survey than Roosevelt supporters, creating what is called a nonresponse bias. By contrast, a poll by George Gallup's organization successfully predicted the result because its sample was more representative of the voting population.

b. In a sample that included both landline *and* cell phone interviews, the Republican lead was only 6 points, 48% to 42%. Based on this and other surveys, the Pew Research Center concluded that landline-only surveys tend to be biased toward Republicans.

c. The sample may have been biased because it contained only the weights of people who were insured with the company.

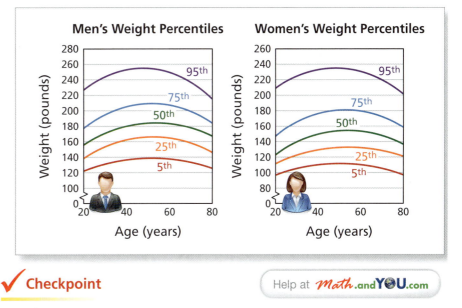

✔ **Checkpoint**

Help at *Math.andY☺U.com*

Give another example of a biased sample.

Sampling and Market Research

The following are five types of sampling used by market research departments.

1. **Surveys:** The wording of the questions is critical. Surveys can be in person, or via telephone, mail, or Internet.

2. **Focus groups:** Focus groups involve a moderator and questions that lead to a discussion among participants.

3. **Personal interviews:** Personal interviews are similar to focus groups and involve open-ended questions.

4. **Observations:** Observations include purchasing habits and product usage of consumers.

5. **Field trials:** Field trials involve selling a product at selected locations to get an idea of how it will perform in the market.

EXAMPLE 5   **Determining a Sample Size**

You work for the market research department of a cosmetics company. You want to identify characteristics of men who are likely to buy men's personal care products. You send a free sample of two of the products to men and ask them to respond to an online survey. How many samples should you send to obtain reliable results?

SOLUTION

Suppose you want a confidence level of 90% and a margin of error of 5%. Using the *Sample Size Calculator* at *Math.andYou.com*, you can determine that you need 271 completed surveys.

Suppose you think that 10% of the men who receive the sample will take the time to use it and respond to the online survey. This means that you should send the sample products to about 3000 men. (You still have to decide how to randomly choose 3000 men to obtain a representative sample.)

 **Checkpoint**

Help at *Math.andYOU.com*

How would you conduct market research on the men's personal care products?

| EXAMPLE 6 | **Analyzing a Historical Example of Sampling** |

Edward Bernays, a nephew of Sigmund Freud, used some of his uncle's ideas about psychology to influence public opinion. In one well-known campaign, Bernays took an assignment from a large bacon producer, Beechnut Packing Company. Rather than try to take away business from Beechnut's competitors, Bernays decided to try to change America's attitude toward breakfast. At the time, Americans tended to eat small breakfasts, often consisting of juice, toast, and coffee.

In the 1920s, Bernays persuaded a well-known physician in New York to write to his colleagues and ask whether they recommended light breakfasts or hearty breakfasts. The result was "hearty." Newspapers spread the message. As a result of Bernays's campaign, the phrase "bacon and eggs" became synonymous with "American breakfast."

Do you think that his claim that doctors recommend a hearty breakfast is valid?

SOLUTION

There are two ways to view this question.

- Is Bernays's implied claim that a big breakfast is healthy true?
- Is the claim statistically valid based on his survey of doctors?

Today, most people in the fitness and health fields recommend that people eat breakfast, but there is no consensus on what size it should be.

Is the claim statistically valid based on his survey? The answer is unknown because you do not know the sample size, the form of the questions, or the responses. However, based on other campaigns that Bernays ran, it is clear that he was not as concerned with the truth as he was with the results.

✓ **Checkpoint**

Help at *Math*.and**Y♥U**.com

Which of the following questions might produce more people saying they believe Australians are more sports minded than Americans? Explain.

a. Do you agree that Australians are more sports minded than Americans?

  ○ Agree   ○ Disagree

b. Do you think that Australians are more sports minded than Americans?

  ○ Yes   ○ No

c. Do you believe that Australians are more sports minded than Americans, less sports minded, or about the same?

  ○ More   ○ Less   ○ About the same

9.4 Exercises

IQ Scores The histogram shows the distribution of the IQ scores of 50 adults. In Exercises 1 and 2, use the histogram. *(See Examples 1 and 2.)*

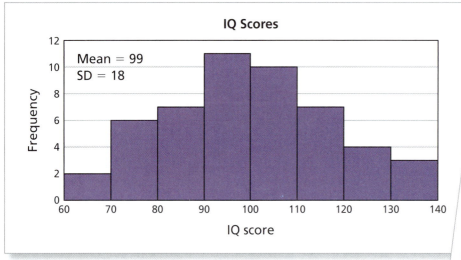

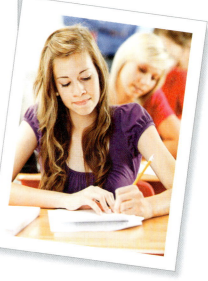

1. Use the *Confidence Interval Calculator* at *Math.andYou.com* and a 95% confidence level to estimate the population mean IQ score.

2. Repeat Exercise 1 using a sample size of 100. Compare the confidence intervals. What happens when the sample size increases? Explain.

Fuel Efficiency The graph shows the distribution of the fuel efficiencies of 60 sedans. In Exercises 3 and 4, use the histogram. *(See Examples 1 and 2.)*

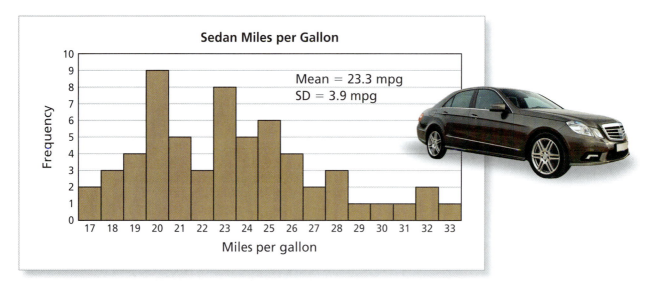

3. Use the *Confidence Interval Calculator* at *Math.andYou.com* and a 90% confidence level to estimate the population mean fuel efficiency.

4. Repeat Exercise 3 using a 99% confidence level. Compare the confidence intervals. What happens when the confidence level increases? Explain.

Biased Samples In Exercises 5–12, explain why the sample may be biased.
Then explain how to find an unbiased sample. (*See Examples 3 and 4.*)

5. A research company wants to determine how many people floss
   their teeth. The company asks a dentist to conduct a survey by
   randomly selecting patients.

6. You want to estimate the number of students in a high school
   who ride the school bus. You randomly survey 60 seniors.

7. You want to estimate the number of defective items produced
   by a factory. You randomly inspect 200 items from one of the
   machines in the factory.

8. A college wants to determine whether to renovate the gym or
   the science lab. The college asks you to conduct a survey. You
   randomly survey 30 students leaving a science club meeting.

9. A radio station wants to determine how many people in the
   listening area support gun control laws. The station asks
   listeners to call in and answer the survey.

10. A city wants to determine whether the residents of the city favor using tax dollars to
    build a new baseball stadium. The city asks you to conduct a survey. You randomly
    survey people entering a sporting goods store.

11. A mayor wants to determine whether the residents of a city support a bill providing
    insurance for nursing home care. The mayor asks you to conduct a survey. You randomly
    survey residents of five nursing homes in the city.

12. A research company wants to determine how many people in the United States spend at
    least 1 week at the beach each year. The company surveys residents of California, Florida,
    and North Carolina.

Alternative Energy You work for a research company. You want to determine whether U.S. adults support more funding for alternative energy. In Exercises 13–15, use the *Sample Size Calculator* at *Math.andYou.com*. *(See Example 5.)*

13. Choose a 95% confidence level and a 3% margin of error. How many people should you survey to obtain reliable results?

14. Repeat Exercise 13 using a 99% confidence level. Compare the sample sizes. What happens when the confidence level increases? Explain.

15. Repeat Exercise 13 choosing a 5% margin of error. Compare the sample sizes. What happens when the margin of error increases? Explain.

Biased Questions In Exercises 16–18, determine which question might produce biased results. Explain. *(See Example 6.)*

16.

    a. Do you think solar panels should be installed at city hall using taxpayer money?

        ◯ Yes   ◯ No

    b. Do you agree with most of your neighbors that it is a waste of taxpayer money to install solar panels at city hall?

        ◯ Agree   ◯ Disagree

17.

    a. Do you agree that the unfair policy of requiring students to do a time-consuming community service project should be changed?

        ◯ Agree   ◯ Disagree

    b. Do you think the policy of requiring students to do a community service project should be kept?

        ◯ Yes   ◯ No

18.

    a. Do you think people should recycle phone books?

        ◯ Yes   ◯ No

    b. Do you think people should recycle old, out-of-date phone books?

        ◯ Yes   ◯ No

▶ Extending Concepts

Confidence Intervals **In Exercises 19–22, use the confidence interval to find the sample mean and the margin of error.**

19. An electronics magazine reports that a 90% confidence interval for the mean price of GPS navigation systems is $178.75 to $211.87.

20. A state agency reports that a 95% confidence interval for the mean annual salaries of employees in Colorado is $45,832 to $47,890.

21. A hospital reports that a 99% confidence interval for the mean length of stay (in days) of patients is 5.1 to 5.9.

22. A company reports that a 95% confidence interval for the mean weight (in ounces) of filled paint cans is 159.97 to 160.03.

Minimum Sample Size **For a 95% confidence level, the minimum sample size** n **needed to estimate the population mean is**

$$n = \left(\frac{1.96s}{E}\right)^2$$

where E **is the margin of error and** s **is the population standard deviation. In Exercises 23–26, find the minimum sample size. If necessary, round your answer up to a whole number.**

23. You want to estimate the mean weight of newborns within 0.25 pound of the population mean. Assume the population standard deviation is 1.3 pounds.

24. You want to estimate the mean number of text messages sent per day by 18- to 24-year-olds within 5 messages of the population mean. Assume the population standard deviation is 30 messages.

25. You want to estimate the mean number of hours of television watched per person per day within 0.1 hour of the population mean. Assume the population standard deviation is 1.5 hours.

26. You want to estimate the mean number of minutes waiting at a department of motor vehicles office within 0.5 minute of the population mean. Assume the population standard deviation is 7 minutes.

9.3–9.4   Quiz

Ambulance Response Times The graph shows the distribution of the ambulance response times for 250 emergency calls in a city. In Exercises 1–6, use the histogram.

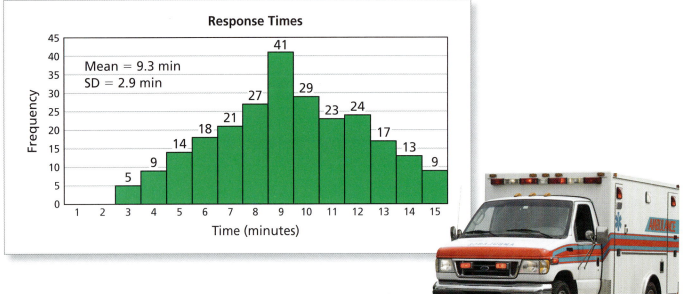

1. Does this data set have a normal distribution? Explain your reasoning.

2. What percent of the response times lie within 1 standard deviation of the mean?

3. What percent of the response times lie within 2 standard deviations of the mean?

4. Compare the percents in Exercises 2 and 3 with the percents given by the normal distribution.

5. Use the *Confidence Interval Calculator* at *Math.andYou.com* and a 95% confidence level to estimate the population mean response time.

6. Use the *Confidence Interval Calculator* at *Math.andYou.com* and a 99% confidence level to estimate the population mean response time.

7. **Fire Station** A city wants to know whether residents will favor a tax increase for the renovation of a fire station. You randomly survey 100 people in the neighborhood around the fire station. Explain why the sample may be biased. Then explain how to find an unbiased sample.

8. **Hospital** You work for a research company. You want to estimate the percent of U.S. adults who have contributed to a hospital fundraiser in the past 12 months. Use the *Sample Size Calculator* at *Math.andYou.com*. Choose a 99% confidence level and a 5% margin of error. How many people should you survey to obtain reliable results?

Chapter 9　Summary

Section Objectives		**How does it apply to you?**

Section 1

Use stacked area graphs to represent the changing parts of a whole.	→	You can see how related variables graphically change over time. *(See Examples 1 and 2.)*
Use a radar graph and an area graph to represent data.	→	There are many creative and unique ways to help people understand and use information. *(See Examples 3 and 4.)*
Graphically represent data sets that have several variables.	→	Not all information is represented with simple types of information design, such as bar graphs, circle graphs, and scatter plots. *(See Examples 5 and 6.)*

Section 2

Use mean, median, and mode to describe the average value of a data set.	→	Measures of central tendency are used to compare and communicate data in real life. *(See Examples 1 and 2.)*
Read and understand box-and-whisker plots and histograms.	→	You can use box-and-whisker plots and histograms to analyze the variability and distribution of data sets. *(See Examples 3 and 4.)*
Understand the effect of outliers on averages.	→	Outliers can significantly change measures of central tendency. This could lead to using bad information for decision making. *(See Examples 5 and 6.)*

Section 3

Use standard deviation to describe the dispersion of a data set.	→	Standard deviation allows you to determine whether data values are clustered around the mean or spread out over a large range. *(See Examples 1 and 2.)*
Use standard deviation to describe a data set that is normally distributed.	→	Many naturally occurring data sets have a normal distribution. You can use standard deviation to determine whether a data set is normal. *(See Examples 3 and 4.)*
Compare different types of distributions.	→	Not all data distributions are normal. There are other types of distributions. *(See Examples 5 and 6.)*

Section 4

Use a randomly chosen sample to describe a population.	→	If a sample is randomly taken from a population, then you can use the sample mean to estimate the population mean. *(See Examples 1 and 2.)*
Determine whether a sample is representative of a population.	→	If your sample is too small, not random, or biased, the statistics from the sample will not represent the population. *(See Examples 3 and 4.)*
Determine a sample size to obtain valid inferences.	→	The minimum sample size depends on the confidence level and the margin of error chosen. *(See Examples 5 and 6.)*

Chapter 9   Review Exercises

Section 9.1

Religiosity **In Exercises 1–8, use the information below.**

The graph shows the data from a Gallup survey relating religiosity to gross domestic product (GDP) per capita. GDP is the market value of the goods and services produced by a country over a certain time period, usually by year. It is often used as a comparative statistic to gauge a country's standard of living.

1. Describe the variables displayed in the graph.

2. Describe the information presented in the graph.

3. Which country has the highest GDP per capita?

4. What is the dominant religion in France?

5. Does the graph support the comment below? Explain your reasoning.

   COMMENT:

   Religiosity is highly correlated to poverty. Richer countries, in general, are less religious.

6. Are there more religious people in Argentina or Russia? Explain your reasoning.

7. Compare the religiosity of the United States with that of other countries with high GDPs per capita.

8. Could you use a stacked area graph to show the information presented in this graph? Explain your reasoning.

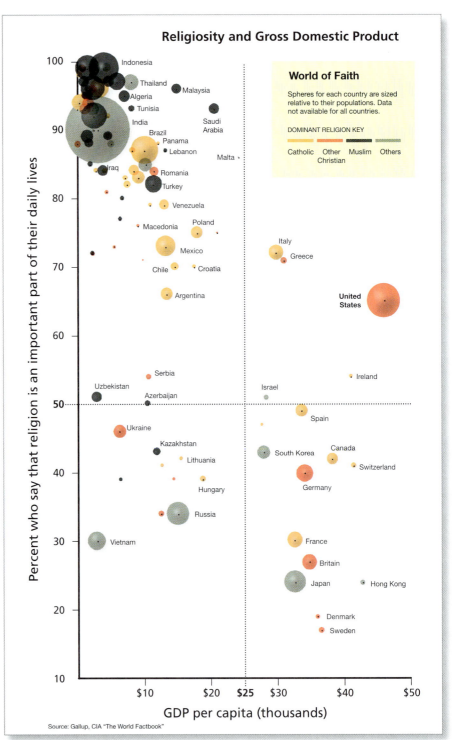

Religiosity and Gross Domestic Product

World of Faith

Spheres for each country are sized relative to their populations. Data not available for all countries.

DOMINANT RELIGION KEY

Catholic   Other Christian   Muslim   Others

Percent who say that religion is an important part of their daily lives

GDP per capita (thousands)

Source: Gallup, CIA "The World Factbook"

Section 9.2

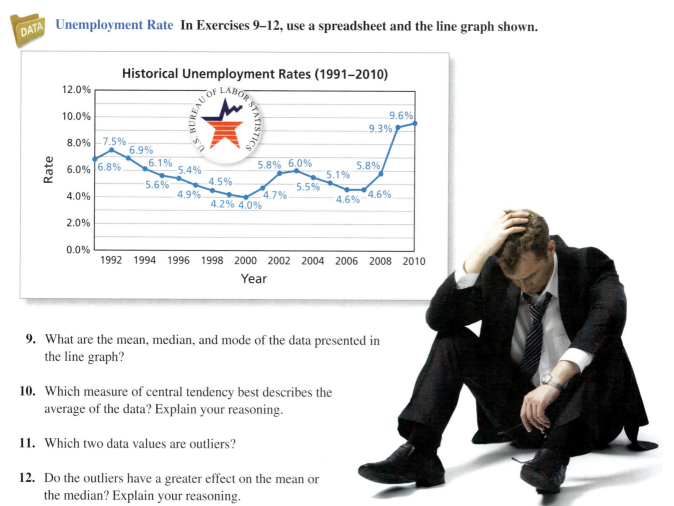

DATA **Unemployment Rate** In Exercises 9–12, use a spreadsheet and the line graph shown.

Historical Unemployment Rates (1991–2010)

U.S. BUREAU OF LABOR STATISTICS

9.6%
9.3%
7.5%
6.9%
6.8%
6.1%
5.4%
5.8% 6.0%
5.6%
4.5%
5.1% 5.8%
4.9%
4.7%
5.5%
4.2% 4.0%
4.6% 4.6%

Rate

Year

9. What are the mean, median, and mode of the data presented in the line graph?

10. Which measure of central tendency best describes the average of the data? Explain your reasoning.

11. Which two data values are outliers?

12. Do the outliers have a greater effect on the mean or the median? Explain your reasoning.

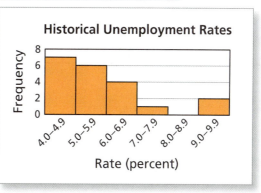

13. **Box-and-Whisker Plot** Use the box-and-whisker plot to analyze the unemployment rates.

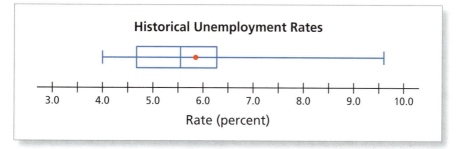

Historical Unemployment Rates

Rate (percent)

14. **Histogram** Explain whether the following statement is supported by the histogram: "The unemployment rate during most years is between 4.0% and 7.0%."

Historical Unemployment Rates

Frequency

Rate (percent)

4.0–4.9 5.0–5.9 6.0–6.9 7.0–7.9 8.0–8.9 9.0–9.9

Section 9.3

Hotel Rooms The histogram shows the distribution of the numbers of hotel rooms occupied for 150 days in a year. In Exercises 15–18, use the histogram.

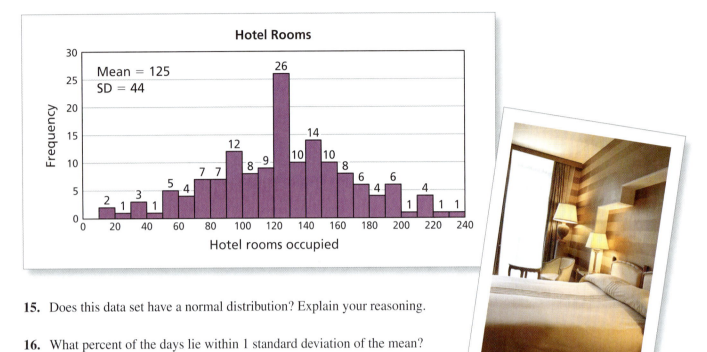

15. Does this data set have a normal distribution? Explain your reasoning.

16. What percent of the days lie within 1 standard deviation of the mean?

17. What percent of the days lie within 2 standard deviations of the mean?

18. Compare the percents in Exercises 16 and 17 with the percents given by the normal distribution.

Television The histograms show the distributions of samples of hours of television watched per day by men and women. In Exercises 19 and 20, use the histograms.

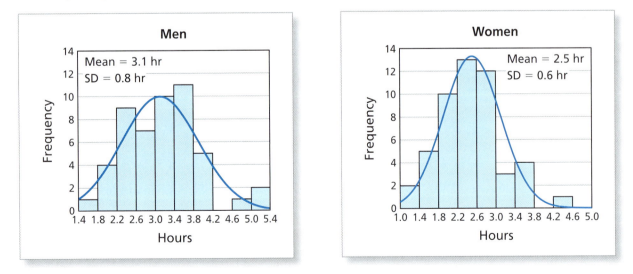

19. What is the significance of the smaller standard deviation for women?

20. Estimate the percents of men and women that watch at least 4 hours of television per day.

Section 9.4

Teacher Salaries The histogram shows the distribution of the salaries of 75 New York public school teachers. In Exercises 21 and 22, use the histogram.

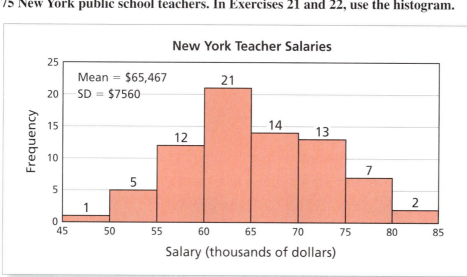

21. Use the *Confidence Interval Calculator* at *Math.andYou.com* and a 95% confidence level to estimate the population mean salary.

22. Repeat Exercise 21 for a 99% confidence level.

Biased Samples In Exercises 23 and 24, explain why the sample may be biased. Then explain how to find an unbiased sample.

23. A politically conservative magazine wants to estimate the percent of U.S. adults who approve of the job the president is doing. The magazine sends the survey to readers on its mailing list.

24. A pollster wants to determine the percent of registered voters who voted for the Republican candidate in an election. The pollster conducts an exit poll in a primarily Republican neighborhood.

25. Political Campaign You work for a political campaign. You want to estimate the percent of registered voters who plan to vote for your candidate. Use the *Sample Size Calculator* at *Math.andYou.com*. Choose a 90% confidence level and a 2% margin of error. How many people should you survey to obtain reliable results?

26. Biased Questions Determine which question could produce biased results. Explain.

> **a.** Do you agree with most of the students that the senator should resign?
>
>   ○ Agree   ○ Disagree
>
> **b.** Do you think the senator should resign?
>
>   ○ Yes   ○ No

10 The Mathematics of Fitness & Sports

10.1 Health & Fitness

▶ Compare a person's weight, height, and body fat percentage.

▶ Interpret and use a person's heart rate and metabolism.

▶ Determine factors for cardiovascular health.

10.2 The Olympics

▶ Analyze winning times and heights in the Summer Olympics.

▶ Analyze winning times in the Winter Olympics.

▶ Understand Olympic scoring.

10.3 Professional Sports

▶ Use mathematics to analyze baseball statistics.

▶ Use mathematics to analyze football statistics.

▶ Use mathematics to analyze statistics in other professional sports.

10.4 Outdoor Sports

▶ Use mathematics to analyze hiking and mountain climbing.

▶ Use mathematics to analyze kayaking and sailing.

▶ Use mathematics to analyze bicycling and cross-country skiing.

Calories Burned During Physical Activities

The graph in Example 6 on page 489 compares the number of calories burned per minute for eight activities. Why does cross-country skiing consume so much energy?

10.1 Health & Fitness

▶ Compare a person's weight, height, and body fat percentage.

▶ Interpret and use a person's heart rate and metabolism.

▶ Determine factors for cardiovascular health.

Comparing Weight, Height, and Fat Percentage

There are many different opinions about the "ideal" weight and body fat percentage for a person of a given height. Some of the opinions are expressed by tables, and some are expressed by formulas.

Study Tip

The Metropolitan Life Insurance Company *Height-Weight Tables* are available at *Math.andYou.com*. When using the tables, you are wearing shoes with 1-inch heels.

EXAMPLE 1   **Analyzing the MetLife Height-Weight Tables**

In a criticism of the MetLife tables, Steven B. Halls, MD, said, "For very tall men and women, the MetLife tables suggest impossibly low weights." As part of the criticism, Halls presented the following graphs. What do Halls's graphs show?

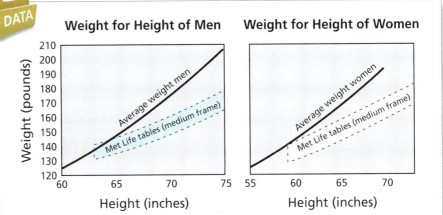

SOLUTION

The graphs are roughly linear. The rate of increase of each graph is about 5 pounds per inch of height, as indicated by the following formulas.

Medium-framed men: Weight = 125 pounds + 5 pounds per inch over 5 feet

Medium-framed women: Weight = 145 pounds + 5 pounds per inch over 5 feet

Compared to average weights, the MetLife tables recommend unrealistically low weights, especially for tall people.

✓ **Checkpoint**   Help at *Math.andYOU.com*

Here is another formula by Dr. G.J. Hamwi that first appeared in 1964. Graphically compare this formula with the average weights of Americans.

Medium-framed men: Weight = 106 pounds + 6 pounds per inch over 5 feet

Medium-framed women: Weight = 100 pounds + 5 pounds per inch over 5 feet

Ideal Body Weight calculators are available at *Math.andYou.com*.

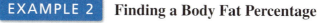

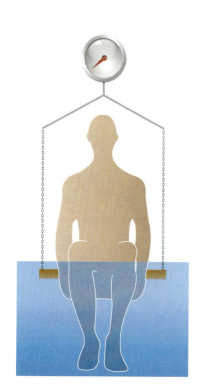

Weight: 145 lb
Wrist: 6.5 in.
Waist: 28 in.
Hips: 36 in.
Forearm: 10 in.

The formulas on this page only give estimates for body fat percentage. One way to find an accurate measurement is to find your weight in and out of water.

EXAMPLE 2   **Finding a Body Fat Percentage**

The human body is composed of many things including muscles, bones, internal organs, and fat. You can find your body fat percentage by dividing your body fat weight by your total body weight. You can use the following formulas to approximate your body fat percentage. (Use pounds and inches.)

Body Fat Formula for Women

Factor 1: (total body weight $\times$ 0.732) + 8.987
Factor 2: wrist circumference (at fullest point)/3.140
Factor 3: waist circumference (at naval) $\times$ 0.157
Factor 4: hip circumference (at fullest point) $\times$ 0.249
Factor 5: forearm circumference (at fullest point) $\times$ 0.434

Lean body mass = factor 1 + factor 2 − factor 3 − factor 4 + factor 5
Body fat weight = total body weight − lean body mass
Body fat percentage = (body fat weight $\times$ 100)/total body weight

Body Fat Formula for Men

Factor 1: (total body weight $\times$ 1.082) + 94.42
Factor 2: waist circumference $\times$ 4.15

Lean body mass = factor 1 − factor 2
Body fat weight = total body weight − lean body mass
Body fat percentage = (body fat weight $\times$ 100)/total body weight

Find the body fat percentage for the woman shown.

SOLUTION

Using a spreadsheet, you can see that the woman's body fat percentage is about 25.4%.

DATA

	A	B	C
1			**Factor**
2	Weight	145	115.13
3	Wrist	6.5	2.07
4	Waist	28	4.40
5	Hips	36	8.96
6	Forearm	10	4.34
7			
8	Lean body mass		108.18
9	Body fat weight		36.82
10	Body fat percentage		25.4%

✓ **Checkpoint**   Help at *Math*.and**YOU**.com

A study was conducted to determine the relationship between body fat percentage (as a percent) and body mass index (BMI), taking age and gender (males = 1, females = 0) into account. The study consisted of 521 males and 708 females, with wide ranges in BMI and age. The researchers determined that the formula for adults is

BMI = (0.83 $\times$ body fat %) − (0.19 $\times$ age) + (9 $\times$ gender) + 4.5.

Find the BMI for the 24-year-old woman in Example 2 and for a 26-year-old man who weighs 210 pounds and has a 36-inch waist.

Heart Rate Levels for a 20-Year-Old

200	**MAXIMUM**
180	
160	
140	**HIGH TARGET**
	LOW TARGET
120	**ABOVE NORMAL**
100	
80	**AT REST**
60	
40	**BELOW NORMAL**
20	**DANGEROUSLY LOW**
0	

Heart Rate and Metabolism

Your **heart rate** is the number of times your heart beats in 1 minute. Your heart rate is lower when you are at rest and increases when you exercise because your body needs more oxygen-rich blood when you exercise. Here is the normal heart rate for a person at rest.

Ages 1–10: 70–120 beats per minute

Ages 11+: 60–100 beats per minute

EXAMPLE 3   **Finding a Target Heart Rate**

An estimate for your maximum heart rate is

Maximum heart rate (MHR) = 220 − (your age).

Many fitness specialists recommend staying within 60–80% of your maximum heart rate during exercise. This range is called your target heart rate zone. Create a table showing the target heart rate zones for different ages.

SOLUTION

DATA

	A	B	C	D
1	Age	Maximum Heart Rate	Low Target Rate	High Target Rate
2	20	200	120	160
3	30	190	114	152
4	40	180	108	144
5	50	170	102	136
6	60	160	96	128
7	70	150	90	120
8	80	140	84	112

For instance, when a 30 year-old exercises, his or her heart rate should be between 114 and 152 beats per minute.

✓ **Checkpoint**    Help at Math.andYOU.com

The 24/5 Complete Personal Training Manual suggests that there is a "fat burning zone." This zone is 60–65% of your maximum heart rate, as indicated in the following table.

Fat burning zone

Estimates for a 130-pound woman during exercise	60–65% MHR	80–85% MHR
Total calories expended per minute	4.86	6.86
Fat calories expended per minute	2.43	2.70

With more fat calories being burned at the higher rate, why is the 60–65% zone called the "fat burning zone"?

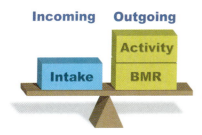

Incoming   Outgoing

If your incoming calories are equal to your outgoing calories, you should maintain your weight. To lose weight, decrease your incoming calories and/or increase your outgoing calories. To gain weight, do the opposite.

Your *basal metabolic rate* (BMR) is the number of calories you expend per day while in a state of rest. BMR decreases with age and increases with the gain of lean body mass.

To determine your BMR accurately, you need a fairly sophisticated test. However, there are several formulas for approximating your BMR. Here is one that was developed by Mark Mifflin and Sachiko St. Jeor in 1990.

Formula for Basal Metabolic Rate (BMR) in Calories Per Day

Factor 1: $4.545 \times$ weight (lb)
Factor 2: $15.875 \times$ height (in.)
Factor 3: $5 \times$ age (yr)
Factor 4: 5 for males and -161 for females

BMR = Factor 1 + Factor 2 − Factor 3 + Factor 4

EXAMPLE 4   **Using a Basal Metabolic Rate**

To determine your daily calorie needs, you must multiply your basal metabolic rate by a number determined by your activity level.

Weight: 198 lb
Height: 72 in.
Age: 32 yr

Outgoing Calories	
Activity level	**BMR multiplier**
Sedentary	1.200
Lightly active	1.375
Moderately active	1.550
Very active	1.725
Extra active	1.900

Incoming Calories	
Food	**Calories per gram**
Carbohydrate	4
Protein	4
Fat	9
Alcohol	7

Assume the man shown is very active and his daily calorie intake is about 3500 calories. Would you expect him to be losing weight or gaining weight? Explain your reasoning.

SOLUTION

DATA

	A	B	C
1			**Factor**
2	Weight	198	900
3	Height	72	1143
4	Age	32	160
5	Gender	M	5
6			
7	BMR		1888
8	Activity level	VA	1.725
9	Outgoing calories		3257

Incoming calories: 3500
Outgoing calories: −3257
Balance: 243

The man should be gaining weight.

 **Checkpoint**

 Help at *Math*.and**Y☺U**.com

Find your daily calorie balance.

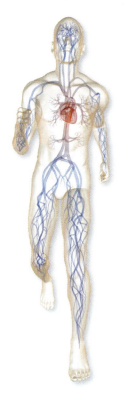

Your cardiovascular system consists of your heart and your blood vessels. Cardiovascular fitness refers to the ability of your heart, lungs, and other organs to transport and use oxygen. The maximum volume of oxygen your body can transport and use is called your *aerobic capacity*. When you exercise regularly, you can increase your cardiovascular fitness as your heart becomes more efficient at pumping blood and oxygen to the body, and as your body becomes more efficient at using that oxygen.

Cardiovascular Health

On page 454, you saw that the 60–65% target of your maximum heart rate is called the "fat burning zone." The 80–85% target is called the "cardio zone." Both of these target zones are classified as *aerobic exercise* because they involve increased oxygen to the lungs. *Anaerobic exercise* (without oxygen) includes activities like weight lifting, which are designed to increase muscle mass.

Cardio zone

	60–65% MHR	**80–85% MHR**
Total calories expended per minute	4.86	6.86
Fat calories expended per minute	2.43	2.70

EXAMPLE 5   Graphically Representing Health Claims

Graphically represent one of the statements.

- Cardiovascular disease accounts for about 34% of all deaths in America.
- Lack of physical activity is a risk factor for cardiovascular disease. About 49% of Americans 18 years old and older are not physically active.
- About 46% of male high school students and 28% of female high school students are physically active.
- Even low-to-moderate intensity activities, such as walking, when done for as little as 30 minutes a day, bring benefits.

SOLUTION

Here is one possibility.

Physically Active High School Students
- Physically active
- Not physically active

28% Females   46% Males

✓ **Checkpoint**   Help at *Math*.and**YOU**.com

Choose a different claim and represent it graphically.

Cholesterol is a fat-like substance that occurs naturally in all areas of the human body. Your body needs some cholesterol to help it work properly. About 75% of the cholesterol in your body is produced by your liver. The rest comes from foods like meats, eggs, and dairy products. The biggest influence on your blood cholesterol level is fats in your diet, not the amount of cholesterol in the food you eat.

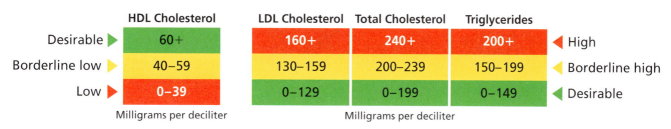

	HDL Cholesterol		LDL Cholesterol	Total Cholesterol	Triglycerides		
Desirable ▶	60+		160+	240+	200+	◀ High	
Borderline low ▶	40–59		130–159	200–239	150–199	◀ Borderline high	
Low ▶	0–39		0–129	0–199	0–149	◀ Desirable	

Milligrams per deciliter   Milligrams per deciliter

EXAMPLE 6   **Describing Insulin and Glucose Interaction**

Use the graph to describe the interaction between glucose and insulin in a typical daily diet.

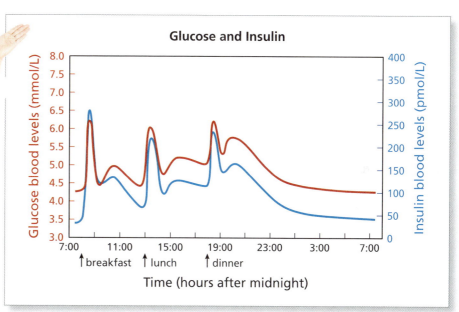

SOLUTION

Food that you eat is broken down into glucose, the simple sugar that is the main source of energy for your body's cells. But, your cells cannot use glucose without insulin, a hormone secreted by the pancreas. Insulin helps the cells take in glucose and convert it to energy.

Throughout the day, each time you eat, your pancreas receives a signal to produce insulin. The graph shows that the amount of insulin a healthy person produces is proportional to the amount of glucose in the blood.

✓ **Checkpoint**   Help at

Use the Internet to describe the relationship between cholesterol, glucose, and cardiovascular health.

Testing your cholesterol and glucose levels is a standard part of a physical examination by your doctor. Both of these tests require you to fast for several hours before the test.

10.1 Exercises

The Robinson Formula **In 1983, Dr. J.D. Robinson published a formula for ideal weight. In Exercises 1–6, use the formula.** *(See Example 1.)*

1. Find the ideal weight of a woman who is 5 feet 2 inches tall.

2. Find the ideal weight of a man who is 5 feet 3 inches tall.

3. Find the ideal weight of a man who is 6 feet tall.

4. Find the ideal weight of a woman who is 6 feet 4 inches tall.

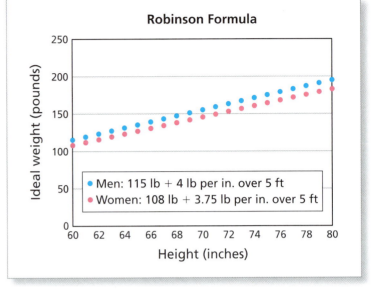

Robinson Formula

Men: 115 lb + 4 lb per in. over 5 ft
Women: 108 lb + 3.75 lb per in. over 5 ft

5. How do you think you should calculate the ideal weight of a person who is shorter than 5 feet tall? Explain your reasoning. Then find the ideal weight of a woman who is 4 feet 8 inches tall.

6. Compare the ideal weights for men and women given by the Robinson formula.

DATA **Body Fat Percentage** **In Exercises 7 and 8, find (a) the body fat percentage and (b) the body mass index for the person shown.** *(See Example 2.)*

7.

8.

Weight: 160 lb
Waist: 34 in.
Age: 21 yr

Weight: 125 lb
Wrist: 4.5 in.
Waist: 24 in.
Hips: 30 in.
Forearm: 7 in.
Age: 18 yr

Heart Rate Zones In Exercises 9–12, use the graph. *(See Example 3.)*

9. Describe any trends in the graph.

10. Is a 25-year-old with a heart rate of 150 beats per minute in the target heart rate zone?

11. Is a 42-year-old with a heart rate of 130 beats per minute in the fat burning zone?

12. Is a 40-year-old with a heart rate of 110 beats per minute in the fat burning zone?

Daily Calorie Balance In Exercises 13 and 14, determine whether you would expect the person to be losing weight or gaining weight. Explain your reasoning. *(See Example 4.)*

13.

Weight: 121 lb
Height: 64 in.
Age: 24 yr
Activity level:
    moderately active
Daily calorie intake:
    1500 calories

14.

Weight: 154 lb
Height: 68 in.
Age: 26 yr
Activity level:
    lightly active
Daily calorie intake:
    3000 calories

Graphical Representation In Exercises 15–18, graphically represent the statement. *(See Example 5.)*

15. About 15% of adults 20 years old and older have total cholesterol greater than or equal to 240 milligrams per deciliter.

16. About 13% of males 20 years old and older and 16% of females 20 years old and older have total cholesterol greater than or equal to 240 milligrams per deciliter.

17. The average total cholesterol of adults 20 years old and older is 198 milligrams per deciliter.

18. The average total cholesterol of males 20 years old and older is 195 milligrams per deciliter, and the average total cholesterol of females 20 years old and older is 200 milligrams per deciliter.

Cholesterol and Coronary Heart Disease The graph shows the results from a study. In Exercises 19–22, use the graph. *(See Example 6.)*

19. Describe any trends in the graph.

20. Compare the information in the graph to the information about total cholesterol in the chart on page 457.

21. What is the 6-year mortality rate for men with total cholesterols between 182 and 192 milligrams per deciliter?

22. How many times greater is the risk of dying from coronary heart disease for a man with total cholesterol greater than or equal to 264 milligrams per deciliter than for a man with total cholesterol less than or equal to 167 milligrams per deciliter?

▶ Extending Concepts

 The Katch-McArdle Formula The Katch-McArdle formula for basal metabolic rate applies to both men and women. In Exercises 23–30, use the formula.

$$\text{Basal metabolic rate (calories per day)} = 370 + 9.8 \times \text{lean body mass (lb)}$$

23. Make a graph for the Katch-McArdle formula. Describe any trends in the graph.

24. How much does one additional pound of lean body mass increase a person's basal metabolic rate?

25. Find the basal metabolic rate for a person with 100 pounds of lean body mass.

26. Find the basal metabolic rate for a person with 130 pounds of lean body mass.

27. Find the basal metabolic rate for a person who weighs 160 pounds and has a body fat percentage of 25%.

28. Find the basal metabolic rate for a person who weighs 175 pounds and has a body fat percentage of 20%.

29. How many calories per day does this person need to take in to maintain his weight?

30. How many calories per day does this person need to take in to maintain her weight?

Weight: 180 lb
Waist: 36 in.
Activity level: sedentary

Weight: 130 lb
Wrist: 4.5 in.
Waist: 26 in.
Hips: 33 in.
Forearm: 8 in.
Activity level: extra active

10.2 The Olympics

▶ Analyze winning times and heights in the Summer Olympics.

▶ Analyze winning times in the Winter Olympics.

▶ Understand Olympic scoring.

The Summer Olympics

The Summer Olympics are held every 4 years. The games feature thousands of athletes with over 30 sports and about 400 different events.

| EXAMPLE 1 | **Analyzing Winning Times** |

The graph shows the winning times for the men's 200-meter freestyle swimming event from 1968 through 2008. Describe the pattern in the graph.

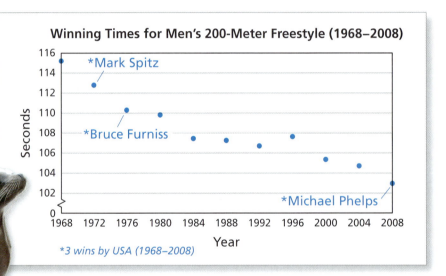

Winning Times for Men's 200-Meter Freestyle (1968–2008)

*Mark Spitz

*Bruce Furniss

*Michael Phelps

*3 wins by USA (1968–2008)

SOLUTION

Except for 1996, the winning time has decreased each year. The pattern for the times is roughly linear. At the 2008 Olympics, the use of full-body suits coincided with many new records, which caused some controversy.

✓ **Checkpoint**

Help at *Math*.and**YOU**.com

Sketch a graph of the winning times (in seconds) for the women's 200-meter freestyle swimming event from 1968 through 2008. Describe any patterns in the graph.

DATA (1968, 130.50), (1972, 123.56), (1976, 119.26), (1980, 118.33), (1984, 119.23), (1988, 117.65), (1992, 117.90), (1996, 118.16), (2000, 118.24), (2004, 118.03), (2008, 114.82)

At the 2008 Olympics, Michael Phelps's speed in the 200-meter freestyle swimming event was about 4.3 miles per hour. A California sea lion, which is larger than a human, can swim up to 25 miles per hour.

EXAMPLE 2   **Analyzing Winning Heights**

Pole vaulting has been an event at the Olympic Games since 1896 for men and since 2000 for women. Early Olympians used solid wood or bamboo poles. Describe the pattern of the winning heights for men's pole vaulting from 1896 through 2008. When do you think fiberglass poles were first used?

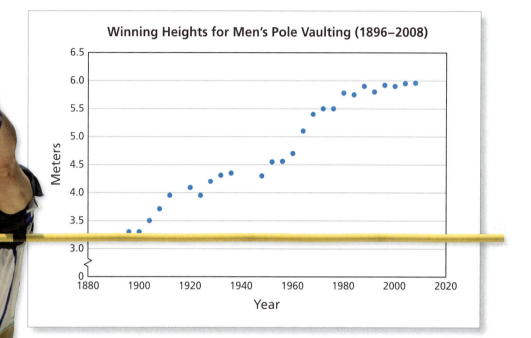

Winning Heights for Men's Pole Vaulting (1896–2008)

SOLUTION

The winning heights increased until around 1950 when it appeared as though they would level off at around 4.25 meters. Then in the 1950s, pole-vault technology improved with fiberglass poles. With that, the heights started rising again. Now they seem to have leveled off at around 6 meters.

Many years ago, pole vaulters went over the bar with their feet pointing downward. Today, they do a complicated gymnastic maneuver, turning upside down as the jump takes place.

 **Checkpoint**   Help at *Math*.and**YOU**.com

Today, high school students can pole vault at heights that would have broken Olympic records in 1950. How is this possible?

The Winter Olympics

Like the Summer Olympics, the Winter Olympics are also held every 4 years. The 2010 Winter Olympics were held in Vancouver, British Columbia.

EXAMPLE 3 **Analyzing Winning Times**

The graph shows the winning times for the men's 1500-meter speed skating event from 1924 through 2010. Describe the pattern in the graph.

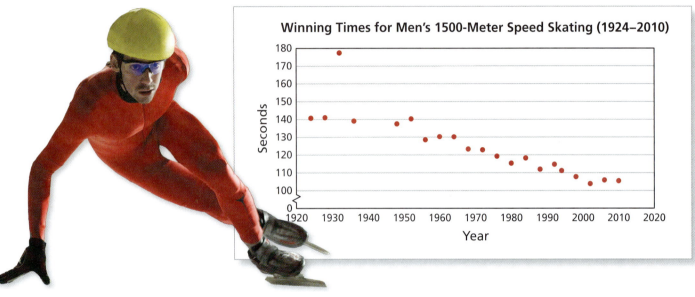

Winning Times for Men's 1500-Meter Speed Skating (1924–2010)

SOLUTION

One thing that stands out in the graph is that the 1932 winning time is an outlier. The 1500-meter speed skating event at the 2010 Olympics (held at Lake Placid) was not typical. Instead of racing against the clock individually, skaters raced against each other in small packs. This accounts for the time that was longer than the other winning times.

From 1948 through 2010, the winning time generally decreased. The biggest change in technology occurred in the 1998 Olympics when the "clap skate" was first allowed. It has allowed speed skaters to reach speeds that skaters on regular skates could not reach.

Regular Skate   Clap Skate

✓ **Checkpoint**

Help at *Math*.and**Y♥U**.com

DATA

Sketch a graph of the winning times (in seconds) for the women's 1500-meter speed skating event from 1960 through 2010. Describe any patterns in the graph.

(1960, 145.2), (1964, 142.6), (1968, 142.4), (1972, 140.85),
(1976, 136.58), (1980, 130.95), (1984, 123.42), (1988, 120.68),
(1992, 125.87), (1994, 122.19), (1998, 117.58), (2002, 114.02),
(2006, 115.27), (2010, 116.89)

| EXAMPLE 4 | **Analyzing Winning Times** |

Downhill skiing has been an event at the Winter Olympics since 1948. Before 1964, the skiers' times were measured to the nearest tenth of a second. Since 1964, the times have been measured to the nearest one-hundredth of a second. Describe the pattern of the winning times.

For downhill skiing, remember that the local skiing and weather conditions may affect skiers' times.

SOLUTION

From 1948 through 1976, the winning times decreased dramatically. Since 1968, the winning times seem to have clustered between 105 seconds and 120 seconds. The one time since 1968 that it was out of this range was 2002, when the winning time by Fritz Strobl (Austria) was only 99.13 seconds. To travel about 2 miles in 100 seconds, Strobl had an average speed of about

$$\frac{2 \text{ mi}}{100 \text{ sec}} = \frac{2 \text{ mi}}{100 \text{ sec}} \times \frac{60 \text{ sec}}{1 \text{ min}} \times \frac{60 \text{ min}}{1 \text{ hr}} = 72 \text{ mph.}$$

✓ **Checkpoint**

Help at *Math*.and**Y●U**.com

In 2010, the men's downhill skiing course had a vertical drop of 2799 feet. The course length was 1.929 miles. The time differences between the medalists were the closest in the history of the event at the Olympics. Graphically represent the top 15 times (shown in seconds).

Olympic Scoring

In both the Summer Olympics and the Winter Olympics, the scores for some events are based on subjective decisions by judges. Some examples are gymnastics, diving, figure skating, and ski jumping. Example 5 describes the judging system for diving.

EXAMPLE 5   Analyzing Olympic Scoring

For a women's diving event, a panel of seven judges evaluate a dive. Each judge awards a score between 0 and 10, as shown in the table. The degree of difficulty of the dive is 1.8. To determine the diver's final score, discard the two highest and two lowest scores awarded by the judges. Then add the remaining scores and multiply by the degree of difficulty. What is the diver's score? Is this a good score?

Judge	Russia	China	Mexico	Germany	Italy	Japan	Brazil
Score	7.5	8.0	6.5	8.5	7.0	7.5	7.0

SOLUTION

Of the 7 scores, discard 8.5, 8.0, 7.0, and 6.5.

7.5   ~~8.0~~   ~~6.5~~   ~~8.5~~   ~~7.0~~   7.5   7.0

       High   Low   High   Low

Add the remaining scores and multiply by the degree of difficulty.

$$1.8(7.5 + 7.5 + 7.0) = 1.8(22) = 39.6$$

The diver's score is 39.6. This is not particularly good.

✓ **Checkpoint**   Help at

Watch videos of five dives at *Math.andYou.com*. Award a score between 0 and 10 to each dive. (You must award a whole number or a half, as in 6.0 or 8.5. You cannot award a score like 8.3.) Use the following criteria when awarding a score.

- Rate the approach, the height above the diving board, and the acrobatics.
- Toes pointed = good   • Feet touching = good
- No splash = high points   • Ripples = lower points
- Diver straight up and down on entry into water = good

EXAMPLE 6   **Analyzing Olympic Scoring**

Olympic ski jumping scores are based on a point system that combines *style points* and *distance points*.

Style points: For style points, a jump is divided into three parts: flight, landing, and outrun. A panel of five judges evaluate the jump. Each judge awards a maximum of 20 points. The highest and lowest scores are discarded, and the remaining three scores are totaled.

Distance points: A "K-point" is marked in the landing area. When the K-point is at 120 meters, jumpers are awarded 60 points plus 1.8 points for each meter that they exceed the K-point, or 60 points minus 1.8 points for each meter that they are short of the K-point.

In the 2014 Winter Olympics, women will be allowed to compete in ski jumping for the first time.

Suppose the K-point is at 120 meters. Find the score for a ski jumper with the following style points and distance.

Style points: 17.0, 18.0, 18.5, 20.0, 19.0   Distance: 125 meters

SOLUTION

Style points: Discard the 20.0 and 17.0.   18.0 + 18.5 + 19.0 = 55.5 points
Distance points: 60 + 1.8(5) = 69 points
Total points: 55.5 + 69 = 124.5 points

 Checkpoint   Help at *Math*.and**Y☺U**.com

Sketch a graph of the winning points for Olympic ski jumping from 1932 through 2010. Describe any patterns in the graph.

(1932, 228.1), (1936, 232.0), (1948, 228.1), (1952, 226.0), (1956, 227.0),
(1960, 227.2), (1964, 230.7), (1968, 231.3), (1972, 219.9), (1976, 234.8),
(1980, 271.0), (1984, 231.2), (1988, 224.0), (1992, 239.5), (1994, 274.5),
(1998, 272.3), (2002, 281.4), (2006, 276.9), (2010, 283.6)

10.2 Exercises

Men's Discus Throw In Exercises 1–4, use the graph. *(See Examples 1 and 2.)*

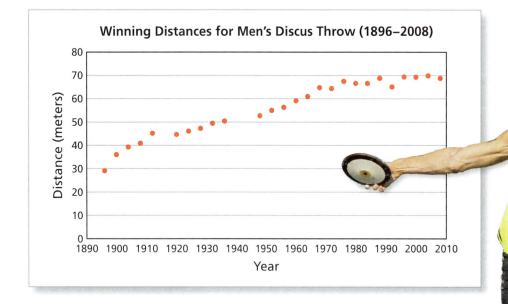

Winning Distances for Men's Discus Throw (1896–2008)

1. Describe any patterns in the graph.

2. The concrete throwing circle was introduced to the Olympic discus throw in 1954. Discuss the impact of the concrete throwing circle on the winning distances.

3. The Olympics were not held in 1940 or 1944 because of World War II. Based on the graph, what do you think the winning distance would have been in 1940 if the Olympics had been held? Explain your reasoning.

4. What is a reasonable expectation for the winning distance in the 2012 Olympics? Explain your reasoning.

DATA **Women's Discus Throw** The winning distances for the women's discus throw are shown. In Exercises 5 and 6, use the data. *(See Example 1.)*

(1928, 39.62), (1932, 40.58), (1936, 47.63), (1948, 41.92), (1952, 51.42), (1956, 53.69), (1960, 55.10), (1964, 57.27), (1968, 58.28), (1972, 66.62), (1976, 69.00), (1980, 69.96), (1984, 65.36), (1988, 72.30), (1992, 70.06), (1996, 69.66), (2000, 68.40), (2004, 67.02), (2008, 64.74)

5. Sketch a graph of the winning distances. Describe any patterns in the graph.

6. Sketch a graph that shows the Olympic record for each of the years in the data set.

Women's 1000-Meter Speed Skating In Exercises 7–10, use the graph. *(See Examples 3 and 4.)*

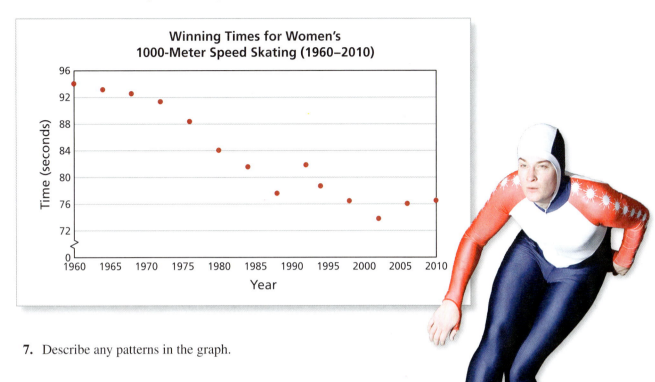

**Winning Times for Women's
1000-Meter Speed Skating (1960–2010)**

7. Describe any patterns in the graph.

8. What is the Olympic record for the women's 1000-meter speed skating event?

9. What percent of the winning times were Olympic records?

 10. Sketch a graph that shows the Olympic record for each year in the data set.

 11. **Finishing Times** Graphically represent the top 12 times for the 2010 Olympic women's 1000-meter speed skating event (shown in seconds). *(See Example 4.)*

76.56,   76.58,   76.72,   76.78,   76.80,   76.94,

77.08,   77.28,   77.37,   77.43,   77.46,   77.53

 12. **Men's 1000-Meter Speed Skating** The winning times (in seconds) for the men's 1000-meter speed skating event are shown. Sketch a graph and describe any patterns. *(See Example 3.)*

(1976, 79.32), (1980, 75.18), (1984, 75.80), (1988, 73.03), (1992, 74.85),
(1994, 72.43), (1998, 70.64), (2002, 67.18), (2006, 68.90), (2010, 68.94)

Diving Scores In Exercises 13–16, find the diver's score. *(See Example 5.)*

13. Degree of difficulty: 1.7

   Judges' scores:

Judge	Russia	China	Mexico	Germany	Italy	Japan	Brazil
Score	7.0	8.5	8.5	9.0	9.5	9.0	8.5

14. Degree of difficulty: 3.6

   Judges' scores:

Judge	Russia	China	Mexico	Germany	Italy	Japan	Brazil
Score	7.0	7.5	7.0	8.5	7.5	8.0	6.5

15. Degree of difficulty: 2.4

   Judges' scores:

Judge	Russia	China	Mexico	Germany	Italy	Japan	Brazil
Score	9.0	7.5	8.0	8.5	6.5	7.0	7.5

16. Degree of difficulty: 2.6

   Judges' scores:

Judge	Russia	China	Mexico	Germany	Italy	Japan	Brazil
Score	8.0	9.5	9.5	9.0	8.5	9.0	10.0

Ski Jumping Scores In Exercises 17–20, find the score for a ski jumper with the given style points and distance. Assume the K-point is at 120 meters. *(See Example 6.)*

17. Style Points: 17.0, 18.5, 17.5, 17.5, 19.0   Distance: 124 meters

18. Style Points: 18.0, 19.0, 20.0, 19.5, 19.5   Distance: 130 meters

19. Style Points: 16.5, 17.0, 18.5, 17.0, 19.0   Distance: 110 meters

20. Style Points: 18.0, 19.5, 19.5, 20.0, 19.5   Distance: 115 meters

▶ Extending Concepts

Men's Super Combined In the super combined, skiers compete in two events. The finishing times for the events are added together to determine the skier with the fastest combined time. The graphs show the finishing times for both events for the skiers with the top 6 combined times in 2010. In Exercises 21–23, use the graphs.

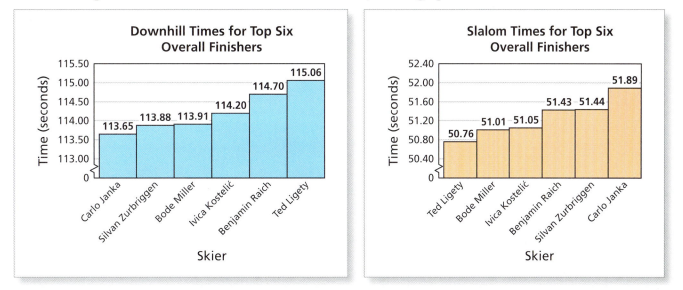

21. Find the combined time for each skier and make a bar graph showing the combined times from least to greatest. Who won the gold medal? the silver medal? the bronze medal?

22. Determine how many seconds the gold medalist finished ahead of each of the other skiers in the top six. Then graphically represent the information.

23. Aksel Lund Svindal had the fastest finishing time in the downhill portion at 113.15 seconds, but he skied off course in the slalom and did not finish that portion. What finishing time in the slalom would have won him the gold medal? Compare this time to the finishing times in the slalom graph.

Women's Super Combined The table shows the finishing times for the 2 events in the women's super combined for the skiers with the top 6 combined times in 2010. In Exercises 24–26, use the table.

24. Find the combined time for each skier and make a bar graph showing the combined times from least to greatest. Who won the gold medal? the silver medal? the bronze medal?

25. Determine how many seconds the gold medalist finished ahead of each of the other skiers in the top six. Then graphically represent the information.

26. Lindsey Vonn had the fastest finishing time in the downhill portion at 84.16 seconds, but she missed a gate and fell in the slalom and did not finish that portion. What finishing time in the slalom would have won her the gold medal? Compare this time to the slalom finishing times for the top six overall finishers.

Times for Top Six Overall Finishers in Women's Super Combined, 2010		
	Time (seconds)	
Skier	**Downhill**	**Slalom**
Julia Mancuso	84.96	45.12
Tina Maze	85.97	44.56
Anja Pärson	85.57	44.62
Maria Riesch	84.49	44.65
Fabienne Suter	85.29	45.56
Kathrin Zettel	86.01	44.49

10.1–10.2   Quiz

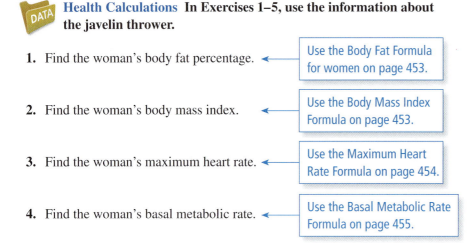

Health Calculations In Exercises 1–5, use the information about the javelin thrower.

1. Find the woman's body fat percentage. ← Use the Body Fat Formula for women on page 453.

2. Find the woman's body mass index. ← Use the Body Mass Index Formula on page 453.

3. Find the woman's maximum heart rate. ← Use the Maximum Heart Rate Formula on page 454.

4. Find the woman's basal metabolic rate. ← Use the Basal Metabolic Rate Formula on page 455.

5. The woman is extra active and has a daily calorie intake of 2400 calories. Would you expect the woman to be losing weight or gaining weight? Explain your reasoning.

Women's Javelin Throw In Exercises 6–8, use the graph.

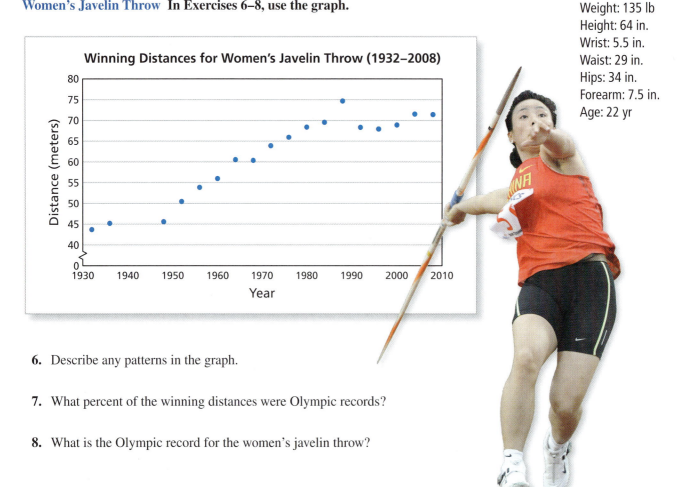

Weight: 135 lb
Height: 64 in.
Wrist: 5.5 in.
Waist: 29 in.
Hips: 34 in.
Forearm: 7.5 in.
Age: 22 yr

6. Describe any patterns in the graph.

7. What percent of the winning distances were Olympic records?

8. What is the Olympic record for the women's javelin throw?

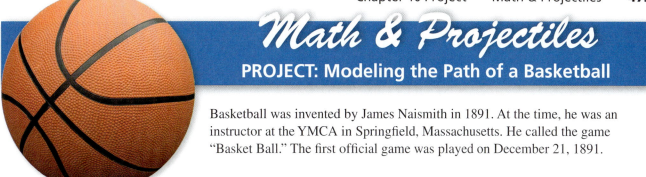

Math & Projectiles
PROJECT: Modeling the Path of a Basketball

Basketball was invented by James Naismith in 1891. At the time, he was an instructor at the YMCA in Springfield, Massachusetts. He called the game "Basket Ball." The first official game was played on December 21, 1891.

1. The path that a basketball takes depends on several things. Here are three of them.

   • The height from which the basketball is tossed
   • The angle at which the basketball is tossed (as measured from the horizontal)
   • The initial speed of the basketball

   Use the *Basketball Simulator* at *Math.andYou.com* to find a height, angle, and speed that successfully make a basket. Set the distance at 25 feet.

Angle: 45 (degrees)   Speed: 5 (ft/sec)   Height: 9 (feet)   Distance: 25 (feet)   **Throw**

2. You are 25 feet from the basket and release the ball from a height of 9 feet. Is it possible to make a basket by shooting with an initial speed of 5 feet per second? Use the simulator to verify your answer.

3. You are 25 feet from the basket and release the ball from a height of only 7 feet. Is it possible to make a basket by shooting with an initial speed of 5 feet per second? Use the simulator to verify your answer. Compare your answer with your answer in Exercise 2.

4. Is it true that "the greater the speed, the smaller the angle you should throw the ball"? Explain your answer.

10.3 Professional Sports

▶ Use mathematics to analyze baseball statistics.

▶ Use mathematics to analyze football statistics.

▶ Use mathematics to analyze statistics in other professional sports.

Baseball Statistics

The statistics of baseball seem to have attracted more interest than the statistics of any other professional sport. There are dozens of different types of statistics, including batting, pitching, fielding, and baserunning statistics.

EXAMPLE 1   Analyzing Batting Statistics

The batting average of a *player* or of a team is the ratio of the number of hits to the number of "at bats." The earned run average (ERA) is a measure of a *pitcher's* performance obtained by dividing the total number of earned runs allowed by the total number of innings pitched, and then multiplying by nine.

The scatter plot compares the batting averages and the ERAs of the American League and the National League from 2000 through 2010. What can you conclude?

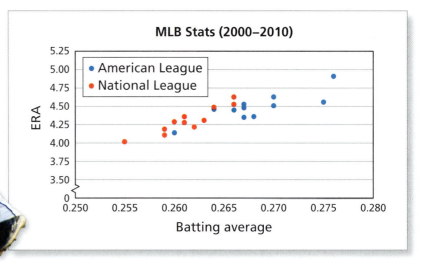

SOLUTION

The American League had better (higher) batting averages. The National League had better (lower) earned run averages.

✓ Checkpoint

Help at *Math*.and**YOU**.com

It is possible for 1 player to have a higher batting average than another player 2 years in a row, but to have a lower batting average when the 2 years are combined. Here is an example. How can you explain this?

	1995		1996		Combined	
Derek Jeter	12/48	.250	183/582	.314	195/630	.310
David Justice	104/411	.253	45/140	.321	149/551	.270

| EXAMPLE 2 | **Analyzing the Strike Zone** |

The strike zone is an imaginary box above home plate. The dimensions of the strike zone vary according to the batter's height and stance. It is defined as "that area over home plate the upper limit of which is a horizontal line at the midpoint between the top of the shoulders and the top of the uniform pants, and the lower level is a line at the hollow beneath the kneecap."

In his book *The Science of Hitting*, Ted Williams figured that for a man of his height, the official strike zone would be 7 baseballs wide by 11 baseballs high, as shown. The numbers inside the circles are Williams's estimates of his batting average on balls pitched in that part of the strike zone.

Which part of the strike zone did Williams consider his weakest? Explain your reasoning.

Ted Williams was voted the American League Most Valuable Player twice. He led the league in batting 6 times and had a career batting average of .344.

SOLUTION

Ted Williams was a left-handed batter. The lowest batting averages are in the lower right corner of the diagram. So, Williams believed that his weakest area was farthest from his body and down, as shown.

This might not be true for all batters, but Williams must have believed that when he had to extend his arms downward, he was not able to obtain the power he wanted.

Weakest area

✓ **Checkpoint**

Help at *Math*.and**Y☻U**.com

Which part of the strike zone did Williams consider his strongest? Explain.

Football Statistics

EXAMPLE 3 Calculating Quarterback Ratings

The NFL uses a complicated formula to rate a quarterback's passing efficiency. It involves the following four categories.

a. Completion percentage: $a = 5\left(\dfrac{\text{completions}}{\text{attempts}} - 0.3\right)$

b. Average yards per attempt: $b = 0.25\left(\dfrac{\text{yards}}{\text{attempts}} - 3\right)$

c. Touchdown pass percentage: $c = 20\left(\dfrac{\text{touchdowns}}{\text{attempts}}\right)$

d. Interception percentage: $d = 2.375 - 25\left(\dfrac{\text{interceptions}}{\text{attempts}}\right)$

If the result of any of these is greater than 2.375, the result is lowered to 2.375. If the result is negative, it is raised to 0. The quarterback's passer rating is

$$\text{Passer rating} = 100\left(\frac{a + b + c + d}{6}\right).$$

On November 25, 2010, Tom Brady completed a game with a perfect passer rating. He had 21 completions, 27 attempts, 341 yards, 4 touchdowns, and 0 interceptions. What was Brady's passer rating in that game?

SOLUTION

$a = 5\left(\dfrac{\text{completions}}{\text{attempts}} - 0.3\right) = 5\left(\dfrac{21}{27} - 0.3\right) \approx 2.39$ Lowered to 2.375.

$b = 0.25\left(\dfrac{\text{yards}}{\text{attempts}} - 3\right) = 0.25\left(\dfrac{341}{27} - 3\right) \approx 2.41$ Lowered to 2.375.

$c = 20\left(\dfrac{\text{touchdowns}}{\text{attempts}}\right) = 20\left(\dfrac{4}{27}\right) \approx 2.96$ Lowered to 2.375.

$d = 2.375 - 25\left(\dfrac{\text{interceptions}}{\text{attempts}}\right) = 2.375 - 25(0) = 2.375$

$\text{Passer rating} = 100\left(\dfrac{2.375 + 2.375 + 2.375 + 2.375}{6}\right) \approx 158.3$

So, Brady's rating was 158.3. This is the highest possible rating in this rating system. That is why it is called a perfect passer rating.

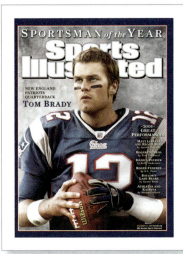

Tom Brady is a quarterback for the New England Patriots. As of 2011, he played in four Super Bowls, winning three of them. In 2005, he was named *Sports Illustrated*'s Sportsman of the Year.

✔ Checkpoint

Help at *Math*.and**Y☺U**.com

Which of the following had perfect passer ratings? Explain.

		CMP	ATT	YDS	TD	INT
Ben Roethlisberger	December 20, 2007	16	20	261	3	0
Kurt Warner	September 14, 2008	19	24	361	3	0
Eli Manning	October 11, 2009	8	10	173	2	0
Drew Brees	November 30, 2009	18	23	371	5	0

EXAMPLE 4   **Describing the Probability of Winning**

Statistics from NFL games have been analyzed to produce the following graph, which compares the probability of winning for the team with possession of the ball, the point difference, and the minutes remaining in the game. The graph is based on all of the regular season NFL games from 2000 through 2007.

At any given time in a football game, a team's probability of winning depends on its lead. When there is not much time left in the game, the team with the lead is likely to win even if it is only ahead by 7 points or less. Surprises do, however, occur. For instance, on November 11, 2010, the Atlanta Falcons scored a touchdown and beat the Baltimore Ravens with only 20 seconds left in the game.

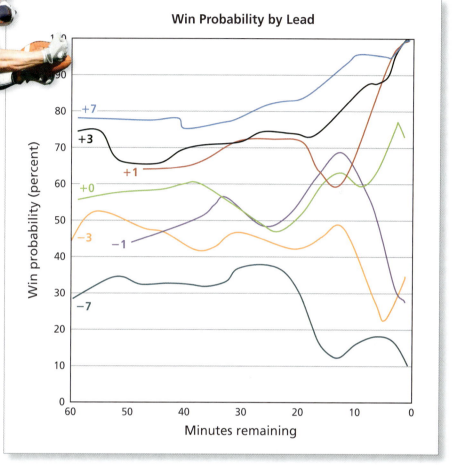

Win Probability by Lead

Describe the implications of the graph for a team that is 3 points ahead.

SOLUTION

A team that is 3 points ahead and has possession of the ball is likely to win, regardless of the number of minutes remaining.

- With 20 to 30 minutes remaining, the team's chance of winning is about 70%.

- With 10 minutes remaining, the team's chance of winning increases to about 85%.

- With 5 minutes remaining, the team's chance of winning is about 90%.

 **Checkpoint**    Help at *Math*.and**Y⊙U**.com

Suppose you are the coach of a football team. Explain how you could use the graph above to make decisions about how your team plays. Describe the different strategies you would use depending on the point difference between your team and your opponent.

Other Professional Sports

| EXAMPLE 5 | Analyzing U.S. Open Wins and Losses |

The graph shows the wins and losses by the "top four seeds" (players with the best rankings prior to the tournament) in the first three rounds of the U.S. Open Tennis Men's Singles Tournament. The **green line** shows straight set wins, the **red line** shows wins when they dropped one or two sets, and the **blue line** shows losses.

In men's tennis, a player must win 3 out of 5 sets. A "straight set win" means that the player won the first 3 sets. (In women's tennis, a player must win 2 out of 3 sets. A "straight set win" means that the player won the first 2 sets.)

In men's tennis, a player can drop 1 or 2 sets and still win the match. This can happen in 9 different orders: L-W-W-W, W-L-W-W, W-W-L-W, L-L-W-W-W, L-W-L-W-W, L-W-W-L-W, W-L-L-W-W, W-L-W-L-W, and W-W-L-L-W. (In women's tennis, a player can drop 1 set and still win the match. This can happen in only 2 orders: L-W-W and W-L-W.)

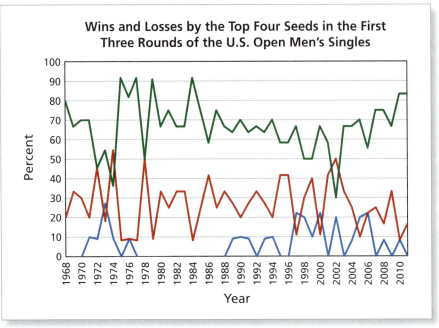

Wins and Losses by the Top Four Seeds in the First Three Rounds of the U.S. Open Men's Singles

What observations can you make from this graph?

SOLUTION

The most obvious observation is that the "seeding system" really works! The players who entered the U.S. Open as 1 of the top 4 ranked players were matched against players with lower rankings, and the top 4 seeds appeared to win easily. Another observation is that the green line always exceeded the red line, except in 1972, 1974, and 2002.

 **Checkpoint** Help at *Math*.and**Y⊕U**.com

a. Do the 3 lines in the graph always total 100%? Explain.

b. In the years in which there were no losses, are the green and red lines mirror images of each other? Explain.

Seeding is a system of ranking players in a tournament draw to avoid the highest-ranked players playing against each other in the early stages of the event.

EXAMPLE 6   **Analyzing Splits and Strikes**

The graphs show that the greater a bowler's skill, the tighter the range of locations where the bowling ball hits the pins.

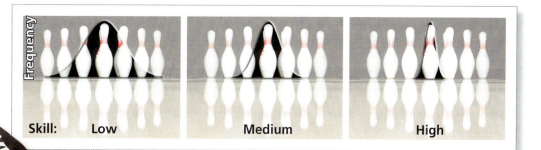

| Skill: | Low | Medium | High |

Professional bowlers almost always get 8, 9, or 10 pins down on their first ball. The box-and-whisker plots compare the number of splits a bowler gets to his or her first-ball average. What does this graph show?

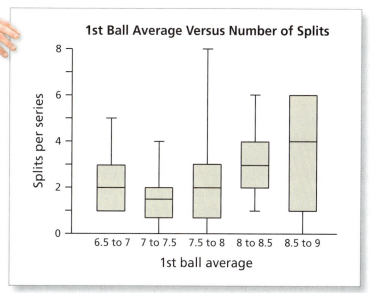

1st Ball Average Versus Number of Splits

SOLUTION

Overall, the box-and-whisker plots show that the median number of splits a bowler gets in a series (3 games) increases as his or her first-ball average increases. This makes sense because you cannot have a high pin average on the first ball unless you are hitting the head pin almost every time. The bad news is that when you hit the head pin, you have an increased likelihood of ending up with a split—a result in which the standing pins are "split" with a space between them.

✓ **Checkpoint**   Help at *Math*.and**YOU**.com

Bowling pins are numbered from 1 through 10. The "7-10 split" is considered to be the worst. List three other possible splits. Include a diagram with your answer.

As of 2010, Pete Weber won 35 Professional Bowlers Association (PBA) Tour events, including 8 major titles. His 35th tour win on April 4, 2010, secured him the position of 3rd place on the all-time PBA Tour titles list.

10.3 Exercises

DATA

Hitting Statistics The 2010 hitting statistics of 10 Major League Baseball players are shown. In Exercises 1–6, use a spreadsheet. *(See Examples 1 and 2.)*

PLAYER	TEAM	POS	G	AB	R	H	2B	3B	HR	RBI	TB	BB	SO	SB	CS	SF	HBP
M Cabrera	DET	1B	150	548	111	180	45	1	38	126	341	89	95	3	3	8	3
S Choo	CLE	OF	144	550	81	165	31	2	22	90	266	83	118	22	7	2	11
P Fielder	MIL	1B	161	578	94	151	25	0	32	83	272	114	138	1	0	1	21
J Hamilton	TEX	OF	133	518	95	186	40	3	32	100	328	43	95	8	1	4	5
J Mauer	MIN	C	137	510	88	167	43	1	9	75	239	65	53	1	4	6	3
J Morneau	MIN	1B	81	296	53	102	25	1	18	56	183	50	62	0	0	2	0
A Pujols	STL	1B	159	587	115	183	39	1	42	118	350	103	76	14	4	6	4
J Thome	MIN	DH	108	276	48	78	16	2	25	59	173	60	82	0	0	2	2
J Votto	CIN	1B	150	547	106	177	36	2	37	113	328	91	125	16	5	3	7
K Youkilis	BOS	1B	102	362	77	111	26	5	19	62	204	58	67	4	1	5	10

1. Which player had the greatest batting average (AVG) in 2010?

2. Slugging percentage (SLG) is a statistic used to measure the power of a hitter. The formula to calculate slugging percentage is shown. Find the slugging percentage for each player.

$$SLG = \frac{TB}{AB}$$

3. On-base percentage (OBP) is a statistic used to measure how often a player reaches base. The formula to calculate on-base percentage is shown. Find the on-base percentage for each player.

$$OBP = \frac{H + BB + HBP}{AB + BB + HBP + SF}$$

4. Considering AVG, SLG, and OBP, which player would you choose as the most valuable player (MVP)? Explain your reasoning.

5. Is it possible for a player to have an OBP that is less than his AVG? Explain your reasoning.

6. Albert Pujols had a stolen base percentage (SB%) of 77.8%. Write a formula using abbreviations to calculate SB%.

Abbreviations

2B	Doubles
3B	Triples
AB	At Bats
AVG	Batting Average
BB	Bases on Balls (Walks)
CS	Caught Stealing
G	Games Played
H	Hits
HBP	Hit by Pitch
HR	Home Runs
OBP	On-base Percentage
R	Runs Scored
RBI	Runs Batted In
SB%	Stolen Base Percentage
SB	Stolen Bases
SF	Sacrifice Flies
SLG	Slugging Percentage
SO	Strikeouts
TB	Total Bases
TPA	Total Plate Appearances

3rd Down The graphs show 3rd down statistics for all of the regular season NFL games from 2000 through 2007. Plays within 2 minutes of the end of a half and all plays within field goal range are excluded. **In Exercises 7–14, use the graphs.** *(See Examples 3 and 4.)*

7. Do NFL coaches call more pass plays or run plays on 3rd down? Explain your reasoning.

8. Describe the pattern in the "3rd Down Success" graph.

9. For what "to go" distances are the conversion rates for running and for passing about equal on 3rd down?

10. With 2 yards to go on 3rd down, should the defense expect a pass play or a run play? Explain your reasoning.

11. With 2 yards to go on 3rd down, should the coach call a pass play or a run play? Explain your reasoning.

12. How might a quarterback's passing efficiency influence a coach's play selection on 3rd down and 4 yards to go?

13. The optimum mix for running and passing occurs when the conversion rates are about equal. When this happens, the overall conversion rate will be greatest. Is the pass/run ratio at an optimum mix on 3rd down and 1 yard to go? Explain your reasoning.

14. Suppose you are the coach of an NFL team. Your team is facing a 3rd down with 5 yards to go. Would you call a pass play or a run play? Explain your reasoning.

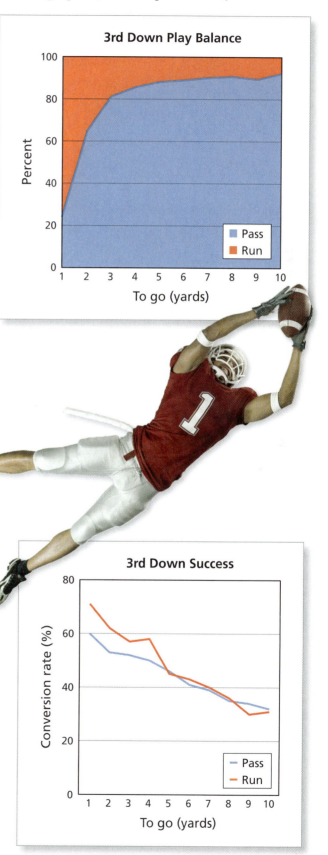

NBA Shot Chart The Dallas Mavericks hosted the Oklahoma City Thunder in game 1 of the Western Conference finals on May 17, 2011. The shot chart shows the final field goal data for Kevin Durant and Dirk Nowitzki, the top scorers on each team. In Exercises 15–21, use the shot chart. *(See Examples 5 and 6.)*

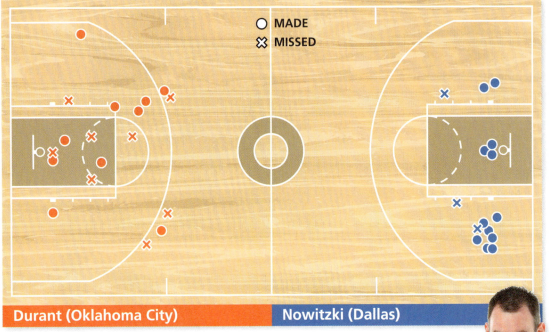

Durant (Oklahoma City)   **Nowitzki (Dallas)**

15. Which player attempted more field goals? Explain your reasoning.

16. Field goal percentage is the number of field goals made divided by the number of field goals attempted. Which player had a greater field goal percentage?

17. Nowitzki had one of the most efficient playoff games in NBA history. Not only did he shoot a high percentage from the floor, he also made 24 out of 24 free throws (an NBA playoff record). How many points did Nowitzki score?

18. The Thunder's final score was 112. Nowitzki accounted for 39.7% of the Mavericks' final score. Which team won the game? Explain your reasoning.

19. What was Durant's 3-point field goal percentage?

20. Durant accounted for 35.7% of the Thunder's 112 points. How many free throws did he make during the game?

21. What observations can you make about the variability of shot selection for each player?

22. **Long Range** Would you rather shoot 40% from 3-point range or 50% from inside the 3-point line? Explain your reasoning.

▶ Extending Concepts

Voting Habits of Sports Fans The bubble graph shows the results of a study about the voting habits of sports fans in the United States. In Exercises 23–28, use the graph.

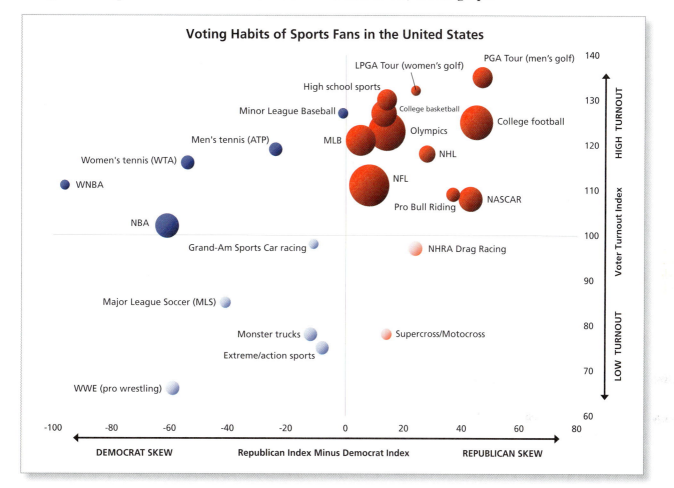

Voting Habits of Sports Fans in the United States

23. What do the size and color of a bubble represent in the graph?

24. Fans of which sport are most likely to vote? least likely to vote?

25. Are sports fans more likely to vote Republican or Democrat? Explain your reasoning.

26. How would you rate the voter turnout of sports fans?

27. Television ad buyers tend to focus on sporting events because sports fans usually watch sporting events live rather than on a DVR machine. This means viewers are unable to skip the ads. Which sporting events should ad buyers target to reach Republican voters? Democratic voters?

28. You are in charge of promoting a Democratic campaign. Would you rather advertise during an NBA game or a women's tennis (WTA) match? Explain your reasoning.

10.4 Outdoor Sports

▶ Use mathematics to analyze hiking and mountain climbing.

▶ Use mathematics to analyze kayaking and sailing.

▶ Use mathematics to analyze bicycling and cross-country skiing.

Hiking and Mountain Climbing

Hiking is an outdoor sport, often done on trails created by the National Forest Service or by a state agency. There are thousands of trails in the United States, such as Glacier Gorge Trail in Estes Park, Colorado, and the Appalachian Trail in the Eastern United States.

Negative Grade

Positive Grade

$$\text{Grade} = \frac{\text{Rise}}{\text{Run}}$$

EXAMPLE 1   Comparing Grade and Speed

The scatter plot shown was created by a hiker in California. The hiker used a GPS to compare the grade of the trail with his speed. Describe the scatter plot. What does the green line represent?

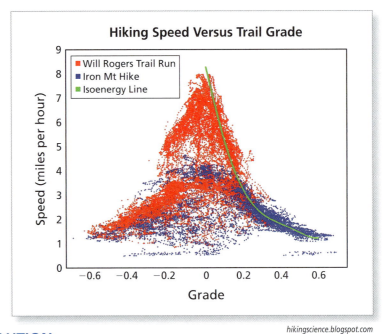

hikingscience.blogspot.com

SOLUTION

As a general observation, you can see that the hiker was traveling faster on level ground. As the grade changed to uphill or downhill, the hiker's speed decreased. In the Internet post about this experiment, the hiker called the green line his "isoenergy line" because he figured he was exerting the same amount of energy at every point on the line. For instance, walking at 2 miles per hour at a grade of 0.4 uses the same energy as running at 5 miles per hour at a grade of 0.1.

The American Hiking Society works "toward ensuring that hiking trails and natural places are cherished and preserved" for all generations.

✓ **Checkpoint**   Help at *Math*.and**YOU**.com

The suggestion for a casual hiking pace on a trail is 1 hour to walk 1.8 miles with a change in elevation of 0.2 mile. The suggestion for a fast hiking pace is 30 minutes. How do these speeds compare to the scatter plot shown in Example 1?

EXAMPLE 2   **Analyzing UV Radiation**

Sunlight consists of visible and invisible light. Ultraviolet (UV) light is invisible and is classified according to its wavelength, measured in nanometers (one-billionth of a meter). UV radiation is dangerous. It causes premature aging of the skin and can also cause various forms of skin cancer.

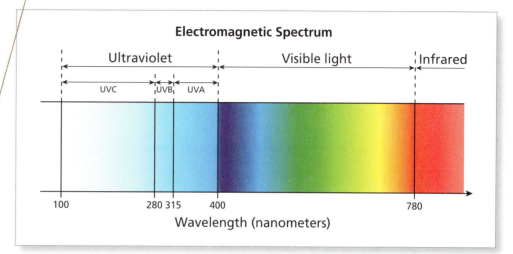

Electromagnetic Spectrum

Wavelength (nanometers)

UV radiation increases with elevation. It increases about 5% for every 1000 feet. Compare the UV radiation at the following elevations.

- Seattle, Washington (0 feet)
- Denver, Colorado (5000 feet)
- Mount McKinley, Alaska (20,000 feet)

SOLUTION

- For the sake of comparison, assume that the amount of UV radiation in Seattle is 1.

- Because Denver has an elevation of 5000 feet, the amount of UV radiation in Denver is

$$(1.05)^5 \approx 1.276 \qquad \text{5000 feet elevation}$$

  or about 28% more than the UV radiation in Seattle.

- The peak of Mount McKinley has an elevation of 20,000 feet. The amount of UV radiation near the peak is

$$(1.05)^{20} \approx 2.653 \qquad \text{20,000 feet elevation}$$

  or about 165% more than the UV radiation in Seattle.

Mountain climbers need special protection for their skin and eyes. They should wear goggles and sunscreen that block both UVA and UVB rays.

✓ **Checkpoint**

Help at *Math*.and**Y⊙U**.com

Compare the UV radiation at the following elevations.

- Reno, Nevada (4000 feet)
- Mount Whitney, California (14,000 feet)
- Mount Everest, Nepal (29,000 feet)

Kayaking and Sailing

A *kayak* is a boat in which a paddler faces forward, legs in front, using a double-bladed paddle. A *canoe*, on the other hand, is a boat in which a paddler faces forward and sits or kneels in the boat, using a single-bladed paddle.

EXAMPLE 3 **Analyzing a Graph**

The graph shows the discharge (amount of water) flowing through Rock Creek in Oregon. You want to kayak in Rock Creek only when the discharge is 50 cubic feet per second or greater. Estimate the percent of days from January 26 through May 26 that the creek meets your criteria.

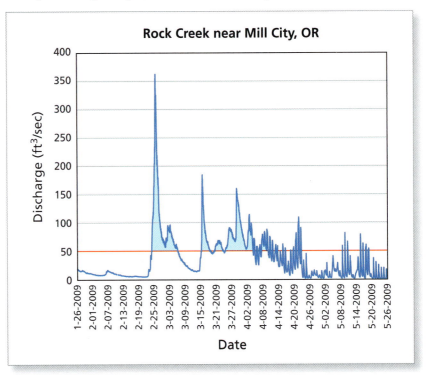

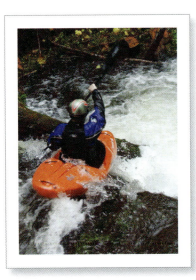

From a blog by a kayaker: "The East Fork of Rock Creek (aka *The Rock Creek* or simply 'The Rock') is a full-on, bare-knuckled brawler and one of my personal favorites. This creek has an outlandish gradient created by a non-stop series of ledges, falls, and boulder gardens that thunder down through Rock Creek Canyon."

SOLUTION

There are only 2 time periods when the creek has a discharge that is comfortably above 50 cubic feet per second. The first is about 11 days from February 23 through March 5. The second is about 23 days between March 14 and April 8. So, the percent of the days from January 26 through May 26 (121 days) that the creek water is high enough is

$$\frac{\text{days above 50 ft}^3/\text{sec}}{\text{total days}} = \frac{34}{121} \approx 0.281.$$

The creek meets your criteria about 28% of the time.

 **✓ Checkpoint** Help at *Math.andYOU.com*

You want to kayak in Rock Creek only when the discharge is 150 cubic feet per second or greater. Estimate the percent of days from January 26 through May 26 that the creek meets your criteria.

| EXAMPLE 4 | **Analyzing Sail Designs** |

The sail design of a modern yacht is quite different from the sail design used during most of the past 5000 years. Why is that?

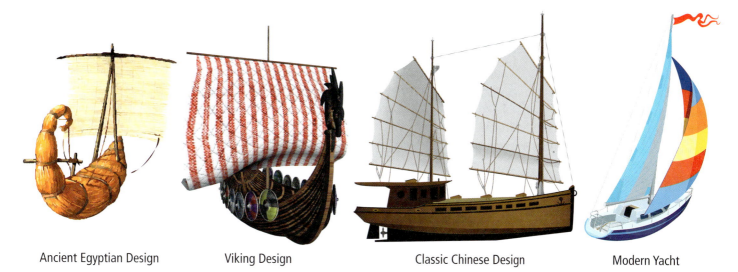

Ancient Egyptian Design   Viking Design   Classic Chinese Design   Modern Yacht

SOLUTION

The sail design of a modern yacht is different from the sail design used during most of the past 5000 years because of advancements in sail design. The following is a summary of some of the advancements.

To sail against the wind, a boat can sail at an angle to the wind and zigzag. This is called "tacking." Early sailboats with square sails were not very effective when sailing against the wind. They were most effective when sailing with the wind. The development of triangular sails enabled boats to sail against the wind more effectively. This was because the boats were able to tack at a smaller angle with the wind than with square sails.

When a triangular sail is positioned correctly, the air flow creates a pressure differential. This differential generates a force called the lift, which pulls a ship forward. A modern yacht like the one shown above has two sails—the mainsail and the jib. This design increases the pressure differential and generates more lift.

✓ **Checkpoint**

Help at *Math*.and**YOU**.com

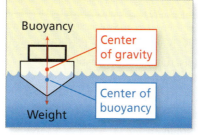

In an answer to the question, "Why doesn't a sailboat tip over?," the following explanation is given. "It's a balancing act between the boat's center of gravity and its center of buoyancy." Explain what this means.

Bicycling and Cross-Country Skiing

Bicycling is one of the most popular outdoor sports in the United States. People not only ride bicycles for exercise and for pleasure, but many commute to work by bicycle.

EXAMPLE 5   Describing a Bicycle Trip

The following graph was created by a cyclist using two devices: one that measured his heart rate and another that measured the elevation. Describe his cycling trip.

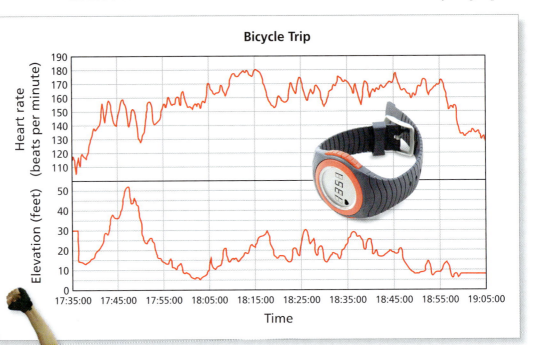

SOLUTION

- The trip lasted from 17:35 (5:35 P.M.) to 19:05 (7:05 P.M.), which is 90 minutes.
- The trip did not start and stop at the same elevation, so the person did not make a round trip—beginning and ending at the same location.
- The elevation is low, at times only 5 feet above sea level. So, the person must have been riding near the ocean. On the other hand, there was 1 point on the trip where the elevation was 50 feet above sea level.
- The person's heart rate was elevated during most of the trip. In fact, during much of the trip, his heart rate was above 160 beats per minute, which is the maximum target heart rate for cardio exercise for a 20-year-old (see page 454).

✓ Checkpoint

Help at

In a survey, bicyclists in the United States were asked why they ride bicycles. Graphically represent the results.

Recreation	26.0%	Visit friend/relative	10.1%
Exercise	23.6%	Go on a bicycle ride	2.3%
Commute to school/work	19.2%	Other	4.9%
Personal errand	13.9%		

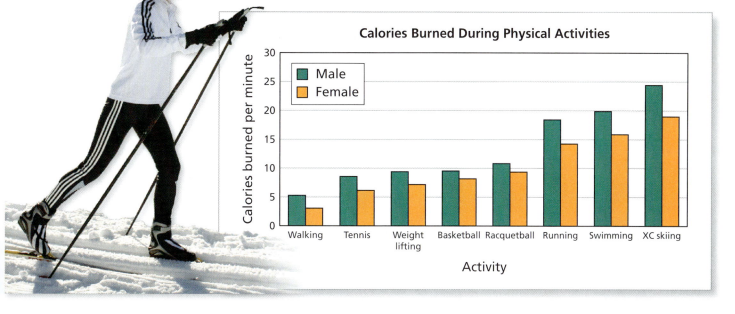

EXAMPLE 6   **Analyzing Energy Use in Outdoor Sports**

Cross-country (XC) skiing is one of the most demanding outdoor sports. The graph compares the number of calories burned per minute for eight activities.

Why does cross-country skiing consume so much energy?

SOLUTION

- Cross-country skiing is 95–100% dependent on aerobic energy output.

- Cross-country skiing consists of repeated contractions of arm and leg muscles.

- Of the total muscle mass of a human body, cycling uses 40%, running uses 60%, and cross-country skiing uses 80%.

- In spite of the fact that skis glide on snow, friction is still present.

- Cross-country skiing is a cold weather sport. Your body uses energy trying to keep you warm.

 **Checkpoint**   Help at *Math*.and**YOU**.com

The graph shows the oxygen uptake for U.S. and Swedish cross-country skiers. Why do you think the uptake is greater for Swedish athletes?

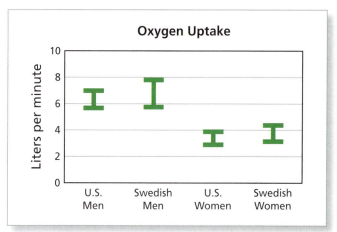

10.4 Exercises

Camping The graph shows the demographics of campers in the United States.
In Exercises 1–6, use the graph. *(See Examples 1 and 2.)*

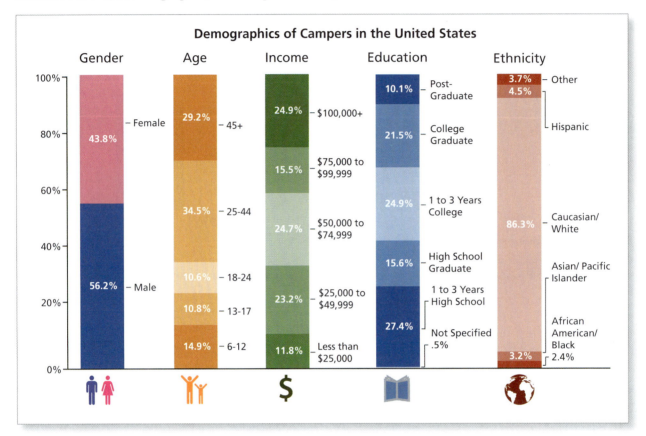

Demographics of Campers in the United States

Gender
- 43.8% — Female
- 56.2% — Male

Age
- 29.2% — 45+
- 34.5% — 25-44
- 10.6% — 18-24
- 10.8% — 13-17
- 14.9% — 6-12

Income
- 24.9% — $100,000+
- 15.5% — $75,000 to $99,999
- 24.7% — $50,000 to $74,999
- 23.2% — $25,000 to $49,999
- 11.8% — Less than $25,000

Education
- 10.1% — Post-Graduate
- 21.5% — College Graduate
- 24.9% — 1 to 3 Years College
- 15.6% — High School Graduate
- 1 to 3 Years High School
- 27.4%
- Not Specified .5%

Ethnicity
- 3.7% — Other
- 4.5% — Hispanic
- 86.3% — Caucasian/White
- Asian/Pacific Islander
- African American/Black
- 3.2%
- 2.4%

1. Is a camper more likely to be younger than 18 years old or older than 44 years old? Explain your reasoning.

2. What is the probability that a random camper is a male between the ages of 25 and 44?

3. The median household income in the United States is about $50,000. On average, do campers earn more or less than the median household income? Explain your reasoning.

4. The graph is based on a survey of 41,500 campers. How many of the campers in the survey are African American?

5. Can you use the graph to determine the likelihood that any random person participates in camping? Explain your reasoning.

6. Explain how the education bar is different from the other four bars in the graph.

Fishing In Exercises 7–13, use the diagram and the information below. *(See Examples 3 and 4.)*

The 28-foot-long fishing boat has twin turbo diesel engines and a 220-gallon fuel tank. The average cruising speed of the boat to its anchoring point is 20 knots. The horizontal distance between the fishing boat and its anchor is 138 feet. (*Note:* 1 knot = 1.15 mph)

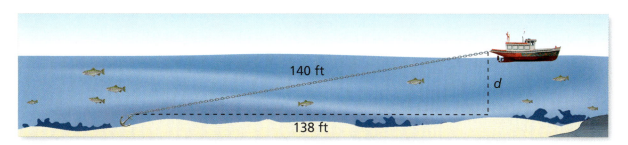

140 ft

d

138 ft

7. What is the average speed in miles per hour of the boat to its anchoring point?

8. The fishing boat takes 48 minutes to cruise straight to its anchoring point from the dock. How far from the dock is the anchoring point?

9. A general rule of thumb is to let out 7 to 10 feet of anchor line for every foot of water depth. The fishing boat in the diagram used this rule when it anchored. What is the range of the depth d of the water?

10. The anchor line ends with a section of chain. This chain provides extra weight and prevents jagged bottom rocks from cutting the anchor line. The recommended length of an anchor chain is one-half foot for every foot of boat length. What is the recommended length for the boat shown?

11. What is the maximum water surface area that the fishing boat can cover when the anchor is stationary?

12. Diesel fuel weighs about 7.1 pounds per gallon. How much does three-fourths of a tank of fuel weigh?

13. Diesel engines use about 1 gallon of fuel per hour for every 18 horsepower used. How many hours can the boat run at 342 horsepower on a half tank of fuel?

14. **Fuel Reserves** A general rule of thumb when fishing is to use one-third of the fuel in your tank to get there, use one-third to get back, and save the last third as an emergency backup. The fuel gauge shows the amount of fuel in a fishing boat after reaching its anchor point. Will the boat have enough fuel reserves according to the rule of thumb? Explain your reasoning.

Mountain Biking The graph shows the data for eight mountain bike trails.
In Exercises 15–20, use the graph. *(See Examples 5 and 6.)*

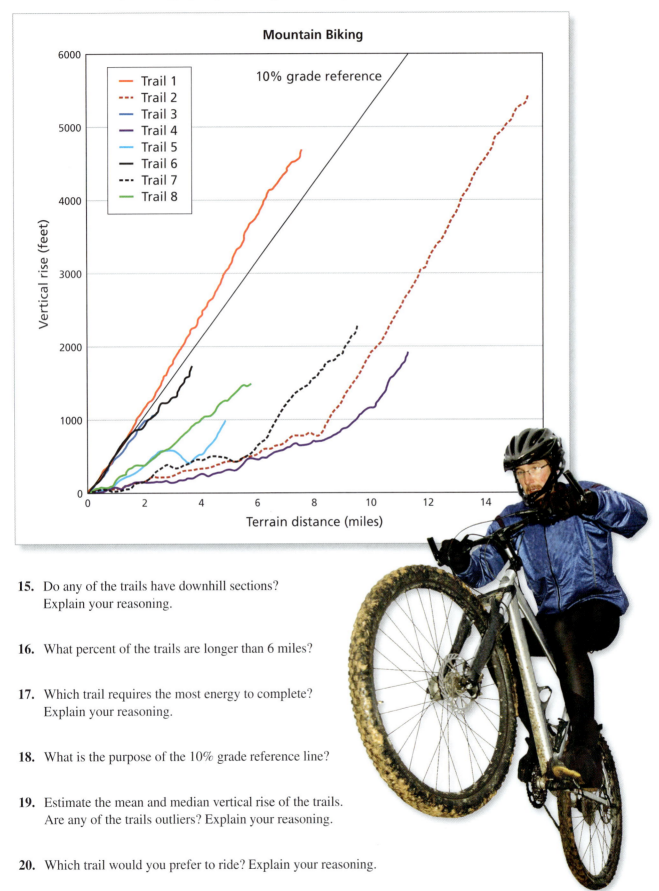

15. Do any of the trails have downhill sections?
    Explain your reasoning.

16. What percent of the trails are longer than 6 miles?

17. Which trail requires the most energy to complete?
    Explain your reasoning.

18. What is the purpose of the 10% grade reference line?

19. Estimate the mean and median vertical rise of the trails.
    Are any of the trails outliers? Explain your reasoning.

20. Which trail would you prefer to ride? Explain your reasoning.

▶ Extending Concepts

Heat Index In Exercises 21–26, use the heat index chart and the information below.

The heat index is the temperature the body feels when heat and humidity are combined. The chart shows the heat index that corresponds to the actual air temperature and relative humidity. This chart is based upon shady, light wind conditions. Exposure to direct sunlight can increase the heat index by up to 15°F. The National Weather Service will issue an excessive heat warning when the heat index is expected to exceed 105°F in the next 36 hours. From 2000 to 2009, heat killed more people in the United States than any other weather-related incident.

Heat Index
Temperature (°F)

Relative humidity (%)	80	82	84	86	88	90	92	94	96	98	100	102	104	106	108	110
40	80	81	83	85	88	91	94	97	101	105	109	114	119	124	130	136
45	80	82	84	87	89	93	96	100	104	109	114	119	124	130	137	
50	81	83	85	88	91	95	99	103	108	113	118	124	131	137		
55	81	84	86	89	93	97	101	106	112	117	124	130	137			
60	82	84	88	91	95	100	105	110	116	123	129	137				
65	82	85	89	93	98	103	108	114	121	126	130					
70	83	86	90	95	100	105	112	119	126	134						
75	84	88	92	97	103	109	116	124	132							
80	84	89	94	100	106	113	121	129								
85	85	90	96	102	110	117	126	135								
90	86	91	98	105	113	122	131									
95	86	93	100	108	117	127										
100	87	95	103	112	121	132										

With Prolonged Exposure and/or Physical Activity

Extreme Danger	Heat stroke or sunstroke highly likely
Danger	Sunstroke, muscle cramps, and/or heat exhaustion likely
Extreme Caution	Sunstroke, muscle cramps, and/or heat exhaustion possible
Caution	Fatigue possible

21. On average, lightning kills 48 people per year. Heat kills 237.5% more people each year than lightning. What is the annual fatality rate of heat?

22. On average, tornadoes kill 100 fewer people per year than heat. What is the annual fatality rate of tornadoes?

23. Suppose tomorrow's high temperature is predicted to be 92°F with a relative humidity of 80%. Should the National Weather Service issue an excessive heat warning? Explain your reasoning.

24. Suppose you are climbing the west side of a mountain on a sunny afternoon. The temperature is 88°F with a relative humidity of 60%. Should you be concerned about the heat index? Explain your reasoning.

25. Holding the relative humidity constant, does the heat index have a linear relationship with the temperature? Explain your reasoning.

26. How might the heat index affect the planning of a hiking trip?

10.3–10.4   Quiz

Competitive Balance The chart shows the average absolute change in win percentage for all the teams in a league for each season shown. For instance, the value for the NBA in 2007 is based on the average absolute change in win percentages between the 2006–2007 season and the 2007–2008 season. In Exercises 1–4, use the chart.

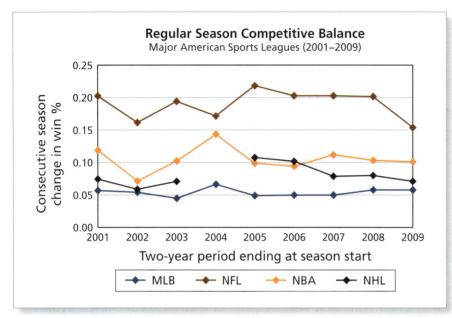

1. Identify and estimate any missing data points.

2. Which major sports league has the most competitive balance? Explain your reasoning.

3. In the 2009–2010 season, an NBA team won 50 out of 82 games. How many games would you expect the team to win in the 2010–2011 season?

4. Do you agree with the following comment? Use the chart to explain your reasoning.

    COMMENT:

    Bad baseball teams remain bad more often than bad football teams.

Frostbite In Exercises 5–8, use the chart to identify the frostbite risk for each activity.

5. Camping in −15°F weather with a 20 mph wind

6. Hiking in 0°F weather with a 5 mph wind

7. Skiing in −30°F weather with a 25 mph wind

8. Climbing in −25°F weather with a 10 mph wind

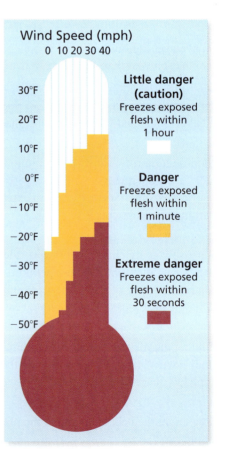

Chapter 10   Summary

Section Objectives		How does it apply to you?

Section 1

Compare a person's weight, height, and body fat percentage.	→	You can identify a healthy weight for a given height and calculate a person's body fat percentage given his or her measurements. *(See Examples 1 and 2.)*
Interpret and use a person's heart rate and metabolism.	→	Use your target heart rate to maximize exercise and your BMR to manage your weight. *(See Examples 3 and 4.)*
Determine factors for cardiovascular health.	→	Understand the indicators of a healthy cardiovascular system, such as exercise, cholesterol, triglycerides, insulin, and glucose. *(See Examples 5 and 6.)*

Section 2

Analyze winning times and heights in the Summer Olympics.	→	Olympic records are consistently broken over time as athletes refine techniques and technology improves. *(See Examples 1 and 2.)*
Analyze winning times in the Winter Olympics.	→	Olympic records are consistently broken over time as athletes refine techniques and technology improves. *(See Examples 3 and 4.)*
Understand Olympic scoring.	→	Understand that the scores for some events are based on subjective decisions by judges. *(See Examples 5 and 6.)*

Section 3

Use mathematics to analyze baseball statistics.	→	You can determine which teams and players are better in various categories. *(See Examples 1 and 2.)*
Use mathematics to analyze football statistics.	→	You can determine a quarterback's passer rating and the probability of winning depending on the score and the time remaining in the game. *(See Examples 3 and 4.)*
Use mathematics to analyze statistics in other professional sports.	→	You can determine how likely it is for a top-seeded tennis player to win a tennis match, or how likely it is for a bowler to throw a split. *(See Examples 5 and 6.)*

Section 4

Use mathematics to analyze hiking and mountain climbing.	→	You can exert the same amount of energy walking on a steep grade as you can running on a lesser grade. As elevation increases, so does UV radiation exposure. *(See Examples 1 and 2.)*
Use mathematics to analyze kayaking and sailing.	→	You should only kayak or canoe when the water level is high enough. Sail designs have evolved over the years to allow ships to travel in any direction. *(See Examples 3 and 4.)*
Use mathematics to analyze bicycling and cross-country skiing.	→	Bicycling can be rigorous exercise that elevates your heart rate. Cross-country skiing uses more energy than most other outdoor sports. *(See Examples 5 and 6.)*

Chapter 10 Review Exercises

Section 10.1

DATA **Body Fat Percentage** In Exercises 1 and 2, find (a) the body fat percentage and (b) the body mass index for the person shown.

1.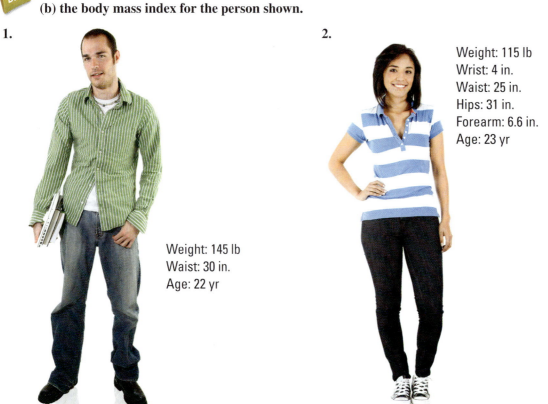

Weight: 145 lb
Waist: 30 in.
Age: 22 yr

2.

Weight: 115 lb
Wrist: 4 in.
Waist: 25 in.
Hips: 31 in.
Forearm: 6.6 in.
Age: 23 yr

DATA **Daily Calorie Balance** In Exercises 3 and 4, determine whether you would expect the person to be losing weight or gaining weight. Explain your reasoning.

3. The man in Exercise 1 is 68 inches tall, very active, and has a daily calorie intake of 3200 calories.

4. The woman in Exercise 2 is 64 inches tall, moderately active, and has a daily calorie intake of 1600 calories.

Heart Rate Zones In Exercises 5 and 6, use the graph.

5. Is a 25-year-old with a heart rate of 160 beats per minute in the cardio zone?

6. Is a 50-year-old with a heart rate of 100 beats per minute in the cardio zone?

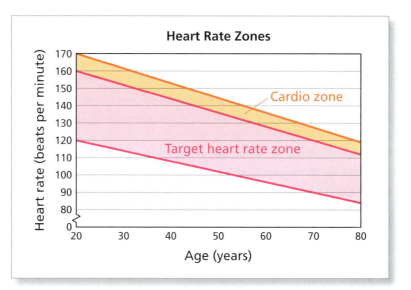

Section 10.2

Men's Hammer Throw In Exercises 7–11, use the graph.

7. Describe any patterns in the graph.

8. What percent of the winning distances were Olympic records?

9. What is the Olympic record for the men's hammer throw?

10. What do you think the winning distance would have been in 1940 if the Olympics had been held? Explain your reasoning.

Winning Distances for Men's Hammer Throw (1900–2008)

DATA 11. Sketch a graph that shows the Olympic record for each year in the data set.

DATA 12. **Distances** Graphically represent the top 8 distances (shown in meters) for the final round of the 2008 Olympic men's hammer throw.

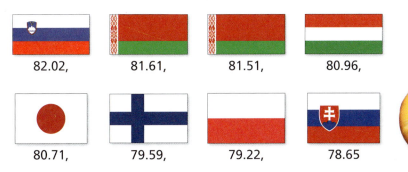

82.02,   81.61,   81.51,   80.96,

80.71,   79.59,   79.22,   78.65

13. **Diving Score** A diver performs a dive with a degree of difficulty of 2.2. The judges' scores are shown. Find the diver's score.

Judge	Russia	China	Mexico	Germany	Italy	Japan	Brazil
Score	7.5	8.0	8.5	8.5	7.0	8.0	9.0

14. **Ski Jumping Score** Find the score for a ski jumper with the following style points and distance. Assume the K-point is at 120 meters.

Style Points: 16.0, 16.5, 19.0, 17.5, 18.0   Distance: 128 meters

Section 10.3

MLB Win Percentage   The scatter plot compares the win percentage on June 1 (about 50 games) to the final win percentage on October 3 (162 games) of every Major League Baseball team for the 2010 season. **In Exercises 15–20, use the scatter plot.**

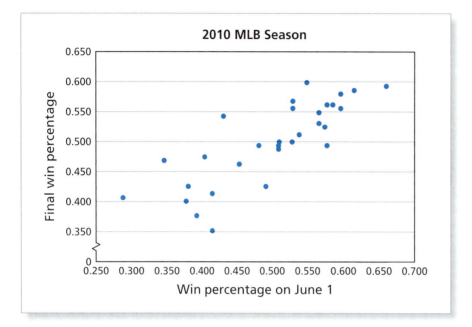

15. What can you conclude from the scatter plot?

16. Identify 2 data points that do not agree with your conclusion in Exercise 15.

17. A team has a 0.600 win percentage on June 1. Predict the team's final win percentage.

18. There is an old cliché that the regular season in baseball is a marathon, not a sprint. Does the scatter plot support this cliché? Explain your reasoning.

19. Of the 8 teams that made the playoffs in 2010, the least win total was 90. Is it possible that a team with a win percentage less than 0.500 on June 1 made the playoffs? Explain your reasoning.

20. Does the scatter plot support the following quote?

"Since 1996, only 9% of teams with a losing record on June 1 finished the season with 90 wins."

Section 10.4

Duck Hunting The bar graph shows the data for an entire duck hunting season.
In Exercises 21–28, use the bar graph.

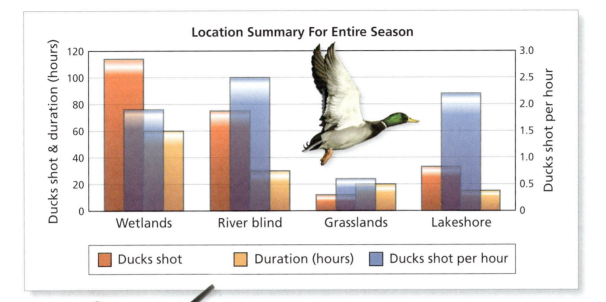

Location Summary For Entire Season

Legend: ■ Ducks shot   ■ Duration (hours)   ■ Ducks shot per hour

Y-axis (left): Ducks shot & duration (hours)
Y-axis (right): Ducks shot per hour

Categories: Wetlands, River blind, Grasslands, Lakeshore

21. Which location had the greatest number of ducks shot per hour?

22. How is the blue bar related to the red and orange bars?

23. How many hours were spent duck hunting for the entire season?

24. Which statistic is used to order the locations from left to right on the bar graph? Explain your reasoning.

25. What percent of the ducks were shot on grasslands?

26. What was the average number of ducks shot per hour for the entire season? Are any of the locations outliers?

27. Hunting the wetlands yields 1.9 ducks per hour. Hunting the lakeshore yields 2.2 ducks per hour. Which of these statistics is a better predictor of future hunts? Explain your reasoning.

28. What other observations can you make from the bar graph?

Answers to Odd-Numbered Exercises

Section 1.1 (page 8)

1. $47.08

3. $39.41

5. **a.** 1361.25 cal **b.** 151.25 cal

7. b; The keystroke sequence in part (a) gives an incorrect total of $3810 because it will multiply 254 by 15. The actual total parking cost is $310.

9. a; Having parentheses around 3×10 and 5×2 clearly indicates the operations being performed, giving a total cost of $40.

11. **a.** 29 bags **b.** $805.91

13. About 499 sec

15. Between about 399 seconds and about 1121 seconds

17. About 36,049 mph

19. **a.** $-$21.49

    b. You have overpaid the electric company, so you have a credit of $21.49.

    c. $80.71; Your average monthly payment in the current year is $80.71. So as long as your monthly usage and the rates stay the same, you should make this your monthly payment next year.

21. **a.** Find the area of the base and the wall of the hot tub. **b.** 13,565 tiles

23. Answers will vary. *Sample answer:* Offer B because the minimum compensation is much greater than the minimum compensation for Offer A, but the maximum compensation is only slightly less than the maximum compensation for Offer A. However, there are many non-financial factors that could also affect your decision.

Section 1.2 (page 18)

1. 3900 billion kWh

3. About 12,829 kWh/person

5. **a.** Top-freezer: $63.59; Side-by-side: $76.21 **b.** $12.62

7. **a.** $588.98 **b.** $445.13 **c.** No, the difference in electricity costs is only $143.85.

9. 1.2×10^{-3} m; meso-plankton

11. 0.0000002 m; femto-plankton

13. In 2009, the population of Texas was about 25 million; It is difficult to determine exact populations.

15. The weight of a bag of sugar is about 5 pounds; You cannot be sure of the accuracy of the weight, plus most people would not care about the extra 0.01 pound.

17. **a.** 297 ft^2 **b.** $1036.53

19. When rounding each item to the nearest dollar, the sum is $31. So the grocery bill total of $31.62 is reasonable.

21. About 200 mi

23. Using a distance of 310 miles, the fuel cost for the trip will be about $42.

25. Answers will vary.

1.1–1.2 Quiz   *(page 22)*

1. b; The keystroke sequence in part (a) gives an incorrect total of 1072 ft^2 because it will multiply 134 by 8. The actual area that you are adding to the patio is 92 ft^2.

3. **a.** About 3.4 people/km^2 **b.** About 3.75 people/km^2

   c. Answers will vary. *Sample answer:* Either the total area or land area can be used, as long as it is specified as to which is being used.

5. **a.** About 3968 people/km^2 **b.** Toronto has about 1000 times the population density of Canada.

Section 1.3   *(page 30)*

1. 67,860,000 men

3. Method I: Add the populations of men and women, and then find 4% of that total.

   Method II: Separately find the number of men and women who have no opinion and add the two results.

   No; Method I gives a result of 9,600,000 people, while Method II gives a result of 8,430,000 people. The difference is probably due to round-off error when summarizing the poll results.

5. About 1,288,219 people   7. Answers will vary.   9. About 73.7%

11. About 4.9%   13. About 51.5%   15. 18.75%

17. 2.00 carats; The price range for a 2.00-carat diamond is $6892 to $16,092. $11,499 is in this range.

19. A 2.00-carat diamond ring could cost anywhere from $0 to $14,066 more than a 1.00-carat diamond ring.

21. When comparing weights, a 0.50-carat diamond is about 67% heavier than a 0.30-carat diamond. However, when comparing diameters, a 0.50-carat diamond (5.2 mm) is not more than 60% larger than a 0.30-carat diamond (4.3 mm).

23. **a.** 100% increase   **b.** 100% increase   **c.** 50% increase

25. False; A loss of 20% of $1250 leaves you with $1000. A 20% increase of $1000 only takes you up to $1200.

27. False; The 1% decrease is being taken from a larger amount than the 5% increase was taken, so the overall increase will be slightly less than 4%.

29. $1530   31. 92%

Section 1.4   *(page 40)*

1. $2.45   3. $10.40

5. **a.** $300.51   **b.** It will cost more to order the treadmill online.   **c.** No; yes

7. b; 72 min   9. 0.625 lb   11. About 5.47 qt   13. 0.8 sec

15. About 20,321.5 ft   17. About 10 ft/min

19. **a.** 156.2°F

   b. Yes; You could boil a potato, but it would take longer then normal to get cooked because it is at a lower boiling temperature than the normal boiling temperature of 100°C.

21. $352.63   23. About $53.97

1.3–1.4 Quiz *(page 44)*

1. 14,040,000 people

3. The percent of the disabled population who are 65 years old and over is 3 times higher than the percent of the general population who are 65 and over.
Answers will vary. *Sample answer:* As people get older, they are more likely and have more opportunity to develop some type of disability.

5. 60 in.

7. 12 furlongs

Chapter 1 Review   *(page 46)*

1. a; The keystroke sequence in part (b) gives an incorrect total cost of $480 because it will multiply 2 by 240. The actual total cost is $300.

3. About 3.5 hr

5. About 234.375 mi/yr

7. a. 19,500,000 barrels/day   **b.** Answers will vary.

9. a. About 0.006 barrels/person/day

   b. The number of gallons of oil consumed by the United States is about 0.064 barrels/person/day, which is about 10 times more than the amount consumed by China per person per day.

11. 189,800,000 barrels; Japan is one of Iran's top imports because Japan does not produce much oil but continues to consume oil.

13. 9,975,000 mi^3

15. About 51.6%

17. About 20.1% increase

19. Answers will vary.

21. Monitor

23. a. 3.45 mi

   b. Territorial waters are now defined to be at most 12 nautical miles from the baseline of a coastal state.

25. 30 ft

Section 2.1   *(page 58)*

1. About $0.17 per oz

3. About $0.05 per fl oz

5. a. $0.98 per lb   **b.** About 38%

7. Brand A: About $0.09 per oz

   Brand B: About $0.05 per oz

   Brand C: $0.06 per oz

   Brand B is the best buy because it has the least unit price.

9. a. About $0.07 per fl oz

   b. Answers will vary. *Sample answer:* Because it results in a lower unit price.

11. $55.73

13. a. $17.55   **b.** $45.75

15. About $0.05

17. Decreasing the amount of product in a package but keeping the same price will increase the unit price, as shown in Exercises 15 and 16.

19. a. $2.34

   b. The unit price in part (a) is $0.35 greater than the unit price on the package.

Section 2.2 *(page 68)*

1. About 122%   **3.** About 64%   **5.** $20   **7.** About 48%

9. a. 3

   b. You can make a profit as long as the total revenue is greater than what you paid for the inventory. If you sell enough items above the wholesale price, then you can afford to sell some below the retail price and still make a profit.

11. About 62%   **13. a.** $298   **b.** $301.75   **15. a.** 25%   **b.** $17.94

17. 40%   **19.** About 21%

21. a. $13.20   **b.** 44%

   c. Answers will vary. *Sample answer:* Having a 30% off sale combined with an additional 20% off makes it appear that the store is having a 50% off sale, when it is actually having a 44% off sale.

2.1–2.2 Quiz *(page 72)*

1. Brand A: About $0.27 per oz
Brand B: About $0.12 per oz
Brand C: About $0.09 per oz
Brand A has the greatest unit price. Brand C has the least unit price.

3. a. $1   **b.** About 101%   **5. a.** About $0.01 per oz   **b.** About 89%

Section 2.3 *(page 80)*

1. $103.87   **3.** 6%

5. Mississippi: $88.13
Alabama: $50.36
The best option would be to drive to Alabama. You would save $88.13 − $50.36 = $37.77 in sales tax. You would not spend nearly that much in gas for the 40-mile round trip.

7. City sales tax: 1.2%   **9.** Wyoming; Alaska   **11.** $0.60
State sales tax: 6%

13. About 19%   **15.** $1380   **17.** $24,380

19. Raw materials: $907.50
Manufacturer: $1320
Dealer: $577.50
The total tax is the same.

21. The gasoline excise tax in the United States is about 9% of the gasoline excise tax in Germany.

23. 1%   **25.** Bottle of cough syrup

27.

	Value-Added Tax	New Value
Raw materials	$90.00	$1,290.00
Manufacturer	$172.50	$3,762.50
Finisher	$37.50	$4,300.00
Retailer	$105.00	$5,805.00
Total	**$405.00**	

7.5%

Section 2.4 *(page 90)*

1. 0.00

3. 3055.00

5. 69.50

7. 250.00

9. $540.26

11. Yes; Check #220

13. $319.01

15. $54,000

17. About 6%

19. No

21. The budget compares well with the guidelines. The only areas of concern are that you are a little high on entertainment and a little low on retirement.

23. $50

25. Answers will vary. *Sample answer:* You could reduce unnecessary expenses, such as entertainment and miscellaneous.

27. If you do not account for the bank fees in your checkbook register, then you will think you have more money in your checking account than you actually do.

2.3–2.4 Quiz *(page 94)*

1. **a.** 4.5%

   b. 1%

   c. Oklahoma

3. **a.** $1.70

   b. About 5.5%

5. $522.35

Chapter 2 Review *(page 96)*

1. 2-Liter bottle: About $0.013 per fl oz
   6-pack: About $0.027 per fl oz
   Case: About $0.026 per fl oz

3. $358.40

5. Bubble A will get larger.

7. Bubble B; It is the smallest bubble, which means it has the least unit price.

9. $60

11. $30

13. 25%

15. $49.50

17. $0.16

19. $6.52

21. Excise taxes are often a fixed dollar amount per sales unit, and a value-added tax is a certain percent added to a product at each stage of its manufacture or distribution.

23. January: $10, February: $19, March: ($5), April: $25, May: $23, June: $17

25. You will need to increase your budgeted amount for cell phone expenses. If you do not want to do this, then you are going to have to find a way to reduce your cell phone expenses.

27. Yes. Even though the ratio for the 36% rule is about 36.1%, most banks would determine that you qualify for the home mortgage.

Section 3.1 *(page 108)*

1.

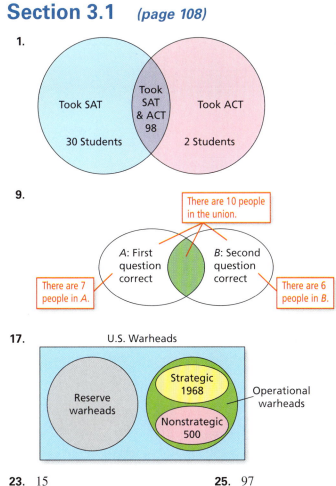

3. 17

5. 7

7. Not likely; Of the 12 colleges with an enrollment of less than 2000, only 3 have tuitions less than $20,000 per year. This is reasonable in real life because small colleges need to charge a substantial amount in tuition to meet their expenses.

9.

There are 10 people in the union.

A: First question correct

B: Second question correct

There are 7 people in A.

There are 6 people in B.

11. 6

13. 4

15. 14

17.

U.S. Warheads

Reserve warheads

Strategic 1968

Nonstrategic 500

Operational warheads

19. 2096

21. **a.** The numbers in each category will be lower.

b. Answers will vary. The graph will continue to decrease.

23. 15

25. 97

27. 47

Section 3.2 *(page 118)*

1.

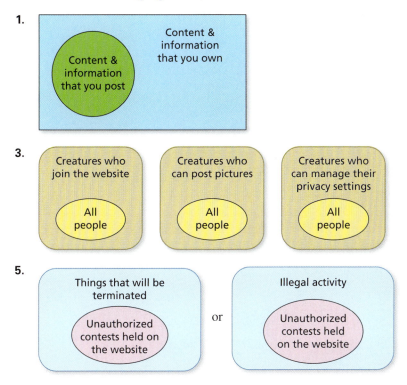

Content & information that you own

Content & information that you post

3.

Creatures who join the website

All people

Creatures who can post pictures

All people

Creatures who can manage their privacy settings

All people

5.

Things that will be terminated

Unauthorized contests held on the website

or

Illegal activity

Unauthorized contests held on the website

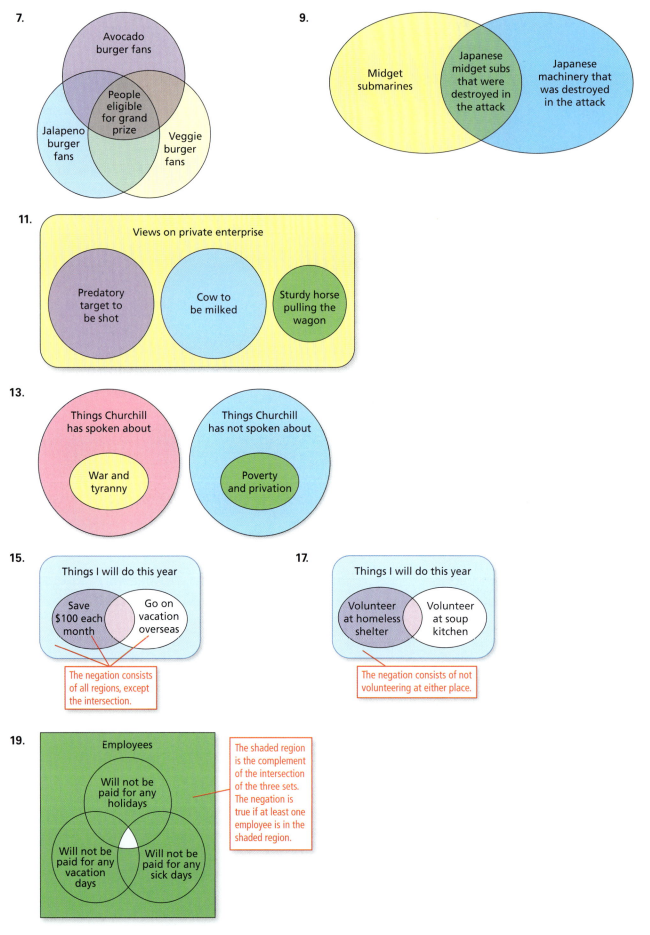

21.

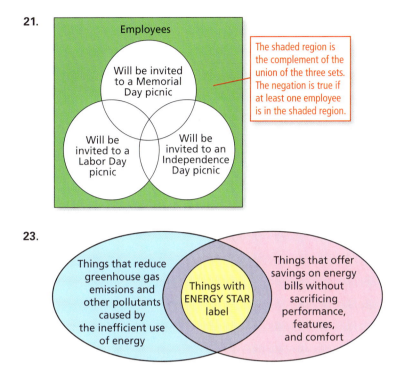

Employees

Will be invited to a Memorial Day picnic

Will be invited to a Labor Day picnic

Will be invited to an Independence Day picnic

The shaded region is the complement of the union of the three sets. The negation is true if at least one employee is in the shaded region.

23.

Things that reduce greenhouse gas emissions and other pollutants caused by the inefficient use of energy

Things with ENERGY STAR label

Things that offer savings on energy bills without sacrificing performance, features, and comfort

25. Answers will vary. *Sample answer:* All of the products shown emit less than 50 grams of CO_2e, while some emit less than 10 grams.

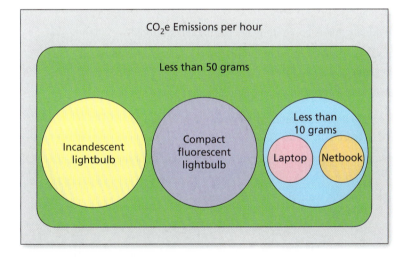

CO_2e Emissions per hour

Less than 50 grams

Incandescent lightbulb

Compact fluorescent lightbulb

Less than 10 grams

Laptop   Netbook

27.

Things I will do

Buy a gas condensing heater

Buy a solar heater

Buy low-flow showerheads

The shaded region is the complement of the intersection of the set where you will buy at least one water heater and the set where you will buy low-flow showerheads. The negation is true if your actions are in the shaded region.

3.1–3.2 Quiz *(page 122)*

1. 311

3. 387

5.

7.

Section 3.3 *(page 130)*

1. All segregated public schools are unconstitutional.

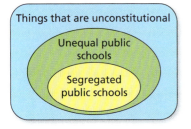

3. The Constitution will govern the case to which they both apply.

5. Premise: All unequal treatment of individuals that is based on gender is unconstitutional.

    Premise: The Virginia Military Institute's male-only admission policy was an unequal treatment of individuals that was based on gender.

    Conclusion: The Virginia Military Institute's male-only admission policy was unconstitutional.

7. Premise: If any resident of a house objects to a police search without a warrant, then it is unconstitutional for the police to search the house.

    Premise: In the case of *Georgia v. Randolph*, one resident objected to a police search of the house without a warrant.

    Conclusion: The police search of the house without a warrant was unconstitutional.

9. "town" and "road"

11. Premise: If there are two towns, then there is a road that passes through them.

    Premise: There are two towns.

    Conclusion: There is a road that passes through the two towns.

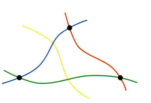

13. **1.** There exists two towns.

    2. There exists a road passing through the two towns.

    3. There exists a town which the road does not pass through.

    4. The town in Step 3 is different from the towns in Step 1.

    5. So, there exists at least three towns.

15. No; The model shown satisfies all three postulates and has an intersection of two roads at which there is no town.

17. *Troodon*: The only places that *Troodon* fossils have been found are Wyoming, Montana, Alberta, and Alaska. Therefore, the only places that *Troodon* lived were Wyoming, Montana, Alberta, and Alaska.

Velociraptor: The only places that *Velociraptor* fossils have been found are China and Mongolia. Therefore, the only places that *Velociraptor* lived were China and Mongolia.

Giganotosaurus: The only place that *Giganotosaurus* fossils have been found is Argentina. Therefore, the only place that *Giganotosaurus* lived was Argentina.

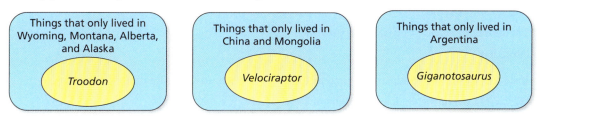

19. All *Giganotosaurus* fossils have been dated to about 100 to 95 million years ago. Therefore, *Giganotosaurus* lived about 100 to 95 million years ago.

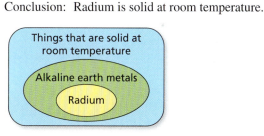

21. It is extremely unlikely that *Velociraptor* encountered *Giganotosaurus* because the two species lived millions of years apart. It is unlikely that *Velociraptor* encountered *Troodon* because they lived in different parts of the world.

23. Premise: All discovered alkali metals react strongly with water.

Premise: Potassium is an alkali metal.

Conclusion: Potassium reacts strongly with water.

25. Premise: All discovered alkaline earth metals are solid at room temperature.

Premise: Radium is an alkaline earth metal.

Conclusion: Radium is solid at room temperature.

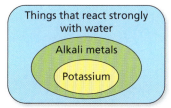

27. All discovered alkali metals react strongly with water. Therefore, all alkali metals (discovered or undiscovered) react strongly with water.

29. Premise: All alkali metals (discovered or undiscovered) react strongly with water.

Premise: Ununennium is an undiscovered alkali metal.

Conclusion: Ununennium reacts strongly with water.

Section 3.4   *(page 140)*

1. Premise: If something is not there, you will not find it.

Premise: We did not find weapons of mass destruction in Iraq.

Conclusion: Therefore, there are no such weapons in Iraq.

Affirming the consequent

3. Premise: If we found weapons of mass destruction in Iraq, it would prove that Iraq had such weapons.

Premise: We have not found weapons of mass destruction in Iraq.

Conclusion: Therefore, Iraq did not have weapons of mass destruction.

Denying the antecedent

5. Premise: If you support the war in Iraq, you support America.

Premise: The senator does not support the war.

Conclusion: Therefore, the senator does not support America.

Denying the antecedent

7. Premise: If we had overthrown Saddam's regime in 1991, then we would not be fighting in Iraq today.

Premise: We did not overthrow Saddam's regime in 1991.

Conclusion: Therefore, we are still fighting in Iraq today.

Denying the antecedent

9.   **11.**   **13.**   **15.**

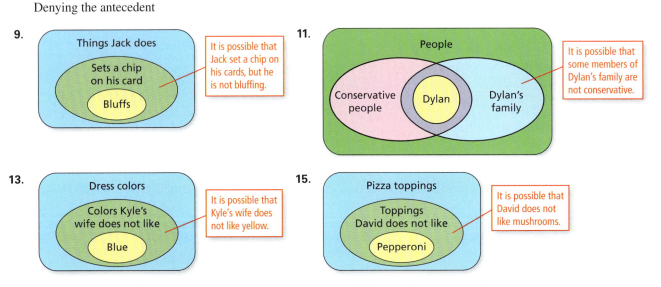

17. The ad implies that if you use the toothpaste, then your teeth will be as white as those shown. It also implies that some type of survey was conducted with dentists.

19. The ad implies that the shoes can make you a good athlete.

21. The ad appeals to emotion by displaying the American flag and by using the colors of the flag. It also contains an appeal to authority by having the candidate pointing and saying "I want you" like Uncle Sam.

23. The ad appeals to celebrity by implying that if President Obama wears this jacket, then you should wear it.

25. Ad hominem   **27.** Self-refuting idea   **29.** Ad populum

3.3–3.4 Quiz   (page 144)

1. **a.** 100°C   **b.** All water boils at 100°C.

c.

Characteristic of all water

Boils at 100°C

3. Premise: If water at sea level is heated to a temperature of 100°C, then it will boil.

Premise: Water at sea level is heated to a temperature of 100°C.

Conclusion: The water boils.

5. Premise: If you move from sea level to a higher elevation, then the boiling point of water will be less than 100°C.

Premise: You move from sea level to a higher elevation.

Conclusion: The boiling point of water is less than 100°C.

7. Premise: If you move from sea level to a higher elevation, then the atmospheric pressure decreases.

Premise: You move from sea level to a higher elevation.

Conclusion: The atmospheric pressure decreases.

Chapter 3 Review *(page 146)*

3. 14

5. 6

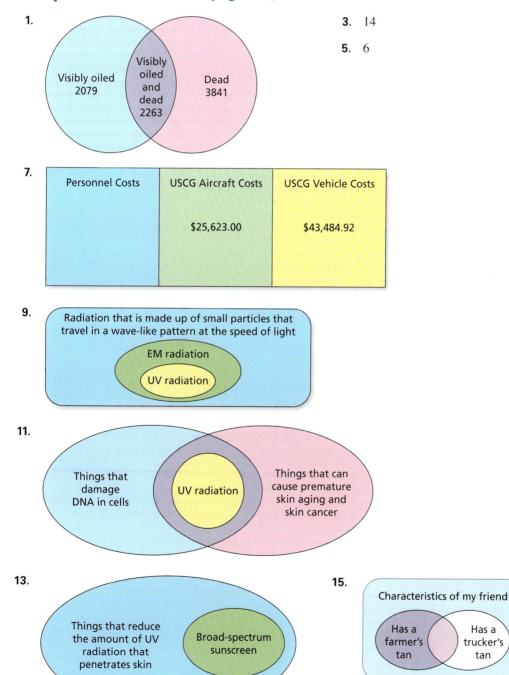

17. Newton observed something falling and noticed that it fell in a path that is perpendicular to the center of Earth; inductive; Newton's conclusion was based on repeated patterns.

19. Inductive; As is true of almost all scientific laws, Newton based his laws of motion on observed patterns.

21. Premise: If an object is in uniform motion and no external force acts on it, then it will remain in uniform motion.

Premise: An object is in uniform motion and no external force is acting on it.

Conclusion: The object will remain in uniform motion.

23. This is an example in which you can reach a false conclusion using inductive reasoning because the reasoning was not based on a large enough and diverse enough sample.

25. 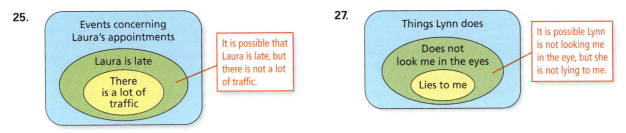 **27.**

29. The ad implies that the golf clubs can make you a good golfer.

31. This ad appeals to emotion by displaying the American flag and by using the colors of the flag. It also begs the question by assuming that we need change.

Section 4.1 *(page 158)*

1. $A = 3(1.25)^n$

3. 635 people

7. 8032 people

5.

	A	B
1	Minutes, n	Informed People
2	0	100
3	3	305
4	6	929
5	9	2,833
6	12	8,638
7	15	26,334
8	18	80,283
9	21	244,753
10	24	746,160
11	27	2,274,763
12	30	6,934,898

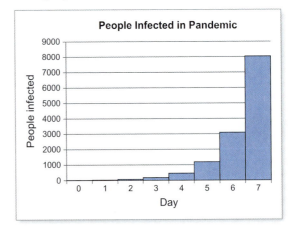

9. 4374 people

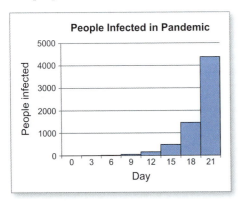

11. The graph of an exponential growth model is a curve that rises more and more rapidly as *n* increases.

13. 1%   **15. a.** $1350.50   **b.** $1461.00

17. $r < 0.732$, or about 73.2%

19. Linear; The balance is increasing by about the same amount each year.

21. Exponential growth continues to increase more and more rapidly as time increases. Logistic growth starts off like exponential growth, but then the rate of growth begins to decrease and eventually approaches zero.

23. The blue part; The growth rate is decreasing.

25. $r < 0.154$, or about 15.4%

Section 4.2   *(page 168)*

1. a. About 39.4%   **b.** About 146.2%
   c. About 64.1%   **d.** About 229.0%

3. 1990–2009; The minimum wage percent increase (90.8%) was higher than the CPI percent increase (64.1%).

5. No; From 1992 to 2010, there was about a 55.5% increase in the Consumer Price Index, but there was only about a 46.6% increase in the company's starting wage.

7. 2005

9.

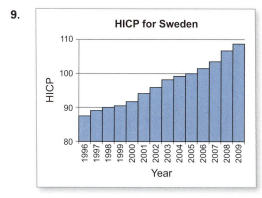

The rate of inflation was steadily increasing.

11. Yes; The HICP increased in all of the countries from year-to-year.

13.

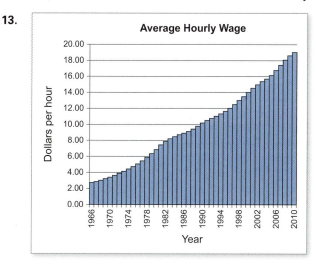

1966–1984: Exponential growth pattern

1984–1998: Linear growth pattern

1998–2010: Steeper linear growth pattern

15. No; From 1970 to 1990, the CPI had about a 237% increase, but the hourly wage only had a 200% increase.

17. $50,000 in 1990; From 1990 to 2010, the CPI had about a 67% increase, but your salary only had a 50% increase.

19.

Year	Gasoline Index	Diesel Index
1995	100.0	100.0
1996	107.0	111.7
1997	107.0	108.1
1998	92.2	93.7
1999	101.7	100.9
2000	131.3	134.2
2001	127.0	126.1
2002	118.3	118.9
2003	138.3	136.0
2004	163.5	163.1
2005	200.0	216.2
2006	225.2	244.1
2007	243.5	260.4
2008	284.3	342.3
2009	204.3	222.5

21. Both gasoline and diesel fuel prices increased at a higher rate than the Consumer Price Index.

23. Answers will vary.

4.1–4.2 Quiz   *(page 172)*

1. $A = 100(1.03)^n$

3.

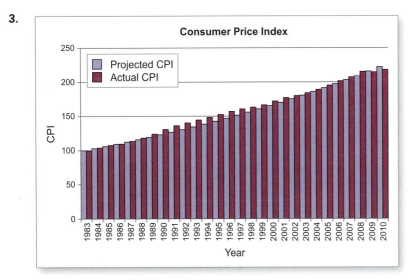

5. About $244 billion

Section 4.3   *(page 180)*

1. $A = 320(0.86)^n$

3.

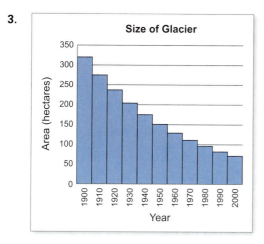

5. 2130

7. 5 mg

9. Saber-toothed cat: 40 hr
Tasmanian tiger: 32 hr

11. 0.94

13. 1995

15. 3.2 billion years

17. a. No; Radon itself has a half-life, so some of the radon will have decayed into polonium.

b. About 4 days; After 2 days, the amount of radon-222 decayed to $34.7/50 = 0.694 = 69.4\%$ of the original amount. After 2 more days, the amount of radon-222 will be $34.7(0.694) \approx 24.1$, which is about half of the original amount.

19. $A = 1024(1.5)^n$

21. $A = 39,366(0.9)^n$

23. Day 30

Section 4.4 *(page 190)*

1.

	A	B	C	D
1	Year	Value before Depreciation	Depreciation	Value after Depreciation
2	1	$1,100.00	$100.00	$1,000.00
3	2	$1,000.00	$100.00	$900.00
4	3	$900.00	$100.00	$800.00
5	4	$800.00	$100.00	$700.00
6	5	$700.00	$100.00	$600.00
7	6	$600.00	$100.00	$500.00
8	7	$500.00	$100.00	$400.00

3. $5000

5. Year 5: $1000
Year 6: $2000
Year 7: $3000
Year 8: $4000

7.

	A	B	C	D
1	Year	Value before Depreciation	Depreciation	Value after Depreciation
2	1	$50,000.00	$20,000.00	$30,000.00
3	2	$30,000.00	$12,000.00	$18,000.00
4	3	$18,000.00	$7,200.00	$10,800.00
5	4	$10,800.00	$800.00	$10,000.00
6	5	$10,000.00	$0.00	$10,000.00
7				

9. $144,675.93

11. $11,628.40

13. $56,275.53

15.

	A	B	C	D
1	Year	Value before Depreciation	Depreciation	Value after Depreciation
2	1	$4,500.00	$1,500.00	$3,000.00
3	2	$3,000.00	$1,200.00	$1,800.00
4	3	$1,800.00	$900.00	$900.00
5	4	$900.00	$600.00	$300.00
6	5	$300.00	$300.00	$0.00
7				

17. Double declining-balance

19. $103,636.36

21. a. Year 4   **b.** Year 7

23.

	A	B	C	D
1	Year	Value before Depreciation	Depreciation	Value after Depreciation
2	1	$4,000.00	$800.00	$3,200.00
3	2	$3,200.00	$1,280.00	$1,920.00
4	3	$1,920.00	$768.00	$1,152.00
5	4	$1,152.00	$460.80	$691.20
6	5	$691.20	$460.80	$230.40
7	6	$230.40	$230.40	$0.00
8				

25.

	A	B	C	D
1	Year	Value before Depreciation	Depreciation	Value after Depreciation
2	1	$24,000.00	$3,428.57	$20,571.43
3	2	$20,571.43	$5,877.55	$14,693.88
4	3	$14,693.88	$4,198.25	$10,495.63
5	4	$10,495.63	$2,998.75	$7,496.88
6	5	$7,496.88	$2,141.97	$5,354.91
7	6	$5,354.91	$2,141.96	$3,212.95
8	7	$3,212.95	$2,141.97	$1,070.98
9	8	$1,070.98	$1,070.98	$0.00
10				

4.3–4.4 Quiz   *(page 194)*

1. a. 54 ppb   **b.** 5 filters

3. 7 filters

5.

	A	B	C	D
1	Year	Value before Depreciation	Depreciation	Value after Depreciation
2	1	$10,000.00	$1,800.00	$8,200.00
3	2	$8,200.00	$1,800.00	$6,400.00
4	3	$6,400.00	$1,800.00	$4,600.00
5	4	$4,600.00	$1,800.00	$2,800.00
6	5	$2,800.00	$1,800.00	$1,000.00
7				

7. Sum of the years-digits; The values are not decreasing by the same amount or by the same percent each year.

Chapter 4 Review   *(page 196)*

1. $A = 415,000(1.115)^n$

5. About 11.8%; The populations are relatively close, while the rate of growth is slightly greater.

7. 1970s; The CPI increased over 40 units in the 1970s but less than 40 units in the 1980s and 1990s.

3.

Year	Population per Representative
2020	797,439
2030	889,144
2040	991,396
2050	1,105,406
2060	1,232,528
2070	1,374,269
2080	1,532,310
2090	1,708,525
2100	1,905,006

11. $1.09 per thousand cubic feet

13. $115.50 in 2004; From 1990 to 2004, the CPI had about a 45% increase but your natural gas bill only had about a 23% increase.

15. $A = 100(0.9)^n$

9.

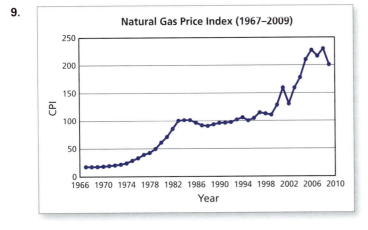

Answers will vary.

17.

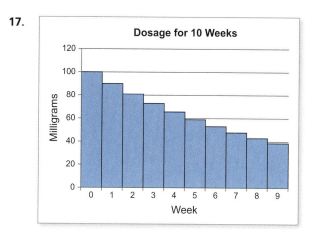

Dosage for 10 Weeks

19. 15 g

21. 6 years; It took 6 years for the amount to decay from 10 grams to 5 grams.

23.

	A	B	C	D	
1	Year	Value before Depreciation	Depreciation	Value after Depreciation	
2	1	$2,500.00	$400.00	$2,100.00	
3	2	$2,100.00	$400.00	$1,700.00	
4	3	$1,700.00	$400.00	$1,300.00	
5	4	$1,300.00	$400.00	$900.00	
6	5	$900.00	$400.00	$500.00	
7					

25. $14,285.71

27. $8413.15

Section 5.1   *(page 208)*

1. $1666

3. $1088

5. Not taxable

7. Taxable

9. Regressive; As the earnings over $106,800 increases, the percent of the Social Security tax paid on the earnings decreases.

11. Flat; The same percentage is paid on all incomes.

13. Sales & Excise Taxes: $2030
Property Taxes: $2436
Income Taxes: $1682

15. Sales & Excise Taxes: Regressive
Property Taxes: Regressive
Income Taxes: Progressive

17. $1.25; The retailer doubles the price to $30. Because $30(0.04166) = \$1.25$, the consumer pays $31.25 for the lei. Of this, $1.25 is the general excise tax.

19.

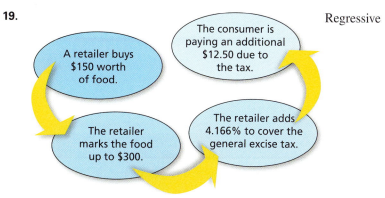

Regressive

21. The state requires a general excise tax of 4% on the entire retail price, including the extra charge to cover the tax. So just charging an extra 4% of the marked-up price would not totally cover the tax. Retailers are allowed to add 4.166% of the marked-up price to compensate for this.

23. $4900

25. $6590

27. Tax credit of $1000; A tax credit saves the taxpayer $1000. A tax deduction of $1000 only saves the taxpayer 1000(0.14) = $140.

29. a. Receives a check; $280   **b.** $42,000

Section 5.2 *(page 218)*

1. $7975; about 13.6%

3. $23,062.50; about 18.8%

5. No; Filing "separately," you would be in the income bracket with a marginal tax rate of 25%. Filing "jointly," you would be in the bracket with a rate of only 15%.

7. South Carolina's state income tax is a graduated income tax system. The top 25% of all wage earners in South Carolina pay 84.9% of all the state income tax paid. This seems equitable because the top 25% of all wage earners in South Carolina earn 79.6% of all the income.

9. About 22%

11. The sources are quite different. The two largest sources of tax revenue for South Carolina are property taxes and sales taxes. The federal government does not collect either of these taxes. The two largest sources of tax revenue for the federal government are Social Security & Medicare taxes and individual income taxes. South Carolina does not collect Social Security & Medicare taxes.

13. California could raise this amount of income tax revenue by using a flat tax rate of 58,000,000,000/1,610,000,000,000 ≈ 0.03602 ≈ 3.6%.

15. $11,251,000,000; 4.3%

17. Income tax: $4321.25

Effective tax rate: About 12.1%

19. a. Income tax: $307,643.75

Effective tax rate: About 30.8%

b. Income tax: $227,643.75

Effective tax rate: About 22.8%

c. As more and more income moves from ordinary income to capital gains, the effective tax rate moves closer and closer to the capital gains rate of 15%.

5.1–5.2 Quiz *(page 222)*

1. Sales & Excise Taxes: $1775
Property Taxes: $275
Income taxes: $575

3. Income tax: $1460
Effective tax rate: About 4.9%

5. Answers will vary.

Section 5.3 *(page 230)*

1. a. $5816.16 **b.** About 2.9%   **3. a.** $10,314 **b.** About 5.2%   **5. a.** $3813 **b.** About 1.9%

7. a. Cohoes; $1295.23

b. Answers will vary. *Sample answer:* The house in Cohoes is the most economical choice. The additional cost of fuel in commuting from Sarasota Springs will overshadow the difference in property tax. Plus, the house in Cohoes costs $10,000 less.

9. $2070; $725

11. Answers will vary. *Sample answer:* No; Fred's assessment is similar to that of his neighbors. Plus, because the one house is assessed at $60,000, there is a chance that his assessment could actually be increased.

13. $1300

15. About 42%

17. $4,398,900,000

19. $1120

21. 12 mills

23. $468.75

25. Town A: 40%

    Town B: 40%

    Town C: 20%

27. Answers will vary. *Sample answer:* It would be unfair because Town C would be paying a disproportionate amount based on market value.

Section 5.4   *(page 240)*

1. $2309.19

3. $1672.71

5. About 22.8%

7. $1657

9. $1410

11. Age 82

13. The primary reason that the total dependency ratio is projected to increase is that people are projected to live longer.

15. The working age population will decrease.

17. 2050; The old-age dependency ratio is projected to be more than twice as much in 2050 than it was in 1965. This means that there will be higher expenditures for both Social Security and Medicare.

19.

Year	Indexed Earnings	Year	Indexed Earnings	Year	Indexed Earnings
1975	$19,160.00	1987	$32,551.68	1999	$98,736.00
1976	$18,569.60	1988	$66,768.00	2000	$98,298.00
1977	$18,079.02	1989	$66,369.08	2001	$101,304.00
1978	$28,537.60	1990	$65,687.68	2002	$105,276.00
1979	$27,057.60	1991	$65,455.00	2003	$105,270.00
1980	$25,548.60	1992	$69,381.00	2004	$101,964.00
1981	$24,006.00	1993	$71,390.57	2005	$100,800.00
1982	$34,662.20	1994	$71,924.64	2006	$100,794.00
1983	$34,370.93	1995	$71,477.67	2007	$99,450.00
1984	$33,717.76	1996	$70,750.40	2008	$102,000.00
1985	$33,667.56	1997	$69,114.21	2009	$106,800.00
1986	$33,715.73	1998	$97,812.00		

21. $1523

23. In principle, the economic dependency ratio is supposed to measure the ratio of "nonworking people" to "working people" by using age groups. The problem is that there are people who work even though they are either under the age of 20, or 65 and over, while there are people who do not work even though they are between the ages of 20 and 64.

5.3–5.4 Quiz   *(page 244)*

1. 68.100 mills

3. About 1.4%

5. Local: $8.45

    State: $25.94

7. About 18%

Chapter 5 Review *(page 246)*

1. $2385

3. Sales & Excise Taxes: $1537
Property Taxes: $1802
Income Taxes: $2014

5. Sales & Excise Taxes: regressive
Property Taxes: regressive
Income Taxes: progressive

7. Answers will vary.

9. $489.40; about 2.7%

11. $2611.73; about 3.5%

13. About 14.8%

15. Arizona: about 48.3%
South Carolina: about 34.9%

17. $256

19. 0.7%

21. $893.20

23. $3596

25. $1473

27. No; A person needs at least 40 credits (10 years of work) to qualify for Social Security retirement benefits.

29. Answers will vary.

Section 6.1 *(page 258)*

1. a. 2 yr   **b.** $1900

3. May 30, 2012

5. May 18, 2012; $20,923.04

7. $16.75

9. $36,825

11. Loan Proceeds: $1750
Other Charges: $0
Amount Financed: $1750
Finance Charge: $41.06
Total Amount Due: $1791.06

13.

t	P $100	$400	$1000
30 days	$3.29	$13.15	$32.88
60 days	$6.58	$26.30	$65.75
180 days	$19.73	$78.90	$197.26
1 year	$40.00	$160.00	$400.00

15. $6.58

17. a. 40%   **b.** 68.6%

19. a. 25%   **b.** 41.7%

21. a. 140 days   **b.** $4345.37

23. Ordinary simple interest; No; For instance, ordinary simple interest would be cheaper on a loan from July 15 to August 15 because 30/360 < 31/365.

25. Lender; The interest rate using the Banker's rule will be greater because when converting the length of the loan to years, the denominator will be 360 instead of 365. Decreasing the denominator gives a greater percentage.

Section 6.2 *(page 268)*

1. a. $660.33   **b.**

	A	B	C	D	E
1	Payment Number	Balance before Payment	Monthly Interest	Monthly Payment	Balance after Payment
2	1	$35,000.00	$145.54	$660.33	$34,485.21
3	2	$34,485.21	$143.40	$660.33	$33,968.28
4	3	$33,968.28	$141.25	$660.33	$33,449.20
5	4	$33,449.20	$139.09	$660.33	$32,927.97
6	5	$32,927.97	$136.93	$660.33	$32,404.56
57	56	$3,261.05	$13.56	$660.33	$2,614.28
58	57	$2,614.28	$10.87	$660.33	$1,964.82
59	58	$1,964.82	$8.17	$660.33	$1,312.66
60	59	$1,312.66	$5.46	$660.33	$657.79
61	60	$657.79	$2.74	$660.52	$0.00

3. No; after 13 months

5. **a.** 2 years: $565.91   **b.** 2 years: $581.84
   3 years: $386.65   3 years: $919.40
   4 years: $297.85   4 years: $1296.80
   5 years: $245.27   5 years: $1716.20
                              The cost of credit increases as the term increases.

7. 53 months; $190.54   **9.** 12 months; $64.24   **11.** 1-year installment loan; $34.20

13. Yes; The rises and falls of the two rates occur at the same times.

15. No; The CPI increased by a slightly greater factor than student loan debt from 2000 to 2004.

17. Yes; Credit card debt increased by a slightly greater factor than the CPI from 2000 to 2008.

19. Yes

21. April's ending balance of $115 must not have been paid in full during the month of May.

23. Gold Card: $127.67
   World Card: $643.55

6.1–6.2 Quiz   *(page 272)*

1. 5 years; $100,000   **3.** $118,101; $18,101   **5.** After 47 payments

Section 6.3   *(page 280)*

1. **a.** $130,558.00   **b.** $195,311.20
   An increase of only 2 percentage points increases the interest that you pay by $64,753.20.

3. **a.** $158,111.60   **b.** $254,843.60
   An increase of only 2 percentage points increases the interest that you pay by $96,732.00.

5. **a.** $116,778.40   **7.** **a.** 34 months sooner
   b. $186,510.40   **b.** $24,912.16
   c. $262,907.20

9. **a.** $215
   b. Total interest for 25-year mortgage: $160,465.00
      Total interest for 25-year mortgage paid off 5 years early: $124,933.82
      Difference: $35,531.18

11. Balloon payment: $111,835.67   **13.** The cost of renting is $50,668.80 less than the
   Total interest paid: $33,007.07   cost of buying.

15. 16.875   **17.** **a.** About $190,000   **b.** About $55,000

19. About 13%

21. **a.** Mortgage A: $760.03   Mortgage B: $805.23
      With Mortgage B, you are paying $45.20 more per month.
   b. About 66 months, or $5\frac{1}{2}$ years

23. **a.** $190,491.80   **b.** $5151.00   **25.** About $112,000

Section 6.4 *(page 290)*

1. a. $1647.01 **b.** $2712.64

3. Your account

5. About $6 trillion

7. $719,536,522.10

9. $57,018.65

11. a. $137,799.63
 b. $413,398.89
 c. The balance in part (b) is three times the balance in part (a).

13. a. $396,767.60
 b. $1,432,633.58
 c. $1,829,401.18

15. a. $750,000
 b. $540,029.59
 c. $790,029.59

17. 35.5 years

19. 59 years old

21. a. $440.50 **b.** $172.50

23. a. 6.18% **b.** 6.17% **c.** 6.14% **d.** 6.09% **e.** 6%

25. Annual; When the interest is compounded annually, the APR is the annual rate of increase, which is the definition of APY. Also, substituting $n = 1$ into the formula for APY gives APY $= r$.

6.3–6.4 Quiz *(page 294)*

1. a. $401.82 **b.** $807.99

3. a. 75 months sooner **b.** $20,293.22

5. $413,398.89

Chapter 6 Review *(page 296)*

1. October 14

3. Loan Proceeds: $255
 Other Charges: $0
 Amount Financed: $255
 Finance Charge: $45
 Total Amount Due: $300
 Annual Percentage Rate: 460%
 Payable in 1 payment with a term of 14 days.

5. a. 460% **b.** 226.54%

7. $487.67

9. $20,224.48

11. $928.80

13. 5%: $158,536.00
 7%: $237,163.60
 An increase of only 2 percentage points increases the interest that you pay by $78,627.60.

15. a. 39 months sooner
 b. $20,343.82

17. The cost of renting is $14,719.00 more than the cost of buying.

19. a. $6155.01 **b.** $12,628.03

21. $1,054,907.98

23. a. $92.10 **b.** $1696.51
 c. $31,249.98 **d.** $575,629.52
 e. $10,603,184.55 **f.** $22,485,504,530,179.20

Section 7.1　*(page 308)*

1. Yes; For every 10 feet of depth, the pressure increases by 4.33 pounds per square inch.

3. 0.433 pounds per square inch; 4.33/10 = 0.433

5.

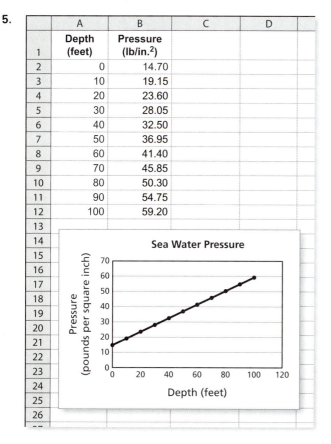

7. 148.2 pounds per square inch

9.

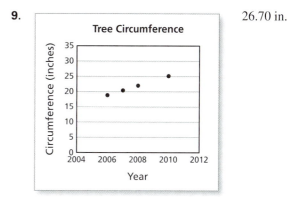

26.70 in.

11. From March 10 through May 20, the migration of the black-and-white warbler appears to be linear.

13. a. For every 1 mg/L of concentration, the absorbance increases by 0.164.

b.

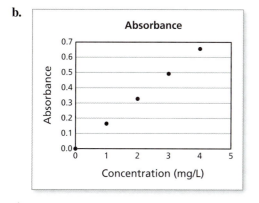

c. 5 mg/L: 0.820

17. 0.265

15. a. For every 0.2 cm of path length, the absorbance increases by 0.088.

b.

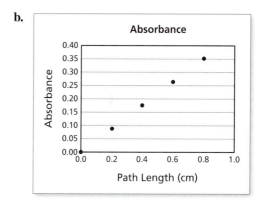

c. 1.0 cm: 0.440

19. a.

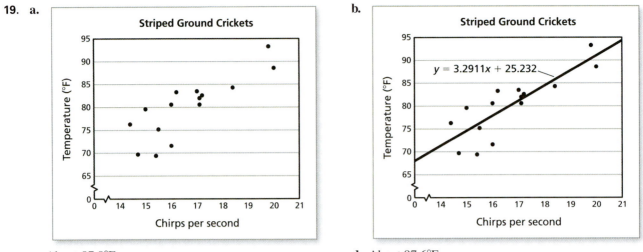

b.

$y = 3.2911x + 25.232$

c. About 87.8°F

d. About 97.6°F

Section 7.2 *(page 318)*

1. No; The difference between the surface area values for consecutive weeks is not constant.

3.

	A	B	C	D
1	**Week**	**Surface area (square feet)**		
2	0	500		
3	1	700		
4	2	980		
5	3	1,372		
6	4	1,921		
7	5	2,689		
8	6	3,765		
9	7	5,271		
10	8	7,379		
11	9	10,331		
12	10	14,463		
13	11	20,248		
14	12	28,347		
15	13	39,686		
16	14	55,560		
17	15	77,784		
18	16	108,898		
19	17	152,457		
20	18	213,439		
21	19	298,815		
22	20	418,341		
23				
24				
25				

Spreading of Water Hyacinth

5.

Week	Surface area (square feet)
0	1,500
1	2,400
2	3,840
3	6,144
4	9,830
5	15,729
6	25,166
7	40,265
8	64,425
9	103,079
10	164,927
11	263,883
12	422,212
13	675,540
14	1,080,864
15	1,729,382
16	2,500,000

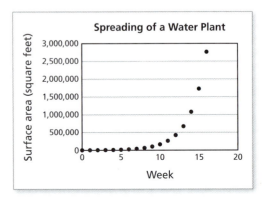

Spreading of a Water Plant

The graph shows exponential growth.

7. The initial population   **9.** 2147 rabbits   **11.** During year 9   **13.** 1291 rabbits

15. About 4000 trout; The population approaches but does not exceed 4000 over time.

17.

Year	Percent change in the number of trout
1	75.0
2	70.0
3	62.4
4	52.1
5	40.0
6	28.1
7	18.2
8	11.1
9	6.5
10	3.7
11	2.1
12	1.2

The percent of increase is decreasing each year.

19. Both species of fish experience a logistic growth pattern. Species A levels off with a population of about 1400 fish. Species B levels off with a population of about 1000 fish.

21. **a.** 39.2%   **b.** 24%   **c.** 4%

23. 0%; When the population reaches its maximum sustainable population, $r = r_0 \times (1 - 1) = r_0 \times 0 = 0$. This is because when a population is at its maximum, it cannot grow any larger.

25.

Year	Exponential population	Superexponential population
0	100	100
1	150	150
2	225	240
3	338	413
4	506	769
5	759	1567
6	1139	3517
7	1709	8768
8	2563	24,477
9	3844	77,101
10	5767	276,012

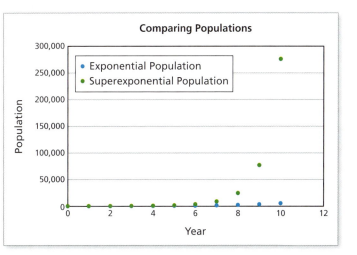

27. Day 29

7.1–7.2 Quiz   *(page 322)*

1. The pattern for Set A is linear. The deer population is projected to grow at a rate of 24 additional deer each year.

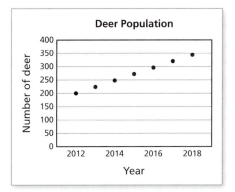

3. The pattern for Set B is exponential. The deer population is projected to grow at a rate of about 12% each year.

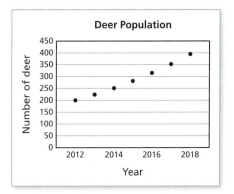

5. 2025

7. Set B; $\frac{200}{1.12^5} \approx 113$, but $200 - (5 \times 24) = 80$.

Section 7.3 *(page 330)*

1. The second differences are constant (-8). The pattern is quadratic.

3. The second differences are constant (-8). The pattern is quadratic.

5.

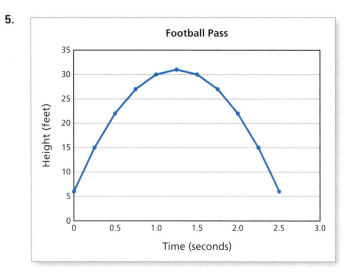

The pattern is quadratic.
The graph is a downward U-shaped curve.

7. Quadratic; The graph is curving upward.

9. About 600 ft

11. Yes; The second differences are constant (20).

13. Quadratic; The second differences are constant $\left(1\frac{1}{3}\right)$.

15. Quadratic; The second differences are constant (-12.4).

17.

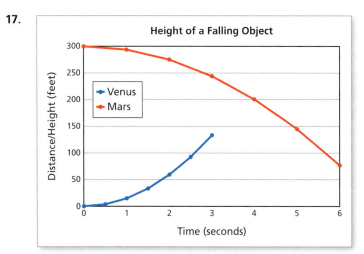

The graph of the data from Venus curves upward. The graph of the data from Mars curves downward. It appears that if the second differences are positive, then the graph curves upward, and if the second differences are negative, then the graph curves downward.

19. Linear; The first differences are about 186.

21. Quadratic; The second differences are about 16.

23.

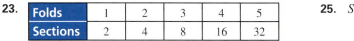

Folds	1	2	3	4	5
Sections	2	4	8	16	32

25. $S = 2^n$

Section 7.4 *(page 340)*

1. The number of petals on a daisy is usually a Fibonacci number.

3. Pineapple scales are patterned into spirals. There are three sets of spirals, and they follow the Fibonacci sequence (5, 8, and 13).

5. The leaves on a sneezewort plant grow in a Fibonacci pattern.

7. Yes, credit cards are in the shape of the golden rectangle.

9. Yes, the ratio of the height of the mirror to the length of the mirror is approximately equal to the golden ratio.

11. The RF mask identifies several prominent dimensions of the human face. Dr. Marquardt believes that if these dimensions are close to the golden ratio, then the person's face will be perceived to be beautiful.

13. Yes; Nefertiti's features align perfectly with the mask.

15. The nth triangular number is the sum of the first n whole numbers ($1, 1 + 2 = 3, 1 + 2 + 3 = 6, 1 + 2 + 3 + 4 = 10$, etc.).

17. No; The first few triangular numbers are 1, 3, 6, 10, 15, 21, 28, 36, 45, 55, 66, 78, 91, 105, and 120.

19. The cactus has 11 spirals in one direction and 18 spirals in the other direction. Both 11 and 18 are Lucas numbers.

21. About 137.51°; The golden angle is formed by dividing the circumference of a circle into two parts so that the ratio of the larger part to the smaller part is equal to the golden ratio.

23. **a.** Yes; The triangle is isosceles and $\frac{8}{5} = 1.6$, which is approximately equal to the golden ratio.

　　b. No; The triangle is not isosceles.

　　c. Yes; The triangle is isosceles and $\frac{55}{34} \approx 1.618$.

　　d. No; $\frac{14}{10} = 1.4$, which is not close to the golden ratio.

25. The ratio of any diagonal to any side is equal to the golden ratio.

7.3–7.4 Quiz   *(page 344)*

1. The second differences are constant (0.16). The pattern is quadratic.

3.

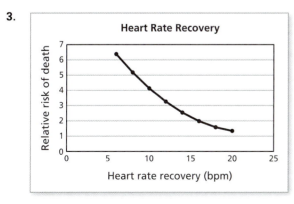

The graph curves downward.

5.

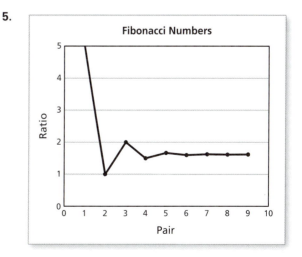

The graph resembles the tail end of the QRS complex shown on the EKG.

Chapter 7 Review   *(page 346)*

1. Yes; For every increase in temperature of 1°C, the resistance increases by 0.38 ohm.

3. 103.80 ohms

5. Yes

7. 45 amperes

9. 100% per week

11. During week 10

13. Logistic

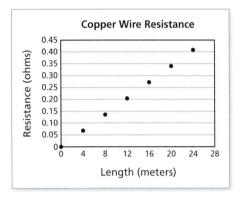

15.

Week	Change in the number of flour beetles
1	33
2	54
3	77
4	88
5	79
6	57
7	35
8	20
9	10
10	5
11	2
12	2

The weekly increase in the number of flour beetles increases until the population is about half the maximum sustainable population. Then the gain in the flour beetle population begins to decrease.

17. The second differences are constant (56). The pattern is quadratic.

19. 708 ft; No; The main towers are only 500 feet above the roadway.

21.

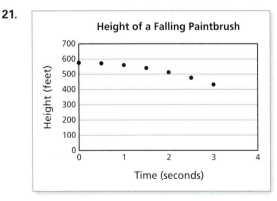

The graph shows a quadratic pattern and curves downward.

23. Each full cycle of a DNA double helix spiral measures 34 angstroms long by 21 angstroms wide. 34 and 21 are numbers in the Fibonacci sequence.

25. The major groove measures 21 angstroms long, and the minor groove measures 13 angstroms long. 21 and 13 are numbers in the Fibonacci sequence.

27. By placing two pentagons together and then rotating one of them, you have the shape of the cross-sectional view of a DNA double helix.

Section 8.1 *(page 358)*

1.

Probability of Fears			
	Fraction	**Decimal**	**Percent**
Dark	$\frac{1}{20}$	0.05	5%
Heights	$\frac{9}{25}$	0.36	36%
Identity theft	$\frac{25}{38}$	0.66	66%
Thunder or lightning	$\frac{11}{100}$	0.11	11%

The probability of a person being afraid of the dark is very unlikely.

The probability of a person being afraid of heights is somewhat unlikely.

The probability of a person being afraid of identity theft is likely.

The probability of a person being afraid of thunder or lightning is very unlikely.

3. Very unlikely

5. Very likely

7. a. Answers will vary.

   b. Industries differ in fatality rates because occupations that are much more dangerous than others tend to have higher fatality rates.

9. Low likelihood, high significance

11. Low likelihood, high significance

13. Low likelihood, high significance

15. The life expectancy for females increased in each country. In Australia, Germany, Spain, and the United States, life expectancy increased at about the same rate. The female life expectancy in Turkey increased at a greater rate than the other countries.

17. 1980: Turkey, Germany, United States, Australia, Spain

   2007: Turkey, United States, Germany, Australia, Spain

   Germany and the United States switched orders. Spain still has the greatest female life expectancy, and Turkey still has the least female life expectancy.

19. In each country in each year, the life expectancy for females is about 5 years greater than for males.

21. a. Event 2; There are 0 numbers less than 1, and there are 5 numbers greater than 1.

   b. Event 2; There are 2 multiples of 3, and there are 3 numbers greater than 3.

23. a. Event 2; There are 2 blue areas on spinner A, and there are 3 blue areas on spinner B.

   b. Event 1; There are 3 multiples of 2 on spinner A, and there is 1 multiple of 6 on spinner B.

Section 8.2 *(page 368)*

1. $\frac{1}{12}$

3. $\frac{2}{3}$

5. $\frac{1}{2}$

7. a. $\frac{1}{1000}$ **b.** $\frac{1}{200}$

9. 64.75%

11. About 58.9%

13. The probabilities are the same because no matter how the data is presented, the overall probability of saying "No" is the same.

15. Underweight: About 14.9%; unlikely

Normal weight: About 23.3%; unlikely

Overweight: About 49.0%; about equally likely to happen or not happen

Obese: About 59.1%; somewhat likely

Morbidly obese: About 62.1%; likely

17. About 52.2%   **19.** About 19.6%   **21.** 3.5   **23.** 25% for all 4 suits

25. The experimental and theoretical probabilities for spades and clubs are about the same. The experimental probability for hearts is greater than the theoretical probability. The experimental probability for diamonds is less than the theoretical probability.

27. About 53.8%

8.1–8.2 Quiz   *(page 372)*

1.

Month	Fraction	Decimal	Percent
November	$\frac{11}{250}$	0.044	4.4%
December	$\frac{25}{74}$	0.338	33.8%
January	$\frac{81}{1000}$	0.081	8.1%
February	$\frac{301}{1000}$	0.301	30.1%
March	$\frac{5}{34}$	0.147	14.7%

3. Likely

5. Nearly impossible

7. 70%

November: very unlikely
December: somewhat unlikely
January: very unlikely
February: unlikely
March: very unlikely

Section 8.3   *(page 380)*

1. −$49.70   **3.** $70.95

5. a. −$100.26   **b.** $100.26   **c.** $1,002,600

7.

	Probability	Profit	Expected Value	
A	10%	$8 million	0.1($8)	The company should develop Laptop B.
	70%	$4 million	0.7($4)	
	20%	−$2 million	+ 0.2(−$2)	
			$3.2 million	
B	30%	$8 million	0.3($8)	
	50%	$4 million	0.5($4)	
	20%	−$2 million	+ 0.2(−$2)	
			$4 million	

9.

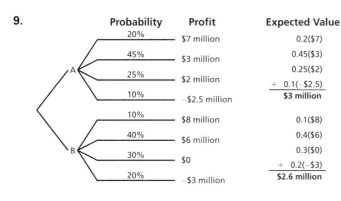

Probability	Profit	Expected Value
20%	$7 million	0.2($7)
45%	$3 million	0.45($3)
25%	$2 million	0.25($2)
10%	–$2.5 million	+ 0.1(–$2.5)
		$3 million
10%	$8 million	0.1($8)
40%	$6 million	0.4($6)
30%	$0	0.3($0)
20%	–$3 million	+ 0.2(–$3)
		$2.6 million

The company should develop E-reader A.

11. Expected value for Option 1: $1000
Expected value for Option 2: $1200
Option 2 has the greater expected value.

13. Expected value for Option 1: $1400
Expected value for Option 2: $1800
Option 2 has the greater expected value.

15. Expected value for speculative investment: $1600
Expected value for conservative investment: $390

17. Expected rate of return for Stock V: 11%
Expected rate of return for Stock W: 7%
Stock V has a greater expected rate of return.

19. 8.72%

21. 8.825%

Section 8.4   *(page 390)*

1. 0.1728%

3. 0.0512%

5. 0.0216%

7. 0.0008%

9. About 44.4%

11. About 98.4%

13. About 45.5%

15. About 77.0%

17. About 33.3%

19. If you do not switch, you win 2 out of 5 times. If you do switch, you win 3 out of 5 times. So, based on probability, you should switch.

21. 20%

23. About 2.0%

25. About 74.7%

8.3–8.4 Quiz   *(page 394)*

1. 2.5 points

3. About 5.6%

5. About 1.9%

7. About 94.4%

Chapter 8 Review   *(page 396)*

1.

Event	Fraction	Decimal	Percent
Being an organ donor	$\frac{4}{15}$	0.267	26.7%
Eats breakfast	$\frac{61}{100}$	0.61	61%
Having a dream that comes true	$\frac{429}{1000}$	0.429	42.9%
Household with television	$\frac{491}{500}$	0.982	98.2%

The probability of being an organ donor is unlikely.

The probability of eating breakfast is about equally likely to happen or not happen.

The probability of having a dream that comes true is about equally likely to happen or not happen.

The probability of a household having a television is almost certain.

3. Almost certain   **5.** Somewhat unlikely

7. Because, for each solid, the probability of landing on any one of the faces is the same.

9. 75%

11. 18–29: About 74.8%; likely

30–49: About 58.3%; about equally likely to happen or not happen

50–64: About 31.1%; unlikely

65 and over: About 9.7%; very unlikely

13. $-$250

15.

	Probability	Profit	Expected Value	The company should develop Toaster A.
A	25%	$500,000	0.25($500,000)	
	65%	$100,000	0.65($100,000)	
	10%	$-$250,000	$+$ 0.1($-$250,000)	
			$165,000	
B	30%	$450,000	0.3($450,000)	
	55%	$50,000	0.55($50,000)	
	15%	$-$250,000	$+$ 0.15($-$250,000)	
			$125,000	

17. Expected value for Option 1: $1000

Expected value for Option 2: $1250

Option 2 has the greater expected value.

19. About 1.56%   **21.** About 86.65%   **23.** 0.81%   **25.** About 91.76%

Section 9.1   *(page 408)*

1. Answers will vary. *Sample answer:* The stacked area graph shows the types and quantities of aircraft that composed the USAF fighter force from 1950 through 2009. In general, the number of aircraft has declined since the peak in the 1950s.

3. About 24%

5. Disagree; The F-16 has decreased slightly in number since 2000 but still makes up about one-half of the fighting force.

7. Company D

9. Company D; Company E

To determine the profit for each company, multiply revenue by profit percent (in decimal form). This shows that Company E makes the most profit.

Company A: $2.5 million

Company B: $18.75 million

Company C: $37.5 million

Company D: $35 million

Company E: $45 million

11. The graph shows the price of a share of company C stock for each week of 2010. In general, the stock price started at about $24 for the first 6 weeks, then significantly decreased to about $15 from weeks 9 through 16, then gradually increased to about $20 by week 31, then significantly increased to about $30 by week 37, and then gradually decreased to about $25 by the end of the year.

13. Earned a profit

15. The graph shows the day of the week (*x*-axis), the type of change between the opening price and the closing price (green if increase, red if decrease), the amount of change from opening to closing (length of candlestick body), the opening price (bottom of candlestick body if green, top of candlestick body if red), the closing price (top of candlestick body if green, bottom of candlestick body if red), the highest price (top of shadow), the lowest price (bottom of shadow), and the range (length of shadow).

17. $22.50; $22.75   **19.** Wednesday; Thursday

21. 12.5%; If you bought the stock at its lowest price ($22 on Monday) and sold it at its highest price ($24.75 on Friday), you would have earned $2.75 per share. This is a return of $\frac{2.75}{22} = 12.5\%$.

23. Answers will vary.

25. The chart in Example 2 shows the changing parts of a whole, but the chart in Example 1 shows data for a total that changes over time.

27. Answers will vary. *Sample answer:* Income, expenses, debt, retirement savings, body weight, blood pressure, cholesterol level, food consumption, exercise routine, study routine, social habits, personal goal accomplishment, etc.

Section 9.2   *(page 418)*

1. Women's team: About 72.5 in.

Men's team: About 79.1 in.

The average player on the men's team is more than 6 inches taller than the average player on the women's team.

3. Women's team: 74 in.

Men's team: 79 in.

The median height of the men's team is 5 inches greater than the median height of the women's team.

5. Median

7. The mean and the median would each be 2 inches greater.

9. About 10 or 11 colleges

11. Uppermost (4th) quartile; lower middle (2nd) quartile; The uppermost quartile has the greatest range of values (41,350 − 24,900 = $16,450), and the lower middle quartile of the data has the least range of values (16,200 − 12,500 = $3700).

13. No; The histogram shows the $5000 interval in which the annual cost for each college occurs, rather than the actual cost. This information cannot be used to determine the sum, middle value, or most common value of the actual annual costs.

15. 13   **17.** 2004, 2005, 2008, and 2010

19. **a.** Mean: About 36,739   **b.** Mean: About 16,898
              Median: 5530   Median: 4000

c. Mean; Because more than half of the data are grouped so closely together, removing the two outliers changes the middle value of the data to a number that is close to the original median. But because their values are so great, removing the outliers has a substantial effect on the mean.

21. Median; The median is close to the majority of the data, but the mean is greater than most of the data.

23. **a.** Mode; Printer A has the greatest mode (32 ppm) of the 3 printers.

b. Mean; Printer B has the greatest mean (30.4 ppm) of the 3 printers.

c. Median; Printer C has the greatest median (31 ppm) of the 3 printers.

25.

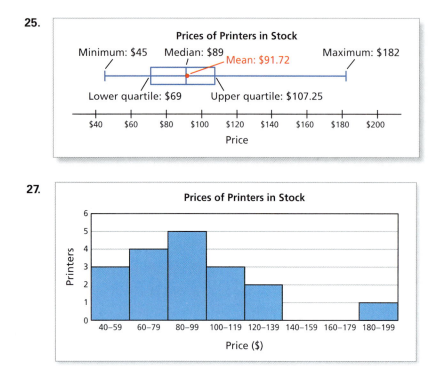

27.

Prices of Printers in Stock

9.1–9.2 Quiz   *(page 422)*

1. The chart shows the ranks and scores for the top 34 countries in reading, math, and science. The scores for each subject are given according to their rank from highest to lowest. The scores for each country are indicated by color coding and lines connecting the scores.

3. No; The overall score for the United States is $500 + 487 + 502 = 1489$, but the overall score for Germany is $497 + 513 + 520 = 1530$.

5. **a.** Reading: The median is 496, and the mode is 500.
   Math: The median is 496.5, and the mode is 487.
   Science: The median is 501, and the modes are 500 and 508.

   b. Reading: 493
   Math: 496
   Science: 501

7. Answers will vary. *Sample answer:* The lowest and uppermost quartiles are more spread out than the middle quartiles. The lowest quartile is the most spread out, and the lower middle quartile is the least spread out.

Section 9.3   *(page 430)*

1. About 78 of the egg diameters lie within 1 standard deviation of the mean.

3. 100%   5. About 51%

7. Yes; The distribution of the data is approximately bell shaped.

9. About 97%

11. The distribution of data values for machine 2 has less dispersion than the distribution of the data values for machine 1.

13. Males tend to have higher mathematics SAT scores than females; females tend to have higher writing SAT scores than males.

15. May and October; The school may have an SAT test in May or June and another SAT test in October or November, so this would account for the higher number of study hours in the months before the tests.

17. Answers will vary.

19.

x	x^2
16	256
12	144
23	529
20	400
18	324
15	225
18	324
19	361
$\sum x = 141$	$\sum x^2 = 2563$

$s \approx 3.3$

Section 9.4 *(page 440)*

1. 99 ± 4.99

3. 23.3 ± 0.83

5. People may lie to their dentists about flossing their teeth; someone other than a dentist should conduct the survey.

7. The one machine you choose could be producing hardly any defective items or could be producing more defective items than the other machines; you should choose items from all of the machines.

9. People who have a very strong opinion will call in; the radio station should randomly survey people in the listening area.

11. People living in the nursing homes are more likely to support the bill; you should survey a random sample of people in the city.

13. 1068 people

15. 385 people; When the margin of error increases, the sample size decreases. This is because the fewer people you survey, the more likely you are to have an error.

17. a; The question could produce biased results because it states that the policy is unfair and the project is time-consuming.

19. Sample mean: $195.31
Margin of error: $16.56

21. Sample mean: 5.5 days
Margin of error: 0.4 day

23. 104 newborns

25. 865 people

9.3–9.4 Quiz *(page 444)*

1. Yes; The distribution of the data is approximately bell shaped.

3. 100%

5. 9.3 ± 0.36

7. People in the neighborhood around the fire station are more likely to want the fire station renovated; you should take a random sample of people in the city.

Chapter 9 Review   *(page 446)*

1. The graph shows the GDP per capita in thousands (*x*-axis), the percent who say that religion is an important part of their daily lives (*y*-axis), the country name (label), the population (size of sphere), and the dominant religion (color of sphere).

3. United States

5. Yes; The majority of the countries with a low GDP per capita have a high percent of the population who say that religion is an important part of their daily lives. The majority of the countries with a high GDP per capita have a low percent of the population who say that religion is an important part of their daily lives.

7. The United States is one of the few countries with a high GDP per capita and high percent of people who say that religion is an important part of their daily lives. The majority of the countries with high GDP per capita have a low percent of people who are religious.

9. Mean: 5.85%

   Median: 5.55%

   Mode: 4.6% and 5.8%

11. 9.3% and 9.6%

13. About 25% of the data is between 4.0% and 4.7%. About 25% of the data is between 4.7% and 5.6%. About 25% of the data is between 5.6% and 6.3%. About 25% of the data is between 6.3% and 9.6%. The uppermost quartile has the greatest range. The other 3 quartiles have similar ranges.

15. Yes; The distribution of the data is approximately bell shaped.

17. About 92%

19. The distribution of data values for women has less dispersion than the distribution of the data values for men.

21. $65,467 \pm \$1710.99$

23. Conservative people are more likely to have a positive opinion of a conservative president and a negative opinion of a liberal president; the magazine should send the survey to a random sample of people.

25. 1691 people

Section 10.1   *(page 458)*

1. 115.5 lb

3. 163 lb

5. For a woman, subtract 3.75 pounds for every inch under 5 feet, and for a man, subtract 4 pounds for every inch under 5 feet; the Robinson formula gives a linear relationship between ideal weight and height; 93 lb

7. **a.** About 21.0%   **b.** About 26.94

9. The upper and lower limits for the target heart rate zone decrease over time. The difference between the upper and lower limits for the target heart rate zone decrease over time. These two trends are also true for the fat burning zone, which is part of the target heart rate zone.

11. No

13. Losing weight; incoming calories − outgoing calories = 1500 − 1992 = −492

15.

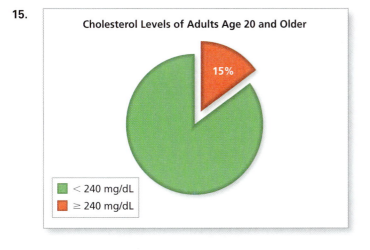

17.

19. As total cholesterol rises, mortality from coronary heart disease rises.

21. About 4 men per 1000

23.

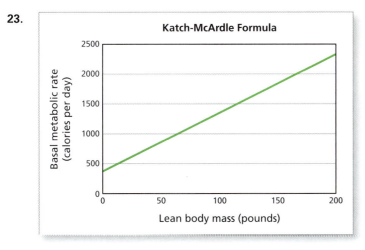

As lean body mass increases, basal metabolic rate increases.

25. 1350 calories per day   **27.** 1546 calories per day

29. About 2088 calories per day

Section 10.2   *(page 468)*

1. The winning distances increased over time and appear to have leveled off around 70 meters.

3. Between 50 and 55 m; The winning distance was increasing each year, and the winning distance in 1936 was about 50 meters.

5.

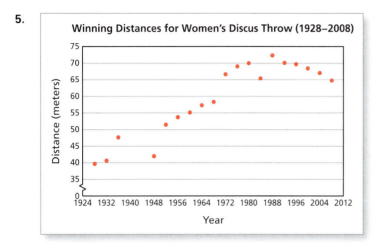

The winning distance increased in every Olympics from 1952 through 1980. The current Olympic record was set in 1988. From 1992 through 2008, the winning distance has decreased in every Olympics.

7. The winning time decreased every Olympics from 1960 through 1988. It increased in 1992, and then decreased through 2002. It increased in 2006 and 2010.

9. About 71.4%

11.

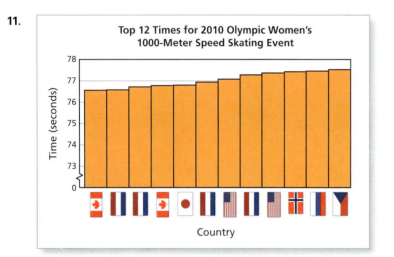

13. 44.2

15. 55.2

17. 120.7 points

19. 94.5 points

21.

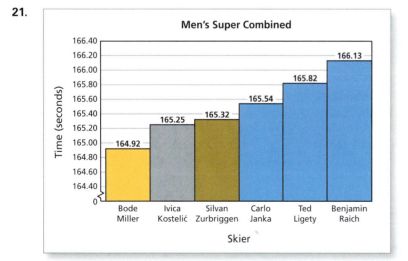

Gold medal: Bode Miller
Silver medal: Ivica Kostelić
Bronze medal: Silvan Zurbriggen

23. 51.76 sec; This would be the sixth best time when compared to the slalom times of the top six overall finishers.

25.

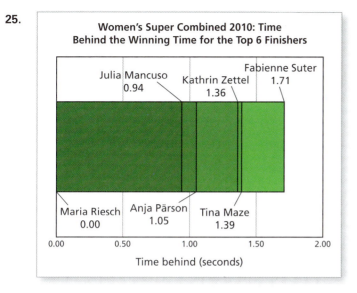

Women's Super Combined 2010: Time Behind the Winning Time for the Top 6 Finishers

Julia Mancuso 0.94

Kathrin Zettel 1.36

Fabienne Suter 1.71

Maria Riesch 0.00

Anja Pärson 1.05

Tina Maze 1.39

Time behind (seconds)

10.1–10.2 Quiz   *(page 472)*

1. About 26.1%

3. 198 beats per minute

5. Losing weight; incoming calories − outgoing calories = $2400 - 2582 = -182$

7. About 66.7%

Section 10.3   *(page 480)*

1. Josh Hamilton

3. Cabrera 0.420, Choo 0.401, Fielder 0.401, Hamilton 0.411, Mauer 0.402, Morneau 0.437, Pujols 0.414, Thome 0.412, Votto 0.424, Youkilis 0.411

5. Yes; This can happen when a player has a low number of bases on balls (BB) and hit by pitches (HBP), and a high number of sacrifice flies (SF).

7. Pass plays; The only time that coaches call more run plays is on 3rd down and 1 yard to go.

9. 5, 6, 7, 8, and 10 yards

11. Run play; The conversion rate is greater for run plays when it is 3rd down and 2 yards to go. Also, the defense is probably expecting a pass play.

13. No; The conversion rate for run plays is about 10% greater than the conversion rate for pass plays on 3rd down and 1 yard to go.

15. Durant; Durant attempted 18 field goals and Nowitzki attempted 15 field goals.

17. 48 points

19. 40%

21. The majority of Nowitzki's shots are from mid-range with a preference for the right side of the court. Durant's shots are more scattered with some close, some mid-range, and some long-range (3-pointers).

23. The size of the bubble represents the relative number of fans of a particular sport. The color of the bubble represents the political party that the fans of the sport are more likely to support.

25. Republican; More of the bubbles are red, and most of the larger bubbles are red.

27. PGA Tour, College football, NASCAR; Men's and women's tennis, WNBA, NBA

Section 10.4 *(page 490)*

1. Older than 44 years old; 29.2% of campers are 45+, and 25.7% of campers are under 18.

3. More; 65.1% of campers earn $50,000+.

5. No; The graph does not tell you what percent of the general population participates in camping.

7. 23 mph

9. 14 ft to 20 ft

11. About 86,526 ft^2

13. About 5.8 hr

15. Yes; A line that decreases from left to right indicates a downhill section. Trail 5 appears to have the biggest downhill section between the 3-mile and 3.5-mile markers.

17. Trail 2; It is the longest trail and it has the steepest incline (after the 8-mile marker).

19. Mean: 2437.5 ft

    Median: 1800

    Yes; Trail 1 and trail 2 are outliers because the vertical rise of both trails is much greater than the rest of the data.

21. 162 fatalities per year

23. Yes; This corresponds to a heat index of 121°F. Any heat index value that exceeds 105°F warrants an excessive heat warning from the National Weather Service.

25. No; The heat index increases at an increasing rate as the temperature increases.

10.3–10.4 Quiz *(page 494)*

1. There is a data point missing for the NHL in 2004. The estimated value is 0.08.

3. 42 to 58 games

5. Danger

7. Extreme danger

Chapter 10 Review *(page 496)*

1. **a.** About 12.5% **b.** About 19.70

3. Gaining weight; incoming calories − outgoing calories = 3200 − 2819 = 381

5. Yes

7. The winning distances increased dramatically between 1948 and 1980. After a slight drop in 1984, the winning distances leveled off between 80 and 85 meters.

9. About 85 m

11.

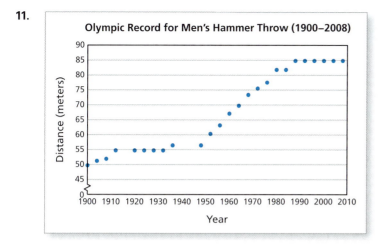

13. 53.9

15. There is a positive correlation between win percentage on June 1 and final win percentage. In general, a team's final win percentage will be slightly less than the team's win percentage on June 1.

17. Between 0.550 and 0.600

19. No; All of the teams that had a win percentage below 0.500 on June 1 finished with a win percentage below 0.550. Because $0.550 \times 162 = 89.1$, which is less than 90, none of these teams could have made the playoffs.

21. River blind   **23.** About 125 hr   **25.** About 5%

27. Wetlands; It is based on more data. The hunting duration in the wetlands is 4 times the hunting duration on the lake shore.

Index

Credits